Stem Cells and
Bone Tissue

Stem Cells and Bone Tissue

Editors

Rajkumar Rajendram

Departments of General Medicine and Intensive Care
John Radcliffe Hospital
Oxford
UK

Victor R. Preedy PhD DSc

Professor of Nutritional Biochemistry
School of Medicine
King's College London
and
Professor of Clinical Biochemistry
King's College Hospital
UK

Vinood B. Patel

Department of Biomedical Science
School of Life Sciences
University of Westminster
London
UK

CRC Press
Taylor & Francis Group
Boca Raton London New York

CRC Press is an imprint of the
Taylor & Francis Group, an **informa** business

A SCIENCE PUBLISHERS BOOK

CRC Press
Taylor & Francis Group
6000 Broken Sound Parkway NW, Suite 300
Boca Raton, FL 33487-2742

First issued in paperback 2019

Cover Illustrations: Reproduced by kind courtesy of the undermentioned authors:
- Figure No. 1 from Chapter 2 by Qisheng Tu and Jake Chen
- Figure No. 1 from Chapter 5 by Agnieszka Arthur et al.

ISBN-13: 978-1-4665-7841-8 (hbk)
ISBN-13: 978-0-367-38039-7 (pbk)

Library of Congress Cataloging-in-Publication Data

Stem cells and bone tissue / editors, Rajkumar Rajendram, Victor R. Preedy.
       p. ; cm.
  Includes bibliographical references and index.
  ISBN 978-1-4665-7841-8 (hardcover : alk. paper)
  I. Rajendram, Rajkumar. II. Preedy, Victor R.
  [DNLM: 1. Bone Diseases--therapy. 2. Stem Cells. WE 225]
  616.02'774--dc23

                                                        2012038882

Visit the Taylor & Francis Web site at
http://www.taylorandfrancis.com

and the CRC Press Web site at
http://www.crcpress.com

Preface

Bone diseases, characterised by either fractures, stunting, loss of mineral composition or strength, are important causes of morbidity and mortality. For example worldwide there are 9 million new cases of fractures due to osteoporosis annually and this has increased by 50% in the past 2 decades. As the population ages, these figures will further increase, imposing burdens on individuals, communities, and health providers. As a result, new treatments and strategies are continually being developed to prevent bone diseases occurring and also to treat such conditions. Stem cells potentially offer a novel therapeutic platform to treat bone disease. They also help the scientist understand the molecular and cellular aetiology of bone disorders. Gaining knowledge on the nature and application of stem cell sciences is a prerequisite for understanding their potential in treating or preventing bone disorders. However, obtaining information on stem cells *per se* and their usage and applications in treating skeletal disorders is currently problematic, as such text-based material is either sporadic or directed towards the expert. This book **Stem Cells and Bone Tissue** is designed to address such limitations. It is multidisciplinary and covers all scientific levels. **Stem Cells and Bone Tissue** has three major sections:

Section 1: Introductory Text and Sources of Stem Cells for Skeletal Tissue
Section 2: Cellular and Molecular Aspects
Section 3: Conditions, Applications, Treatments and Repairs

The simplistic nature of the main section headings, however, should not detract from the fact that individual chapters are highly detailed. Coverage includes for example, general aspects of stems cells, sources of stems cells, isolation and purification, applications in regeneration, nanoscale topography, myostatin (GDF-8) signalling, c-Jun, Lnk, cell-derived Factor 1/CXCR4 signalling, chromatin remodelling, osteoporosis, osteoarthritis, hypophosphatasia, osteopetrosis, osteogenesis, and many other areas of merit too numerous to mention.

There are three distinguishing features of **Stem Cells and Bone Tissue** that sets it apart from other scientific texts. As well as an Abstract, each chapter has one or more Key Facts which expands areas of interest for the novice. There are also at least five definitions and explanations of key terms or words used in each chapter. Finally, each chapter is summarised with at least five bullet points. The book is well illustrated with numerous figures and tables and the entire book has an excellent index. As a consequence of these features the readership is designed to be broad, from the novice to leading expert.

Contributors of **Stem Cells and Bone Tissue** are all either international or national experts, leading authorities or are carrying out ground breaking and innovative work on their subject. The book is an essential text for orthopaedic specialists, doctors and clinicians, surgeons, pathologists, research scientists, molecular biologists, biochemists, health care professionals and general practitioners as well as those interested in disease or bone tissues in general.

The Editors

Contents

Section 2: Cellular and Molecular Aspects

Section 3: Conditions, Applications, Treatments and Repairs

Section 1
Introductory Text and Sources of Stem Cells for Skeletal Tissue

CHAPTER 1

Endothelial Progenitor Cell-based Therapy for Orthopaedic Regenerative Medicine

Ru Li,[1,*] Aaron Nauth,[2,a] Nizar Mahomed,[1] Darrell Ogilvie-Harris[1] and Emil Schemitsch[2]

ABSTRACT

Bone is a highly vascularized tissue that requires a close coupling between blood vessels and osteogenic cells to maintain skeletal integrity. Segmental bone defects after severe trauma, infection, and surgical removal of tumours remain a major clinical problem to be addressed. Vascular in-growth at the fracture site plays a critical role in fracture healing. Endothelial progenitor cells have the potential to enhance bone repair and regeneration. Endothelial progenitor cells have been shown

[1]Division of Orthopaedic Surgery, Toronto Western Hospital, UHN, University of Toronto, 399 Bathurst Street, Toronto, ON M5T 2S8, Canada.
Email: ru.li@utoronto.ca
[2]Division of Orthopaedic Surgery, St. Michael's Hospital, University of Toronto, 30 Bond Street, Toronto, ON M5B 1W8, Canada.
[a]E-mail: aaron.nauth@utoronto.ca
[b]E-mail: nizar.mahomed@uhn.on.ca
[c]E-mail: leafdoc1@rogers.com
[d]E-mail: schemitsche@smh.ca
*Corresponding author

List of abbreviations given at the end of the text.

to have prominent effects in promoting bone regeneration in several animal model studies. Evidence indicates that endothelial progenitor cells promote bone regeneration by stimulating both angiogenesis and osteogenesis through a differentiation process towards endothelial cell lineage and formation of osteoblasts. Moreover, endothelial progenitor cells increase vascularization and osteogenesis by increasing secretion of growth factors and cytokines through paracrine mechanisms. It has been suggested that endothelial progenitor cells mobilized by signals from sites of injury may contribute to neovascularization and thus new bone formation in native fracture healing. Endothelial progenitor cells have a paracrine effect by which capillary and osteoblast density increase in a dose-dependent manner. We have previously evaluated the effects of local delivery of endothelial progenitor cells on the stimulation of angiogenesis and promotion of osteogenesis to enhance bone regeneration in a rat segmental bone defect model. We have shown that local endothelial progenitor cells therapy significantly increased angiogenesis and osteogenesis to promote fracture healing; restored the biomechanical properties of the fractured bone closer to that of the intact bone; and increased the local vascular endothelial growth factor mRNA expression. These results and other recent investigations highlight the promise of EPCs as a potential cell-based therapy for fracture healing.

Introduction

Severe fractures damage blood vessels and disrupt circulation at the fracture site, which can lead to an increased risk of poor fracture healing. Angiogenesis, the formation of new blood vessels and vessel networks, plays a critical role in bone regeneration and fracture repair. Bone is a biologically advantaged tissue, in that it has the capacity to undergo regeneration as part of a repair process. Fracture healing is the most common and recognizable form of bone regeneration. The unique process of bone regeneration requires a coordinated coupling between osteogenesis and angiogenesis. Although the regeneration of the bone can be completed with minimal scarring, a significant percentage of fractures fail to heal adequately. Non-union or delayed union of fractures and the repair of large segmental bone defects after tumour removal, infections or trauma remains a highly challenging and clinically important problem in orthopaedic surgery. Although new technologies and advances have substantially enhanced fracture healing and surgical outcomes, there remains a subset of fractures and related conditions that continue to be present with impaired healing. Consequently, new strategies to optimize bone regeneration are being developed. Current approaches that aim to promote bone healing such as the use of bone morphogenetic proteins (BMPs) and mesenchymal stem cells (MSCs) are lacking angiogenic activity for bone regeneration.

Endothelial progenitor cells (EPCs) are bone marrow (BM)-derived cells with the ability to differentiate into endothelial cells and to participate in the establishment of neovasculature (Asahara et al. 1997). EPCs represent a novel population of precursor cells that enter the circulation in response to trauma, and home to sites of tissue ischemia, where they have been shown to affect functional blood flow recovery in ischemic tissues (Laing et al. 2007; Kudo et al. 2003; Matsumoto et al. 2008). Moreover, EPCs have been shown to be capable of differentiating into osteogenic cells *in vitro*, and have been shown to be up-regulated in response to orthopaedic trauma in humans (Bick et al. 2007; Laing et al. 2007). EPCs mobilized by signals from bone regeneration sites may contribute to neovascularization and thus new bone formation in fracture healing (Lee et al. 2008). The therapeutic application of BM-derived EPCs has been shown to augment fracture healing and local angiogenesis in segmental defect models in rat femurs (Atesok et al. 2010; Li et al. 2011). These findings suggest significant promise for the use of EPCs for cell based therapy directed at bone regeneration.

The purpose of this chapter is to review the available literature and evaluate the use EPCs as a cell-based tissue engineering strategy in the musculoskeletal system. EPCs offer the potential to emerge as a new strategy among other cell-based therapies directed at bone regeneration.

Historical Overview and Sources of EPCs

One of the earliest known speculations on the existence of an EPC was published in 1932 describing the formation of capillary-like structures in leukocyte cultures (Hueper and Russell 1932). Continuing evidence of the existence of an EPC-like cell was collected as stents and aortic grafts gained increased popularity in cardiovascular repair. However, the pivotal step towards our current understanding of EPCs was by Asahara, who defined a cell population that expressed many endothelial-associated markers (e.g., cluster of differentiation molecule: CD34) and demonstrated vascular incorporation *in vivo*. The authors reported for the first time that purified hematopoietic progenitor cells from adults can differentiate into an endothelial phenotype, and the authors named these cells "endothelial progenitor cells" (Asahara et al. 1997).

Identified sources of cells with vascular potential can be divided into four categories: marrow, tissue, blood and embryo (Doyle et al. 2006). The embryonic hemangioblast, derived from the mesodermal layer, is a precursor of both blood and blood vessels. EPCs are generally considered to be a downstream progenitor of hemangioblasts. Irrespective of the exact lineage pathway, the working definition of EPCs maintains that they are bone marrow-derived cells, circulating within the blood stream, that are functionally and phenotypically distinct from mature endothelial cells (EC). These cells contribute to *in vivo* vasculogenesis and/or vascular homeostasis. Although there are various descriptions in the literature with respect to the origin and the surface markers of these cells, EPCs can further be defined as bone-marrow derived precursor cells with the ability to differentiate into endothelial cells and to participate in the formation of new blood vessels.

Phenotypic Characterization of EPC Types

EPCs have been studied and described as two types (early and late types) by researchers. EPC was described as a spindle-shaped cell derived from peripheral blood and cultured for less than 3 wk (Asahara et al. 1997). This form of EPC showed limited proliferating potential for long-term culture and disappeared 4 to 6 wk later during *in vitro* conditions. Other reports suggest the existence of another type of EPC that originates from bone marrow, circulates in peripheral blood, and shows a different morphology and proliferation pattern compared to that described by Asahara. In these studies, prolonged incubation of peripheral mononuclear cells, in the presence of vascular endothelial growth factor (VEGF), produced cells with striking proliferation potential that formed colonies and had a different phenotype from the early spindle-shaped cells. Moreover, another study reported two different types of EPCs according to their time dependent appearance: early EPCs with spindle-shape, and late EPCs with cobblestone-shape (Hur et al. 2004). Both types of cells equally contributed to neovasculogenesis *in vivo* as the early EPCs secreted angiogeneic cytokines, whereas late EPCs supplied a critical population of endothelial cells. These studies suggest two distinct populations of EPCs based on their behaviour in cell culture and their *in vivo* effects. EPCs have more recently been categorized into early and late cellular subsets, termed early endothelial progenitor cells (eEPCs) and late outgrowth endothelial progenitor cells (OEPCs) (Smadja et al. 2008). It is generally accepted that the subsets are defined *in vitro* based on the time in culture. eEPCs are adherent cells which have been under culture conditions for 7 to 14 d, while OEPCs are generally described to occur after 4 wk in culture. Clear phenotypic differences have been described between these two EPC populations, and they have different mechanisms of influencing angiogenesis. OEPCs have been shown to contribute to the formation of perfused vasculature when co-implanted with other cell types (Fuchs et al.

2009), while eEPCs influence angiogenesis in a paracrine manner. Molecular profile of eEPCs and OEPCs isolated from peripheral blood shows that eEPCs have an mRNA fingerprint resembling that of a monocyte or a cell of hematopoietic lineage, with expression of CD45, CD14, Runx, and WAS. OEPCs have multiple endothelial markers, but lack haematopoietic markers such as CD45 and CD14. We have conducted experiments to confirm that the isolated EPC cell population expresses appropriate surface markers with CD34, CD133, Flk1 and vWF by fluorescent immunocytochemistry staining and a reliable flow cytometric assay. We have also found that the cultured EPCs at 1 and 3 wk appeared to have both a cobblestone shape with strong positive CD133, and spindle shape with strong positive CD34 and Flk1, and mixture shapes (cobblestone and spindle) with weak positive vWF staining (Fig. 1.1). Though no apparent differences in the surface appearance of the cells were observed between 7 and 21 d, there was a significant increase in the number of cells.

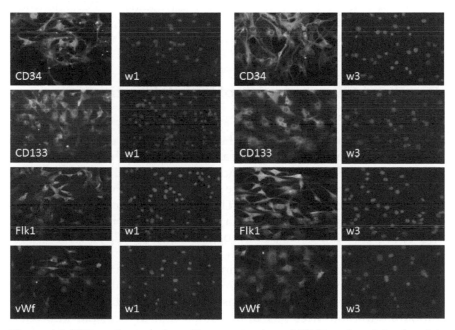

Figure 1.1 EPC surface marker. Characterization of EPCs at 1 and 3 wk using immunohistochemistry staining. The cultured EPCs appeared to have a cobblestone shape with strongly positive CD133, and spindle shape with strongly positive CD34 and Flk1, with both shapes having weakly positive vWF staining at 1 and 3 wk (unpublished data).

Color image of this figure appears in the color plate section at the end of the book.

In vitro, co-culturing of EPCs with other cell types seems to benefit the vascularization process. Co-culture of OEPCs and human primary osteoblasts could lead to angiogenic activation of OEPCs without supplemental stimulatory growth factors (Fuchs et al. 2009). When primary human osteoblasts and OEPC co-cultured cells were then subcutaneously implanted, it was observed that co-cultured cellular implants had higher vascular potential. OEPCs formed functional vascular structures, connected to host vascular supply and demonstrated erythrocytes in the lumen of the structures. These vascular structures closely associated with the scaffold and embedded in an extracellular matrix which was produced by the osteoblasts. The authors hypothesized that the *in vitro* co-culture of OEPCs and osteoblasts helped to prime the OEPCs. These important controversies make the field of EPCs intriguing, but at the same time complicated. Therefore, further efforts should be focused on the development of a standard assay, or set of assays, that are accessible to all investigators, to specifically define and validate the function of (candidate) EPCs. This would provide investigators with a benchmark for comparison, and a rationale for the examination and clinical translation of selected cell subsets in targeted clinical disorders.

Functional Characterization of EPC Mobilization and Paracrine Mechanisms

The mobilization of stem cells in the bone marrow is determined by the local microenvironment, or the so-called "stem cell niche," which consists of fibroblasts, osteoblasts, and endothelial cells (EC). Physiologically, ischemia is believed to up-regulate mobilizing cytokines such as VEGF or stromal-derived factor (SDF)-1. This in turn influences the interactions between EPCs and stromal cells via a matrix metalloproteinase (MMP)-9 dependent mechanism (Capobianco et al. 2010), and EPCs are mobilized from the bone marrow into circulation.

Biological cascades resulting in new vessel formation are initiated following an injury to the vascular or musculoskeletal system. Previous studies have documented the increase in EPC mobilization into peripheral blood circulation after vascular trauma and myocardial infarction. Mobilization of EPCs in response to musculoskeletal trauma has been studied in patients with closed diaphyseal tibia fractures (Laing et al. 2007). The authors demonstrated that circulating CD34 positive mononuclear cell levels increased seven-fold by d 3 post-injury. These cells were identified as EPCs (bound UEA-1 and incorporated fluorescent DiI-Ac-LDL) which suggests that a systemic provascular response is initiated in response to musculoskeletal trauma. The authors suggested that in addition to the inflammatory response and cytokine release, a competent, responsive endothelial cell population is required to obtain angiogenic effects and healing at a fracture site following musculoskeletal trauma.

Distraction osteogenesis (DO) is a procedure by which controlled displacement of a bone fragment is used to generate large volumes of new bone that have been lost due to trauma, infection or tumour resection (Li et al. 2006). The change in the proportion of EPCs during fracture healing and DO was evaluated (Lee et al. 2008). The authors investigated whether the mobilized EPCs were transferred to the bone regeneration site to participate in re/neo-vascularization. They used rodent tibia fracture and DO models in their study. In the fracture model, EPC levels in peripheral blood circulation increased approximately 6-fold at post-fracture d 3 compared with non-operated control animals. In the DO model, the proportion of EPCs increased significantly on post-osteotomy d 3 compared with controls, and decreased gradually to baseline levels. Interestingly, in the DO group, the EPC count increased significantly during the distraction and consolidation periods compared with animals where no distraction was performed after the osteotomy. The authors reported that the relative blood flow of the fragment proximal to the osteotomy gap began to increase towards the end of the distraction period, and peaked during the consolidation period which was after EPC mobilization induced by the distraction strain. As a part of the study, *ex vivo*-expanded and tagged EPCs were transplanted via an intravenous route to the DO model animals to assess the distribution of the cells. Although only a few of the transplanted cells were found at the bone regeneration site, a distinct dose-dependent relationship was observed between the number of infused transplanted cells and the number of cells identified at the distraction gap (Lee et al. 2008). The authors detected no tagged EPCs in the contralateral non-operated tibia after transplantation. They concluded that the proportion of EPCs increases by signals from bone regeneration sites and that this appears to contribute to re/neo-vascularization and thus to increased blood flow during fracture healing and DO in rodent models.

EPCs appear to increase vascularization and osteogenesis by the secretion of growth factors and cytokines through paracrine mechanisms. Fluorescence-activated cell sorting (FACS) analysis demonstrated that the frequency of BM and peripheral blood (PB) EPCs significantly increased post-fracture (Matsumoto et al. 2008). The EPC-derived re/neo-vascularization at the fracture site was confirmed by double immunohistochemistry for CD31 and Sca1 (stem cell antigen-1). The authors showed that the EPCs contributing to formation of new blood vessels at the fracture site were specifically derived from BM. Systemic administration of PB Green Fluorescent Protein (GFP) (Table 1.1) positive EPCs further confirmed incorporation of the mobilized EPCs into the fracture site for fracture

Table 1.1 Key Features of Green Fluorescent Protein. This table lists the key factors of green fluorescent protein including its origin, role, function, and application.

1. Green fluorescent protein (GFP) is a spontaneously fluorescent protein isolated from coelenterates, such as the Pacific jellyfish (Aequorea Victoria).
2. Its role is to transduce blue chemiluminescence of aequorin into green fluorescent light by energy transfer.
3. GFP can be incorporated into a variety of biological systems in order to function as a marker protein.
4. GFP as a transgenic reporter is a very useful tool to view the expression of proteins in living organisms.
5. GFP has become a significant contributor to the research of monitoring gene expression, localization, mobility, interactions between various membrane and cytoplasmic proteins.

healing. Their findings indicated that fracture may induce mobilization of EPCs from BM to PB and recruitment of the mobilized EPCs into the fracture sites, thereby augmenting re/neo-vascularization during the process of bone healing. In their study, the authors utilized a reproducible animal model of femur fracture and noted severe decrease in local blood flow following the fracture by investigation with laser doppler perfusion imaging. The natural history of this model was found to be relevant to the clinical situation of the common fracture. However, histological results demonstrated that part of neovascularization at the fracture site is independent of vasculogenesis by BM-derived EPCs, suggesting other mechanisms such as paracrine effects of the BM-derived EPCs on resident EPCs and ECs. It has been suggested that EPCs mobilized by signals from sites of injury may contribute to neovascularization and thus new bone formation in fracture healing (Lee et al. 2008). EPCs demonstrate a paracrine effect by which capillary and osteoblast density increase in a dose-dependent manner (Matsumoto et al. 2008). An interesting finding that has been consistent in most of the studies was that the incorporation rate of EPCs in an ischemic tissue model was quite low, or at least not enough to explain the observed increase in re/neo-vascularization. The challenge is to illuminate how such a low number of endothelial stem cells can improve re/neo-vascularization. One possible explanation is that the effectiveness of EPC therapy may be due to both the incorporation of EPCs into newly formed vessels and the release of proangiogenic factors in a paracrine manner.

EPCs for Therapeutic Angiogenesis for Bone Healing

Vascularity at the fracture site and soft tissues surrounding the fracture is a critical component of fracture healing. There is a paucity of pre-clinical data on therapies which promote vascularity at sites of desired bone healing and effective clinical therapies for enhancing fracture vascularity are lacking. As the biology of fracture healing has become better understood and the importance of vascularity recognized, there has been significant interest in angiogenic therapies for fracture healing. The incorporation of strategies for therapeutic angiogenesis may produce more effective therapies for the treatment of bone defects secondary to trauma or non-union, given the impaired vascularity that often accompanies these problems.

EPC's major role in new vessel formation and their ability to proliferate and differentiate into endothelial cells present them as an ideal therapeutic alternative for *ex vivo* expansion and transplantation into ischemic areas. Studies have shown that infusion of peripheral blood–derived EPCs, bone marrow mononuclear cells, or purified CD34+ cells may all demonstrate potential to improve re/neo-vascularization and myocardial function in a variety of animal models with an acute myocardial infarct (Kocher et al. 2001). Their results showed that the degree of re/neo-vascularization induced in the ischemic myocardium was directly related to the numbers of CD34 positive angioblasts homing to the ischemic site. In parallel with growth of larger-sized capillaries accompanying injection of high concentrations of human EPCs, ischemic rat hearts developed prominent islands of regenerating myocytes around the infarct region.

In addition to direct differentiation into endothelial cells, paracrine mechanisms resulting in agniogenic growth factor secretion may be a further mechanism of action for EPC therapy. We demonstrated that EPC-based therapy for a segmental bone defect results in increased VEGF mRNA expression during the early period of fracture repair (Fig. 1.2). These findings demonstrate that EPCs may promote fracture healing by increasing VEGF level and thus stimulate angiogenesis and osteogenesis, a process that is essential for early callus formation and bone regeneration. VEGF is of particular interest because it is a critical regulator of both angiogenesis and osteogenesis during fracture healing. We previously applied fibroblast cell-based, non-viral vector, hVEGF gene therapy to a segmental bone defect in a rabbit model (Li et al. 2009). Our results showed improved fracture healing and increased new blood vessel formation in the VEGF treated defects, suggesting that VEGF secretion alone can stimulate fracture healing. We have also demonstrated that hVEGF gene transfer in rat osteoblasts and fibroblasts not only results in hVEGF production but also increases endogenous rat VEGF production, and osteoblast proliferation *in vitro* (Li et al. 2010). These results demonstrate that VEGF is effective at producing the targeted gene locally, impacts positively on important precursor cells, and ultimately enhances both fracture healing and angiogenesis.

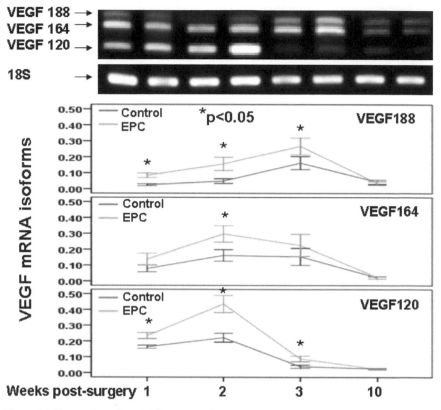

Figure 1.2 Expression of rat VEGF mRNA (RT-PCR). Expression of VEGF mRNA isoforms during endothelial progenitor cells for fracture healing. VEGF120 and VEGF164 levels peaked at two weeks, while VEGF188 levels peaked at 3 wk. All three VEGF isoforms levels were low at 10 wk (unpublished data).

Effects of EPCs and MSCs on Fracture Healing

Recent reports suggest that EPCs derived from peripheral blood contribute to osteogenic differentiation by MSCs *in vitro*, and that MSCs support the proliferation of EPCs and stabilize the formed cellular networks. Physical and biochemical interactions between BM-derived EPCs and MSCs were studied in an *in vitro* co-culture system (Aguirre et al. 2010). Their data suggested that cross-talk occurs between BM- derived EPCs and MSCs through paracrine and direct cell contact mechanisms leading to modulation of the angiogenic response. These interactions between MSCs and EPCs appear to further strengthen the capacity of EPCs to enhance bone healing at a fracture site.

Effects of EPCs alone or in combination with MSCs on early neovascularization and bone healing in a critically-sized defect using a rat model were studied (Seebach et al. 2010). The authors suggested that there is a synergistic effect between EPCs and MSCs, and that the initial stage of neovascularization by EPCs is considered to be crucial for complete bone regeneration. In addition, recent pre-clinical investigation has demonstrated that EPCs resulted in significantly greater evidence of bone healing and blood vessel formation than the MSCs in a rat bone defect model (Nauth et al. 2010).

Musculoskeletal tissue engineering strategies have typically utilized MSCs due to their known ability to differentiate into bone and cartilage-forming cells. Despite demonstrating some success in pre-clinical studies, MSC-based strategies have failed to translate into clinical application. A potential reason for this is that MSC-based constructs fail to address the impaired vascularity at the bone defect or non-union site. The use of EPCs, alone or in combination with MSCs, would seem to address this deficit.

Therapeutic Application of EPCs in Orthopaedic Regenerative Medicine

The investigation and therapeutic application of EPCs in orthopaedic trauma is relatively novel. Our research group and others have very recently reported success with the application of EPCs to critical sized bone defects and non-union models in animal studies (Atesok et al. 2010; Li et al. 2011; Matsumoto et al. 2008; Rozen et al. 2009). Therapy with EPCs has been shown to augment fracture healing and local angiogenesis in animal models of non-union. Moreover, EPCs have been shown to be capable of differentiating into osteogenic cells *in vitro*, and have been shown to be up-regulated in response to orthopaedic trauma in humans (Laing et al. 2007; Bick et al. 2007).

In our previous published works, we assessed the effects of cell-based therapy with EPCs on the healing of a segmental bone defect in the rat femur (Li et al. 2011). In the treatment group, EPCs cultured from the bone marrow of syngeneic rats were impregnated on gelfoam and implanted in the defect. A control group received gelfoam alone in the defect. New bone formation in the EPC treated defects was evident radiographically at 2 to 4 wk after surgery, and osseous union was achieved by 6 to 10 wk. In contrast, the control animals demonstrated very little bone formation and osseous union was not observed in any of the animals (Fig. 1.3). MicroCT evaluation showed significantly more bone formation in the EPC group versus controls (Fig. 1.4). EPC treatment also significantly enhanced the biomechanical properties of critical sized defects in a rat femur model. Biomechanical

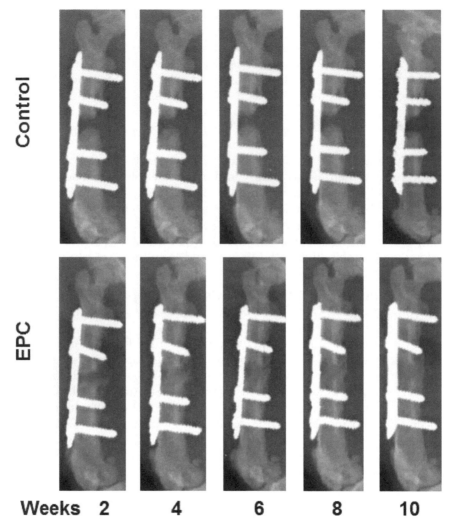

Figure 1.3 Radiographs showing bone healing. Plain radiographs showing the bone healing in EPC treated and control group animals at 2, 4, 6, 8 and 10 wk post fracture. Used with permission from publisher Lippincott Williams and Wilkins.

testing of the samples showed that the EPC treated defects had significantly higher torsional strength and stiffness compared to the control group. The EPC treatment group had complete bridging of the fracture gap and showed a similar stiffness to intact specimens. In contrast, the control group exhibited a weak and compliant behaviour, and was indicative of a fibrous non-union (Fig. 1.5). Quantitative characteristics of fracture healing with the biomechanical properties of the callus were correlated. Pearson's correlation

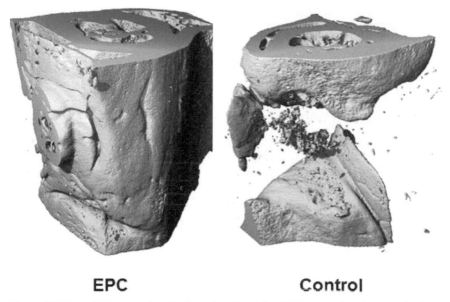

EPC   **Control**

Figure 1.4 Micro-CT images showing bone healing. MicroCT images showing superior bone healing in the EPC treated group compared to negligible bone formation in the saline control group. Used with permission from publisher Lippincott Williams and Wilkins.

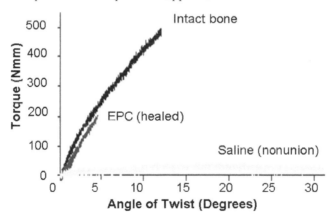

Figure 1.5 Torsional strength. Graph showing the maximum torque versus angle of twist curves for intact bone, EPC treated samples, and saline control samples (the data after maximum torque not being shown). Used with permission from publisher Lippincott Williams and Wilkins.

analysis showed that a number of micro-computed tomographic (micro-CT) parameters were strongly correlated with the biomechanical properties of torsional strength and stiffness (Table 1.2). This finding indicates that, as expected, reconstruction of the bony microarchitecture directly affects

Table 1.2 Correlation between bone architecture microCT parameter and biomechanical properties. Pearson's correlation analysis of strong correlation between micro-CT parameters and biomechanical properties of torsional strength and stiffness: Bone volume (BV), bone surface area (BS), bone surface to bone volume ratio (BS/BV) and trabecular number (Tb.N) had a positive correlation with torsional strength and stiffness. Trabecular spacing (Tb.Sp) was inversely correlated with torsional strength and stiffness. Used with permission from publisher Lippincott Williams and Wilkins.

MicroCT	Torsional Strength (Nmm)		Torsional Stiffness (Nmm/deg)	
parameters	r	*p-value*	r	*p-value*
BV (mm³)	0.817	*0.000*	0.844	*0.000*
BV/TV (%)	0.463	*0.095*	0.564	*0.036*
Conn D (1/mm³)	0.339	*0.236*	0.362	*0.204*
BS (mm)	0.908	*0.000*	0.853	*0.000*
BS/BV (1/mm)	0.780	*0.001*	0.614	*0.019*
Tb.N (1/mm)	0.586	*0.028*	0.625	*0.017*
Tb.Sp (mm)	−0.624	*0.017*	−0.646	*0.013*

the biomechanical properties of healing fractures. In addition, histological analysis showed increased angiogenesis and osteogenesis at the fracture gap in the EPC treated group at 1, 2, 3, and 10 wk (Atesok et al. 2010). Recently, we have used Laser Dopple Flowmetry (LDF) and Laser Doppler Perfusion Imaging (LDPI) to demonstrate enhanced blood flow in both the bone and soft tissues surrounding a segmental bone defect in a rat model (Nauth et al. 2010). These early results suggest that EPCs may be ideally suited for cell-based therapy for fracture healing and angiogenesis.

In a rat tibia fracture model, *ex vivo*-expanded EPCs collected from the spleen in large quantities were compared with those that were homed to the fracture site following intravenous application (Lee et al. 2008). Hence, the local use of *ex vivo* expanded EPCs at a fracture site to enhance new vessel formation and augment bone healing has appeared as a more potent approach. Effects of local treatment with *ex vivo*-expanded EPCs on healing of a critical sized bone defect have been reported recently. *Exvivo*-expanded autologous EPCs were implanted into a wedged-shaped gap platform in sheep tibiae and compared to a control group treated with sham operation (Rozen et al. 2009). Radiographic micro-CT analysis at 12 wk after the procedure revealed complete bridging of the gap in six out of seven animals with better parameters of bone formation in the EPC-transplanted group compared with sham-treated animals where the new bone formation was minimal. Histological analysis of the gap tissue at 12 wk showed dense and massive woven bone formation all throughout the defect in the EPC-transplanted group compared to the control group where the defect was mostly filled with fibrotic scar tissue.

EPC Therapy for other Potential Orthopaedic Applications

Bone tissue engineering is a novel way to repair osseous lesions with cell-free devices or scaffolds seeded with cells. Despite extensive evidence from proof-of-principle studies, bone tissue engineering, particularly the use of scaffolds seeded with cells, has not translated to clinical practice. Similar to other cell-based therapies, EPCs require further study to justify their use as a valuable tissue engineering strategy in orthopaedic surgery. Vascularization remains one of the main obstacles that need to be overcome before large tissue-engineered constructs can be useful in clinical settings. Inability to provide sufficient blood supply in the initial phase after implantation can lead to improper cell integration or cell death in tissue-engineered constructs. EPC therapy represents a promising strategy to increase the success of tissue healing and implant integration.

EPC therapy could be also used for acceleration of graft revascularization and enhancement of tendon-bone osteointegration following Anterior Cruciate Ligament (ACL) reconstruction. This may hasten the revitalization of the tendon graft and integration to the bone which eventually allows earlier functional recovery and return to sport. Matsumoto et al. recently demonstrated that CD34 and CD146 expressing vascular cells exist in human ACL tissues, have the potential for multi-lineage differentiation, and are recruited to the rupture site to participate in the intrinsic healing of injured ACL (Matsumoto et al. 2011).

In a rodent model, Tei et al. studied the effects of locally transplanted human peripheral blood CD34+ cells on the healing of medial collateral ligament (MCL) injury (Tei et al. 2008). Macroscopic, histological, and biomechanical assessments showed significantly enhanced ligament healing in CD34+ cell transplantation group compared with control group. The authors suggested that local transplantation of circulating human CD34+ cells may augment the ligament healing process by promoting a favourable environment through neovascularization.

Articular cartilage is a remarkably durable tissue, it is avascular and thus has a very limited capacity to repair once injured. In fact, cartilage healing requires chondroprogenitor cells from either blood or marrow to enter the damaged region, and thus the reparative process at the chondral level requires angiogenesis and osteogenesis to occur at the level of the subchondral bone. The development of EPCs therapy strategies that can enhance both angiogenesis and osteogenesis mechanisms may be critically important for the prevention and treatment of conditions where the cartilage healing process is impaired. EPC potential in healing of cartilage tissue still needs to be explored.

Clinical applications using EPCs as a cell-based strategy directed to accelerate re/neo-vascularization at a fracture site and to improve bone healing could result in a reduced incidence of delayed or non-union. Endothelial progenitor cells, with their unique features, such as ability to differentiate into endothelial cells and participate in the formation of new blood vessels, and high plasticity, may offer therapeutic alternatives for repair of osteochondral defects, the treatment of avascular necrosis, meniscal tears, and the augmentation of tendon-to-bone healing and ligament repair. Consequently this could bring benefits in terms of health care costs and improved quality of life.

The current literature demonstrates that EPC therapy enhances fracture healing and angiogenesis in animal models. These results add to a growing body of recent literature that suggests that these cells may be an effective therapeutic modality in the treatment of bone defects and non-unions.

Summary Points

- Severe fractures damage blood vessels and disrupt circulation at the fracture site, which can lead to an increased risk of poor fracture healing.
- The unique process of bone regeneration requires a coordinated coupling between osteogenesis and angiogenesis.
- Endothelial progenitor cells are mobilized from bone marrow by growth factors and cytokines from sites of injury through paracrine mechanisms.
- Endothelial progenitor cells promote bone regeneration by stimulating both angiogenesis and osteogenesis through a differentiation process towards endothelial cell lineage and formation of osteoblasts.
- Local endothelial progenitor cell therapy significantly increases fracture healing, restores the biomechanical properties, and increases the local vascular endothelial growth factor mRNA expression.
- There is a synergistic effect between endothelial progenitor cells and mesenchymal stem cells, and the initial stage of neovascularization by endothelial progenitor cells is considered to be crucial for complete bone regeneration.
- Endothelial progenitor cell therapy may offer therapeutic alternatives for regeneration and repair of tissue injury in musculoskeletal system.

Dictionary

- *Endothelial progenitor cell (EPC)*: Progenitor cells from bone marrow to differentiate into endothelial cells, which form blood vessels.
- *Angiogenesis*: new blood vessel formation.
- *Osteogenesis*: new bone formation.
- *Cytokine*: Signalling proteins secreted by cells to play part in cell communication.
- *Vascular Endothelial Growth Factor (VEGF)*: A signalling protein to promote angiogenesis. Recent studies show that it also promotes osteogenesis.
- *Mesenchymal stem cell (MSC)*: A type of multipotent progenitor cell that can self-renew, as well as to differentiate into bone, adipose and cartilage tissues.

List of Abbreviations

ACL	:	Anterior Cruciate Ligament
BM	:	Bone marrow
BMP	:	Bone morphogenetic protein
CD	:	Cluster of differentiation molecule
LDL	:	Acetylated low density lipoprotein
Dil	:	Dioctadecyl– 3,3,3\',3\'-tetramethyl-indocarbocyanine perchlorate
DO	:	Distraction osteogenesis
EC	:	Endothelial cell
eEPC	:	Early endothelial progenitor cell
EPC	:	Endothelial progenitor cell
FACS	:	Fluorescence activated cell sorting
Flk1	:	Fetal liver kinase 1 (VEGF receptor 2)
GFP	:	Green fluorescent protein
LDF	:	Laser Doppler Flowmetry
LDPI	:	Laser Doppler Perfusion Imaging
MCL	:	Medial collateral ligament
Micro-CT	:	Micro-computed tomography
MMP	:	Matrix metalloproteinase
MSC	:	Mesenchymal stem cell
OEPC	:	Outgrowth endothelial progenitor cell
PB	:	Peripheral blood
PCR	:	Polymerase chain reaction

Runx-2	:	Runt-related transcription factor-2
Sca-1	:	Stem cell antigen-1
SDF	:	Stromal-cell derived factor
UEA-1	:	Ulex europaeus agglutinin-1
VEGF	:	Vascular endothelial growth factor
vWF	:	Von Willebrand factor

References

Aguirre, A., J.A. Planell and E. Engel. 2010. Dynamics of bone marrow-derived endothelial progenitor cell/mesenchymal stem cell interaction in co-culture and its implications in angiogenesis. Biochem Biophys Res Commun. 400: 284–291.

Asahara, T., T. Murohara, A. Sullivan, M. Silver, R. van der Zee, T. Li, B. Witzenbichler, G. Schatteman and J.M. Isner. 1997. Isolation of putative progenitor endothelial cells for angiogenesis. Science. 275: 964–967.

Atesok, K., R. Li, D.J. Stewart and E.H. Schemitsch. 2010. Endothelial progenitor cells promote fracture healing in a segmental bone defect model. J Orthop Res. 28: 1007–1014.

Bick, T., N. Rozen, E. Dreyfuss, M. Soudry and D. Lewinson. 2007. Osteogenic differentiation of circulating endothelial progenitor cells. J Bone Min Res. 22: S143–143.

Capobianco, S., V. Chennamaneni, M. Mittal, N. Zhang and C. Zhang. 2010. Endothelial proenitor cells as factors in neovascularization and endothelial repair. World J Cardiol. 2: 411–420.

Doyle, B., P. Metharom and N.M. Caplice. 2006. Endothelial progenitor cells. Endothelium. 13: 403–410.

Fuchs, S., S. Ghanaati, C. Orth, M. Barbeck, M. Kolbe, A. Hofmann, M. Eblenkamp, M. Gomes, R.L. Reis and C.J. Kirkpatrick. 2009. Contribution of outgrowth endothelial cells from human peripheral blood on in vivo vascularization of bone tissue engineered constructs based on starch polycaprolactone scaffolds. Biomaterials. 30: 526–534.

Hueper, W. and M. Russell. 1932. Capillary-like formations in tissue culture of leukocytes. Arch Exp Zellforsch. 12: 407–424.

Hur, J., C.H. Yoon, H.S. Kim, J.H. Choi, H.J. Kang, K.K. Hwang, B.H. Oh, M.M. Lee and Y.B. Park. 2004. Characterization of Two Types of Endothelial Progenitor Cells and Their Different Contributions to Neovasculogenesis. Arterioscler Thromb Vasc Biol. 24: 288–293.

Kocher, A.A., M.D. Schuster, M.J. Szabolcs, S. Takuma, D. Burkhoff, J. Wang, S. Homma, N.M. Edwards and S. Itescu. 2001. Neovascularization of ischemic myocardium by human bone marrow–derived angioblasts prevents cardiomyocyte apoptosis, reduces remodeling and improves cardiac function. Nat Med. 7: 430–436.

Kudo, F.A., T. Nishibe, M. Nishibe and K. Yasuda. 2003. Autologous transplantation of peripheral blood endothelial progenitor cells (CD34+) for therapeutic angiogenesis in patients with critical limb ischemia. Int Angiol. 22: 344–348.

Laing, A.J., J.P. Dillon, E.T. Condon, J.T. Street, J.H. Wang, A.J. McGuinness and H.P. Redmond. 2007. Mobilization of endothelial precursor cells: systemic vascular response to musculoskeletal trauma. J Orthop Res. 25: 44–50.

Lee, D.Y., T.J. Cho, J.A. Kim, H.R. Lee, W.J. Yoo, C.Y. Chung and I.H. Cho. 2008. Mobilization of endothelial progenitor cells in fracture healing and distraction osteogenesis. Bone. 42: 932–941.

Li, R., M. Saleh, L. Yang and L. Coulton. 2006. Radiographic classification of osteogenesis during bone distraction. J Orthop Res. 24: 339–347.

Li, R., D.J. Stewart, H.P. von Schroeder, E.S. Mackinnon and E.H. Schemitsch. 2009. Effect of cell-based VEGF gene therapy on healing of a segmental bone defect. J Orthop Res. 27: 8–14.

Li, R., C. Li, A. Nauth, M.D. McKee and E.H. Schemitsch. 2010. Effect of human vascular endothelial growth factor gene transfer on endogenous vascular endothelial growth factor mRNA expression in a rat fibroblast and osteoblast culture model. J Orthop Trauma. 24: 547–551.

Li, R., K. Atesok, A. Nauth, D. Wright, E. Qamirani, C.M. Whyne and E.H. Schemitsch. 2011. Endothelial progenitor cells for fracture healing: a microcomputed tomography and biomechanical analysis. J Orthop Trauma. 25: 467–471.

Matsumoto, T., Y. Mifune, A. Kawamoto, R. Kuroda, T. Shoji, H. Iwasaki, T. Suzuki, A. Oyamada, M. Horii, A. Yokoyama, H. Nishimura, S.Y. Lee, M. Miwa, M. Doita, M. Kurosaka and T. Asahara. 2008. Fracture induced mobilization and incorporation of bone marrow-derived endothelial progenitor cells for bone healing. J Cell Physiol. 215: 234–242.

Matsumoto, T., S.M. Ingham, Y. Mifune, A. Osawa, A. Logar, A. Usas, R. Kuroda, M. Kurosaka, F.H. Fu and J. Huard. 2011. Isolation and Characterization of Human Anterior Cruciate Ligament-Derived Vascular Stem Cells. Stem Cells Dev. Aug 17 [Epub ahead of print].

Nauth, A., R. Li and E.H. Schemitsch. 2010. Endothelial progenitor cells for fracture healing and angiogenesis: A comparison with Mesenchymal Stem Cells. The Annual Meeting of the American Academy of Orthopaedic Surgeons. Las Vegas, NV.

Rozen, N., T. Bick, A. Bajayo, B. Shamian, M. Schrift-Tzadok, Y. Gabet, A. Yayon, I. Bab, M. Soudry and D. Lewinson. 2009. Transplanted blood-derived endothelial progenitor cells (EPC) enhance bridging of sheep tibia critical size defects. Bone. 45: 918–924.

Seebach, C., D. Henrich, C. Kähling, K. Wilhelm, A.E. Tami, M. Alini and I. Marzi. 2010. Endothelial progenitor cells and mesenchymal stem cells seeded onto beta-TCP granules enhance early vascularization and bone healing in a critical-sized bone defect in rats. Tissue Eng. 16: 1961–1970.

Smadja, D.M., I. Bieche and J. Silvestre. 2008. Bone morphogenic proteins 2 and 4 are selectively expressed by late outgrowth endothelial progenitor cells and promote neoangiogenesis. Arteriosclerosis, Thrombosis, and Vascular Biology. 28: 2137–2143.

Tei, K., T. Matsumoto, Y. Mifune, K. Ishida, K. Sasaki, T. Shoji, S. Kubo, A. Kawamoto, T. Asahara, M. Kurosaka and R. Kuroda. 2008. Administrations of peripheral blood CD34-positive cells contribute to medial collateral ligament healing via vasculogenesis. Stem Cells. 26: 819–830.

CHAPTER 2

Transplanted Bone Marrow Stromal Cells and Bone Tissue Regeneration

Qisheng Tu[a] and Jake Chen[b,*]

ABSTRACT

Bone defects, especially those in the oral and craniofacial region, represent serious public health issues and are a major health concern worldwide. This reality has stimulated the development of bone tissue engineering, which aims at supplying novel and effective bone replacement materials and achieving the ultimate therapeutic goal of regenerating the damaged bone tissue to a normal or pre-disease state. For successful bone tissue-engineered regeneration, several key elements are required, including mesenchymal stem cells (MSCs, or progenitor cells), bioactive factors, and supporting materials such as scaffolds that occupy the wound boundaries. In the last several decades, various stem cell treatments have been applied in bone tissue engineering. These tissue-engineering approaches attempt to heal bone lesions above a critical size, by using resorbable scaffolds supplemented with regeneration-competent cells, such as embryonic stem (ES) cells derived from the inner cell mass of blastocyst stage embryos or adult bone-marrow derived mesenchymal stem cells (BMSCs). Due to ethical

Division of Oral Biology, Tufts University School of Dental Medicine, One Kneeland Street, Boston, MA, 02111, USA.
[a]E-mail: qisheng.tu@tufts.edu
[b]E-mail: jk.chen@tufts.edu
*Corresponding author

List of abbreviations given at the end of the text.

controversy regarding the use of embryonic ES cells, the research focus has shifted towards *ex vivo* expansion of adult BMSC for its use in bone regeneration studies.

BMSCs are adherent cells of non-hematopoietic origin that proliferate and exhibit many of the characteristics attributed to bone marrow stroma *in vivo*. They are capable of self-renewal and can differentiate into several phenotypes including bone, cartilage, and adipocytes. Evidence shows that BMSCs not only serve as an essential stem cell source for bone renewal and remodeling, but also actively participate in local bone regeneration. Moreover, BMSCs can be readily isolated, manipulated and reproduced *in vitro*, creating an ideal strategy to regenerate defected bone. In this chapter, we discuss these new discoveries and strategies obtained from previous published results and from our own laboratory, which were adopted to facilitate the therapeutic use of BMSCs.

Introduction

Bone defects, especially those in the oral and craniofacial region, represent serious public health issues and are a major health concern worldwide. Trauma, surgical treatments for tumors, osteodegenerative diseases, infections and congenital diseases cause bone damage and injury that interfere with normal function and are often disfiguring, which may affect the patients' self-esteem and cause them to withdraw from social and public life. An estimated 1,600,000 bone grafts are performed every year in the United States to regenerate bone tissue lost to trauma and disease, of which 6 percent (96,000) are craniomaxillofacial in nature (Einhorn 2003). Globally, these numbers are believed to be significantly greater, which result in a shortage in the musculoskeletal donor tissue traditionally used to reconstruct the damaged bone. This reality has stimulated the development of bone tissue engineering, which aims to supply novel and effective bone replacement materials and achieve the ultimate therapeutic goal to regenerate the damaged bone tissue to a normal or pre-disease state. Such therapeutic goals require differentiation of reparative cells in a precise temporal and spatial manner to form specific types of hard and soft connective tissues (Mao et al. 2006). A successful reconstruction is characterized by the restoration of both form and function, minimizing patient morbidity from the treatment, and integrating the patient back into society.

Cell Sources Applied in Bone Tissue Engineering

For successful bone tissue-engineered regeneration, several key elements are required, including MSCs (or progenitor cells), bioactive factors, and implanted materials that occupy the wound boundaries. In the past several decades, various stem cells have been applied in bone tissue engineering including embryonic stem cells (ESCs), bone marrow stromal cells (BMSCs), umbilical cord blood-derived mesenchymal stem cells (UCB-MSCs), adipose tissue-derived stem cells (ADSCs), muscle-derived stem cells (MDSCs), dental pulp stem cells (DPSCs), dental follicle cells (DFCs), and the recently invented induced pluripotent stem cells (iPS cells) (Li et al. 2008; Tu et al. 2007; Xu et al. 2009; Ye et al. 2011; Zhang et al. 2011). However, obtaining and using human ESCs is ethically controversial, while the quality of ADSCs and MDSCs regarding their proliferation rates or differentiation potential may be lower than that of BMSCs. In addition, the practical use of UCB-MSCs and dental stem cells may be problematic, as the availability of these cells are restricted to specific points in time and can be isolated only under specific circumstances. For example, dental follicle cells (DFCs) are available only from patients during wisdom tooth eruption, usually between 15 and 28 yr of age. Unfortunately, during adolescence, sophisticated treatments are performed rarely. Special storage facilities are needed to store young cells for future use when the patient gets old. Although iPS cells reprogrammed from somatic cells are an excellent option for regenerative medicine, the research on the application of these cells in bone tissue engineering is still at the beginning stage. In contrast, the use of BMSCs in bone tissue engineering does not bring up any ethical concerns, and BMSCs are frequently accessible for treatment. Moreover, numerous independent research groups have successfully applied BMSCs in bone tissue engineering.

Biological Features of Bone Marrow Stromal Cells

In addition to suspended hematopoietic cells, whole bone marrow cells contain a population of plastic-adherent cells after placed on plastic culture dishes. These plastic-adherent components, with an approximate concentration of 10^{-5} in bone marrow, form round-shaped colony units composed of fibroblastoid cells, and were designated as Colony Forming Unit–fibroblasts (CFU-f) by the researchers. CFU-f derived from bone marrow can be induced to differentiate towards bone forming cells in the presence of transitional epithelium, or undergo spontaneous bone formation in diffusion chambers (Friedenstein et al. 1976). Further investigations demonstrated these cells to have multipotential to differentiate into osteoblasts, chondrocytes, adipocytes, and myoblasts, which all the cell lineages were derived from embryonic mesenchymal components (Phinney et al. 1999). For this reason, these nonhematopoietic, multipotent cells are referred to as mesenchymal stem cells (MSCs) by some researchers. More frequently, these cells are named as bone marrow stromal cells (BMSCs) based on the fact that they are obtained from the supporting structures found in bone marrow.

After their initial adherence to plastic, BMSCs are heterogeneous. Although most of the hematopoietic cells disappear as the primary cultures are maintained for 2 or 3 wk, long-term BMSC cultures still produce a complex mixture of adherent stromal elements including fibroblasts, adipocytes, smooth muscle cells, and macrophages. These bone marrow stromal elements are positive for the following genes: type IV collagen, laminin, vimentin, CD10, muscle actin, STRO-1, yet are negative for CD45, Mac-1, and HLA-DR. Distinct from the primitive hematopoietic progenitors, BMSC precursors that can proliferate and are capable of multi-lineage differentiation have the following phenotypes: positive for CD105, LNGFR, HLA-DR, CD10, CD13, CD90, STRO-1, and bone morphogenetic protein receptor type IA (BMPRIA), and are negative for CD14, CD117, and CD133. However, researchers found that BMSCs from different strains of inbred mouse models required special media for optimal growth, rates of propagation, and presence of surface epitopes, with the major differences being in the expression of CD34. In 2006, the following minimal criteria for identification of human multipotent mesenchymal stromal cells was proposed by the International Society for Cell Therapy: plastic-adherent in standard culture conditions; showing the following expression pattern of surface molecules: CD73+, CD90+, CD105+, CD34-, CD45-, HLA-DR-, CD14- or CD11b-, CD79α- or CD19-; *in vitro* differentiation into osteoblasts, adipocytes and chondroblasts (Dominici et al. 2006).

Multipotency of Bone Marrow Stromal Cells

BMSCs have been considered adult stem cells based on the fact that they maintain the capacity for self-renewal and are able to differentiate, when induced by the appropriate chemical or mechanical signals, into chondrocytes, osteoblasts, adipocytes, myoblasts and possibly neuron-like cells (Fig. 2.1).

Osteogenic differentiation of BMSCs

Osteogenic differentiation in human BMSCs, characterized by morphological transformation of these cells from an elongated to a more cuboidal shape and increased alkaline phosphatase (ALP) activity, can be induced by dexamethasone treatment. Expression of osteocalcin, another molecular marker of osteogenic differentiation, can be enhanced by the co-addition of 1,25-dihydroxyvitamin D3 or bone morphogenetic protein-2 (BMP-2) with dexamethasone. Currently, the classic method for osteogenic induction in BMSCs *in vitro* involves culture medium supplemented with dexamethasone,

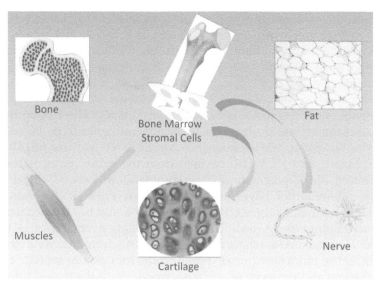

Figure 2.1 Schematic drawing of BMSCs differentiation. (The figure is from Dr. Chen's laboratory, Tufts University).

ascorbic acid, and beta-glycerophosphate. After osteogenic induction using this method, the cells demonstrated the cuboidal shape of osteoblasts and demonstrated expressions of alkaline phosphatase and osteocalcin. Four weeks later, the BMSCs formed mineral nodules which were positively stained by alizarin red and von Kossa techniques (Tu et al. 2006).

Adipogenic differentiation of BMSCs

Interestingly, although both adipogenic and osteogenic differentiation can be induced in BMSCs by the presence of fetal calf serum and dexamethasone (10^{-8} M), the time points when the dexamethasone is added to the culture determine which differential pathway would predominate in the BMSC cultures. The differentiation of BMSCs into adipocytes or osteoblasts is also affected by the inductive chemicals. For example, adipogenic differentiation predominates over osteogenic differentiation in BMSCs when dexamethasone is present in the secondary culture but not in the primary culture. However, if 1,25-dihydroxyvitamin D3 (10^{-8} M) is combined with dexamethasone in the secondary culture, the differentiation of BMSCs into adipocytes is inhibited, whereas osteogenic differentiation is enhanced. Therefore, adipocytes and osteoblasts not only share common precursor cells, but also show an inverse relationship between their differentiation pathways. The molecular basis for the inverse relationship between osteogenic differentiation and adipogenic differentiation in BMSCs has not been fully investigated. It was reported that glucocorticoid receptor signaling supports adipogenesis but inhibits osteogenesis by reducing c-Jun expression and BMSC proliferation. In contrast, platelet-derived growth factor (PDGF) signaling increases JNK/c-Jun activity, which in turn favors osteogenic differentiation instead of adipogenesis (Carcamo-Orive et al. 2011).

Chondrogenic differentiation of BMSCs

Chondrogenesis in BMSCs can be induced in culture medium supplemented with 100 nM dexamethasone and 10 ng/ml transforming growth factor-β (TGF-β). Within 14 d, cells secreted chondrogenic specific genes including type II collagen, aggrecan, and anionic proteoglycans. The cells develop a matrix-rich morphology, which can be positively stained with toluidine blue, indicating an abundance of glycosaminoglycans. BMSCs can be further differentiated to hypertrophic chondrocytes by the addition of 50 nM thyroxine, the withdrawal of TGF-β, and the reduction of dexamethasone concentration to 1 nM (Mackay et al. 1998).

Myogenic differentiation of BMSCs

After being exposed to 5-azacytidine, a compound that can convert rat embryonic fibroblastic cells into myoblasts, BMSCs showed myogenic phenotypes characterized by long, multinucleated myotubes (Haghani et al. 2011). Exposure of mouse BMSCs to amphotericin B also produced a network of multinucleated myotubes which rhythmically contract when exposed to heat.

Neurogenic differentiation of BMSCs

Besides osteogenic, chondrogenic, adipogenic and myogenic differentiation, BMSCs are reported to be induced to overcome their mesenchymal commitment and differentiate into neuron-like cells. In an *in vivo* study, BMSCs were injected into the lateral ventricle of neonatal mice. The exogenous BMSCs migrated throughout the forebrain and cerebellum without disruption to the host brain architecture. Moreover, some BMSCs expressed glial fibrillary acidic protein and differentiated into mature astrocytes. These injected cells may also differentiated into neurons. *In vitro* neurogenic differentiation in rat and human BMSCs were induced using neuronal induction media composed of DMEM/1–10 mM β-mercaptoethanol (BME) or DMEM/2% dimethylsulfoxide (DMSO)/200 mM butylated hydroxyanisole (BHA). Shortly after the exposure of BMSCs to the neuronal induction media, the cells began to exhibit a neuronal phenotype including neuronal morphological characteristics and expressions of neuron-specific enolase, NeuN, neurofilament-M, nestin, trkA, and tau. However, these neuron-like cells were found to lack voltage-gated ion channels which are necessary for generation of action potentials and may not be defined as true neurons. Nevertheless, the researchers found that BMSCs could form physical, nerve fiber-permissive tissue bridges across spinal cord defect areas which may significantly benefit long-term functional improvement (Hofstetter et al. 2002). In a recent study, scaffolds with interconnected pores of inverted colloidal crystal were prepared with the pore surfaces grafted with two laminin-derived peptides. The controlled topography of the scaffold structure and surface LDP considerably promoted the survival/ proliferation of the BMSCs. In addition, induction with neuron growth factor was found to promote the differentiation of BMSCs towards mature neurons in these scaffolds (Kuo and Chiu 2011).

In vivo Functions of Bone Marrow Stromal Cells

Stem cell niche for hematopoietic stem cells (HSCs)

To maintain a stable hematopoietic system, some of the hematopoietic stem cells (HSCs) serve as a deeply quiescent reserve pool, some proliferate to maintain their numbers, and others differentiate into all the lineages of the blood and immune system to replenish the short-lived mature cells. The crucial decision among these biological pathways is balanced by the regulatory molecules provided by the particular microenvironment in which HSCs reside (Lymperi et al. 2010). This microenvironment, which is highly organized and controls HSC homeostasis was named as the stem cell "niche" by Schofield. Anklesaria et al. (1987) reported that hematopoietic recovery from total body irradiation could be achieved after transplantation of a bone marrow stromal cell line. Moreover, BMSCs have been considered as a prerequisite for *in vitro* expansion of HSCs because HSCs may differentiate rapidly *in vitro* and lose self-renewal capacity without the presence of BMSCs. As a heterologous cell group, BMSCs are located in a specific location in close proximity to the HSCs and are essential in supporting hematopoiesis. Therefore as the stem cell niche for HSCs, BMSCs secrete physiological molecules to regulate the stem cell function, and provide structural support and nutritional support for the HSCs.

Cell reservoir to replace mature cells undergoing apoptosis

Another biological function of BMSCs is to serve as a continuing source of progenitor cells for a variety of mesenchymal tissues. After being injected into irradiated mice, BMSCs accounted for 1.5–12 percent of the cells in bone, cartilage, and lung in addition to marrow and spleen, demonstrating that BMSCs can serve as long-lasting precursors for mesenchymal cells in bone, cartilage, and lung. *In vivo* experiments have also demonstrated that after intravenous infusions into nonablated recipients, BMSCs engraft and differentiate into osteoblasts capable of bone matrix formation. These exogenous BMSCs-derived osteoblasts were then encapsulated within the bone lacunae and terminally differentiated into osteocytes. Based on the fact that osteocytes are terminally differentiated cells surrounded by the bone matrix, the researchers believed that these osteocytes could not have been transplanted during the original infusion but must be the progeny of the engrafted BMSCs. Transplanted cells were also detected as flattened bone lining cells on the periosteal bone surface (Nilsson et al. 1999). All of these results strongly indicated that BMSCs are actively involved in the renewal of local bone cells. In another study, the researchers infused BMSCs into recipient mice and found that systemically transplanted BMSCs could differentiate into mature osteoblasts and osteocytes, and support reconstitution of hematopoiesis in radiation-ablated mice (Hou et al. 1999). Although these results did not establish that BMSCs are the only source of progenitor cells for mesenchymal tissues, they demonstrated that BMSCs can make important contributions.

Tissue repair and regeneration

The possibility that nonhematopoietic stem cells residing in bone marrow participate in local wound healing were first suggested by the German pathologist Cohnheim (1867). In this study, Cohnheim injected an insoluble analine dye into the veins of animals and then observed the appearance of dye-containing cells in wounds they created at a distal site. The stained cells included not only inflammatory cells but also cells that had a fibroblast-like morphology and were associated with thin fibrils. He concluded that most of the stained cells appearing in the wounds came from the bloodstream, and, by implication, from bone marrow. Although Cohnheim's work lacked convincing evidence, these findings raised the possibility that bone marrow may be the source of stem cells, which are indispensable in the normal wound repair process.

To answer the question whether the infused BMSCs or their progeny participate in local wound healing and bone regeneration, BMSCs were isolated from a unique transgenic mouse model generated in our laboratory. This mouse model, named BSP-Luc/ACTB-EGFP mice, were double-labeled with a luciferase reporter gene driven by a mouse bone sialoprotein (BSP) promoter and enhanced green fluorescent protein (EGFP) driven by a β-actin promoter and a cytomegalovirus enhancer (Li et al. 2008). The integration of both reporters, luciferase and EGFP, made it feasible to distinguish transplanted BMSCs and their progeny from host cells residing in the wound sites. In short, EGFP staining could be used to track the fate and migration of the transplanted BMSCs, whereas luciferase staining could serve as a marker for osteogenic differentiation of the transplanted BMSCs because the luciferase expression was switched on as these cells differentiated into osteoblasts.

In this study, we provided direct evidence that transplanted BMSCs engrafted from peripheral circulation, repopulated the bone defect sites and participated in the regeneration process (Li et al. 2008). Moreover, the reporter genes were first detected 2 wk after the bone defects were established (7 wk after transplantation) and persisted throughout the experimental period (11 wk after transplantation) (Fig. 2.2). Our results also suggested that under certain local conditions, systemically recruited BMSCs may play a major role in bone regeneration when compared with other local cell populations. Furthermore, during prolonged regenerative processes, there may be a requirement for recruitment of additional blood-borne BMSCs to compensate for a limitation of available local stem cells.

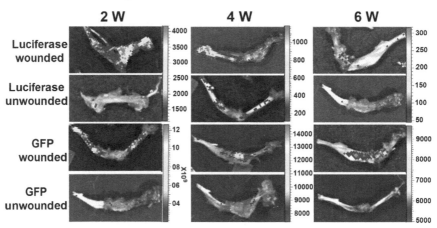

Figure 2.2 Imaging of the kinetic expressions of luciferase and GFP by Xenogen IVIS imaging system in the mouse, long bone defects of an irradiated mouse with BMSCs transplantation which survived until 6 wk after surgery. Bones were isolated from both wounded and unwounded sides of recipient mice sacrificed at 2, 4, and 6 wk after surgery. (The figure is from Dr. Chen's laboratory, Tufts University).

Application of Bone Marrow Stromal Cells in Bone Tissue Engineering

Evidence shows that BMSCs not only serve as an essential stem cell source for bone renewal and remodeling, but also actively participate in local bone regeneration. Moreover, BMSCs can be readily isolated, manipulated, and reproduced *in vitro*, making these cells an ideal strategy to be applied in bone tissue engineering to regenerate bone defects. Indeed, when transplanted subcutaneously into immunocompromised mice, BMSCs formed lamellar bone containing osteocytes and surface-lining osteoblasts, surrounded by a fibrous vascular tissue with active hematopoiesis and adipocytes. However, when BMSCs are applied in bone tissue engineering, appropriate strategies should be adopted to facilitate the therapeutic use of BMSCs. These strategies include local transplantation of BMSCs seeded in supporting biomaterial scaffolds, local transplantation of gene-engineered BMSCs and systemic infusion of BMSCs.

Local Transplantation of BMSCs Seeded in Supporting Biomaterial Scaffolds

In earlier studies, the application of BMSCs in bone tissue engineering was mostly in the form of site-directed delivery of BMSCs in an appropriate carrier vehicle to promote healings of local bone defects. Given the increasing clinical need, biomaterials intended for bone tissue engineering were no longer designed to be "bio-inert". In contrast, the research efforts of the scientists have focused on the development of "bioactive" materials that can integrate with stem cells. Ideally, the biomaterial carrier vehicle applied in bone tissue engineering should be biodegradable, osteoinductive (capable of promoting osteogenic differentiation of the seeded progenitor cells), osteoconductive (capable of supporting bone growth and encouraging in-growth of the surrounding bone), capable of osseointegration, and provide mechanical strength and integrity until newly formed bone tissues can maintain function. Thus far, various biomaterials have been investigated, which can be roughly divided into three categories: natural materials, bioactive inorganic materials, and synthetic polymers.

Natural materials

Natural materials that can be used as scaffolds for supporting BMSCs in bone tissue engineering include collagen, gelatin, albumin, cellulose, hyaluronate, chitin, silk, and alginate. These materials provide innate biological information to guide cell attachment. In one study, BMSCs were dispersed in a type-I collagen gel and transplanted into a large, full-thickness defect in the weight-bearing surface of the medial femoral condyle created in rabbits. The researchers found that the transplanted BMSCs differentiated into articular cartilage and subchondral bone, and the repair tissues appear to undergo the same developmental transitions that originally led to the formation of articular tissue in the embryo. Twenty-four weeks after transplantation, the subchondral bone was completely repaired, without loss of overlying articular cartilage (Wakitani et al. 1994). However, most of the natural materials lacked mechanical strength to provide the necessary resistance for the bone tissue to function normally. Furthermore, these materials have the potential risk of immunogenicity and disease transmission.

Bioactive inorganic materials

Having a similar composition to the mineral phase of bone, bioactive inorganic materials include tricalcium phosphate, hydroxyapatite, bioactive glasses, etc. After implantation, these bioactive materials can rapidly produce a bioactive hydroxycarbonated apatite layer and bind to the surrounding biological tissues. Moreover, these materials can release bioactive ions to enhance cell differentiation and osteogenesis by activating gene transduction pathways. Bioactive glasses can also significantly decrease the plasma membrane stiffness of BMSCs and stimulate BMSCs spreading. *In vivo* experiments further demonstrated that a porous bioactive inorganic scaffold using BMSCs as seed cells is a promising procedure for treatment of large bone defects by enhancing osteogenesis and osteointegration (Khadka et al. 2011). Bioactive inorganic materials show various resorption rates *in vivo* which can be adjusted by changing their composition. While crystalline hydroxyapatite can persist for years after implantation, other calcium phosphates show a faster resorption rate and less resistance to mechanical loading. The brittle nature of bioactive inorganic materials limits their application in load-bearing defect sites.

Synthetic polymers

Synthetic polymers include polyfumarates, polylactic acid (PLA), polyglycolic acid (PGA), copolymers of PLA and PGA (PLGA), and polycaprolactone. They can be easily processed to form three-dimensional scaffolds with different pore network and surface characteristics using various techniques such as porogen leaching, gas foaming, supercritical fluid processing, and three-dimensional printing. It was reported that BMSCs penetrated into the center of scaffolds and began proliferating shortly after seeding into the composite scaffolds composed of synthetic and natural materials, suggesting that fibrous scaffolds made of biomimetic blends of natural and synthetic polymers are useful for tissue engineering (Li et al. 2006). Another research group reported that BMSC viability remains high up to 14 d after encapsulated in a synthetic, degradable poly[poly(ethylene glycol)-co-cyclic acetal] (PECA) hydrogel. Furthermore, BMSCs embedded in PECA hydrogels undergo osteogenic differentiation, which indicate that the PECA hydrogels can be utilized as scaffolds for regeneration of bone tissues (Kaihara et al. 2009).

Local Transplantation of Gene-engineered BMSCs

A second strategy for the application of BMSCs in bone engineering is to introduce osteogenic genes into the BMSCs before transplanting these cells into bone defect sites. Overexpression of these osteoinductive factors enhances and maintains a robust osteoblastic differentiation capacity in BMSCs, and therefore drastically promotes bone regeneration.

Bone morphogenetic proteins (BMPs)

BMPs are a member of the transforming growth factor-β 1 (TGF-β1) superfamily, and BMP-2, BMP-4, and BMP-7 are reported to promote robust new bone formation both *in vitro* and *in vivo*. In a recent study, BMSCs transduced with adenovirus AdBMP-2 were combined with premineralized silk scaffolds and transplanted into mandibular bone defects. Eight weeks after surgery, bone defects treated with BMP-2 gene modified BMSCs demonstrated new enhanced bone formation and higher bone mineral densities (BMD) (Jiang et al. 2009). In another study, BMSCs infected with adenoviral vector containing BMP-7 gene (AdBMP7) were used to enhance bone regeneration in critical-sized femoral defects created in a goat model. The researchers found that new bone formation in bone defects transplanted with BMP-7 overexpressing BMSCs was more prominent. In addition, the mechanical property of the newly formed bone was restored 3 mon earlier in the BMP-7 group than in the control group.

Runt-related gene 2 (Runx2)

Runx2 is a transcription factor identified as a "master gene" required for osteoblastic differentiation. In Runx2 gene knockout mice, there is an almost complete lack of mineralized bone, and mutations in the human Runx2 gene locus have been identified to correlate with cleidocranial dysplasia (CCD), an autosomal dominant disorder characterized by a short stature, varying degrees of bone loss in the clavicles, open cranial fontanels, etc. (Otto et al. 1997). *In vitro* studies showed that Runx2 can upregulate expression levels of bone matrix genes. To investigate the effect of Runx2 on bone formation, adenoviruses encoding mouse Runx2 gene was used to transduce primary cultured BMSCs. The transduced BMSCs showed increased levels of bone matrix proteins and enhanced matrix mineralization *in vitro*. Furthermore, BMSCs transduced with Runx2 adenoviruses significantly promoted wound healing and bone regeneration after being transplanted into the critical-sized calvarial bone defects created in BALB/c mice (Zheng et al. 2004). Another study further investigated the roles of two major N-terminal isoforms of Runx2, Pebp2alphaA and Til-1, in bone regeneration. In this study, overexpression of the two isoforms both induced rapid and marked osteoblast differentiation, with Til-1 being more effective *in vitro*. After being transplanted subcutaneously or into bone defect sites in Fischer rats, Til-1 overexpressing cells showed marked osteogenic capacity *in vivo* (Kojima and Uemura 2005).

Osterix (Osx)

Osx is a zinc-finger-containing transcription factor which is critical in osteoblast differentiation and bone formation. In Osx-null mice, no bone formation occurs as cells in the periosteum and in the condensed mesenchyme of membranous skeletal elements fail to differentiate into osteoblasts. However, these cells do express Runx2 while Osx is not expressed in Runx2 knockout mice (Nakashima et al. 2002). Therefore, Osx acts downstream of Runx2 and induces differentiation of preosteoblasts into fully functional osteoblasts. Overexpression of Osx has been shown to be sufficient to guide differentiation of murine embryonic stem cells towards the osteoblastic lineage *in vitro*. We also showed that Osx enhances *in vitro* proliferation and osteogenic differentiation of BMSCs (Tu et al. 2006). In an *in vivo* study, overexpression of Osx in mouse BMSCs was achieved using retroviral infection. These Osx-overexpressing BMSCs were then transplanted into 4-mm calvarial bone defects in adult mice using type I collagen sponge as a carrier vehicle. The results showed that implantation of Osx-transduced BMSCs resulted in 85 percent healing of calvarial bone defects as detected using radiological analyses (Fig. 2.3). Histological examination also demonstrated that the amount of new bone formation in bone defects transplanted with osx-overexpressing BMSCs is five times higher than that in bone defects transfected with control BMSCs. These results indicate that *ex vivo* gene therapy of Osx is a useful therapeutic approach in regenerating adult bone tissue (Tu et al. 2007).

Special AT-rich sequence-binding protein 2 (SATB2)

As a nuclear matrix protein, SATB2 activates gene transcription through binding to nuclear matrix-attachment regions (MARs), modifies the chromatin structure, and thus plays an important role in integrating genetic and epigenetic signals. In 2003, *SATB2* was identified as the cleft palate gene on 2q32–33, and haploinsufficiency of *SATB2* was reported to affect multiple systems in humans. An individual with a *de novo* germline

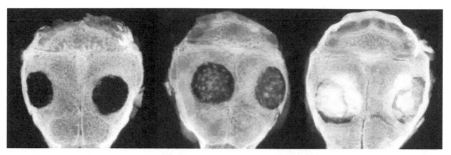

Figure 2.3 Osx overexpression in bone marrow stromal cells stimulates healing of critical-sized defects in murine calvarial bone. Representative images of X-ray radiograph of calvarial defects 5 wk after BMSCs implantation were shown. The defect treatment from the left to the right: Collagen alone; Collagen + RACS vector transduced BMSCs; Collagen + RCAS-OSX transduced BMSCs. (The figure is from Dr. Chen's laboratory, Tufts University).

nonsense in *SATB2* was reported to exhibit generalized osteoporosis, profound mental retardation, and craniofacial dysmorphism including cleft palate, mandibular hypoplasia, and protruding incisors (Leoyklang et al. 2007). Animal studies confirmed the essential role of SATB2 in proper facial patterning of the embryo and in normal bone development, and *Satb2-/-* embryos showed multiple craniofacial defects including a significant truncation of the mandible and a cleft palate (Dobreva et al. 2006). *Satb2-/-* mice also exhibited defects in osteoblast differentiation and function, which consequently delayed bone formation and mineralization (Dobreva et al. 2006).

In a recent study, we transduced SATB2 into murine BMSCs, and found that SATB2 significantly increased expression levels of bone matrix proteins, osteogenic transcription factors, and a potent angiogenic factor, vascular endothelial growth factor (VEGF). We also transplanted SATB2-overexpressing BMSCs into mandibular bone defects, and observed that more transplanted BMSCs undergo osteogenic differentiation in bone defects transplanted with SATB2-transduced BMSCs. New bone formation was consequently accelerated by SATB2-overexpressing BMSCs (Fig. 2.4). Together with previous studies demonstrating that SATB2 plays pivotal roles in both skeletal patterning and osteoblast differentiation, our results suggested that SATB2 can be used as an ideal bioactive factor to overcome the hurdles in craniofacial and dental regeneration by providing functional bone tissue with natural morphology and physiological properties (Zhang et al. 2011).

We also used SATB2-modified BMSCs to accelerate the process of osteogenesis and improve osseointegration of dental implants. To this end, SATB2-overexpressing BMSCs were locally administrated before titanium implants were inserted into the femurs of mice. Histological analysis showed elevated bone mineral density and increased active osteoprogenitor cells

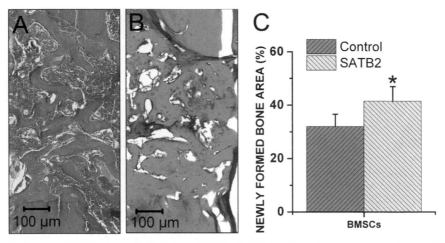

Figure 2.4 SATB2-overexpression in BMSCs facilitates bone tissue regeneration and mineralization in bone defects. H&E stained sections showing bone defects transplanted with control BMSCs (A) or SATB2-overexpressing BMSCs (B). (C) Histomorphometric analysis showed an elevated newly formed bone area in bone defects treated with SATB2-overexpressing BMSCs. Data were represented as mean±SEM. *p<0.05, SATB2 group *vs* control group. (The figure is from Dr. Chen's laboratory, Tufts University).

at the bone-implant contact area in the SATB2 group when compared with the control group. These results showed that overexpression of SATB2 in transplanted BMSCs can accelerate osseointegration of titanium dental implants through up-regulating expression levels of osteogenic transcriptional factors and promoting osteogenic differentiation (unpublished data; Fig. 2.5).

Systemic Infusion of BMSCs

Alternatively, BMSCs, with or without osteogenic gene modification, can be infused systemically to enhance bone regeneration. This represents an important therapeutic strategy especially for the treatment of genetic bone disorders or other bone diseases affected multiple body parts such as osteoporosis and multiple myeloma.

After being infused systemically, BMSCs engraft not only to the bone marrow of the recipients, but also in multiple sites such as bone, cartilage, and lung. In a study, three children with severe osteogenesis imperfecta were treated with allogeneic bone marrow transplantation. Three months after transplantation, new bone formation was observed in the trabecular bone with an increased body bone mineral content. It was suggested that the improvement in the bone formation possibly resulted from the engraftment of functional mesenchymal progenitor cells.

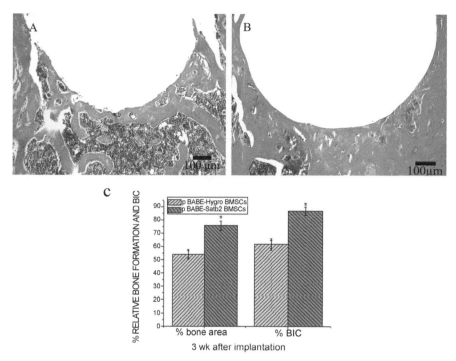

Figure 2.5 SATB2 transduced BMSCs enhances osseointegration after implantation. Histological analysis of H&E stained tissue sections demonstrated that woven bone was replaced by better organized lamellar bone in pBABE-Satb2 BMSCs group (B) compared to the control, pBABE-hygro BMSCs group (A) at 3 wk after implantation. Histomorphometric analysis showed that both the newly formed bone area and the percentage of bone-to-implant contact were increased in the pBABE-Satb2 BMSCs group when compared with those in the pBABE-hygro BMSCs group (C). Data were expressed as mean±SEM (n=6–8). *$p<0.05$, pBABE-hygro BMSCs vs. pBABE-*Satb2*BMSCs. (The figure is from Dr. Chen's laboratory, Tufts University).

In the above-mentioned study performed by Dr. Hou and colleagues (Hou et al. 1999), the systemically transplanted BMSCs can retain competency to engraft and differentiate into mature osteoblasts and osteocytes. More importantly, using a reporter gene driven by the osteocalcin gene promoter, the researchers validated the strategy to use the bone tissue specificity of the osteocalcin gene promoter to specifically deliver therapeutic genes to skeletal cells. One of our studies also provided evidence that systemically transplanted BMSCs could engraft into the whole body, participating in and promoting bone defect regeneration (Li et al. 2008). In another study, double-labeled BMSCs were obtained from the above mentioned BSP-Luc/ACTB-EGFP mice and systemically infused into 4-wk-old male nude mice via intracardiac injection. Five weeks later, titanium implants were inserted into the femurs of these recipient mice. Using this model, we observed that systemically transplanted BMSCs can be recruited from circulation to the implantation sites and participate in the establishment of osseointegration in response to long-range signals originating from the surgical wound of implant placement (Fig. 2.6). In a recently published study, a gene encoding the chemokine receptor largely responsible for stromal-derived factor-1 (SDF-1)-mediated bone marrow homing and engraftment of hematopoietic stem cells (HSCs), CXCR4, was transduced into mouse MSCs by adenovirus infection to enhance the migration and engraftment of these cells (Lien et al. 2009). Higher bone marrow retention and homing of CXCR4-expressing MSCs were observed after these cells were systemically transplanted into immunocompetent mice. Interestingly, a full recovery of bone mass and a partial restoration of bone formation in glucocorticoid-induced osteoporotic mice were also observed 4 wk after a single intravenous infusion of one million CXCR4-expressing MSCs, suggesting the potent therapeutic effects of systemic transplantation of genetically manipulated MSCs in ameliorating bone diseases affecting multiple body parts.

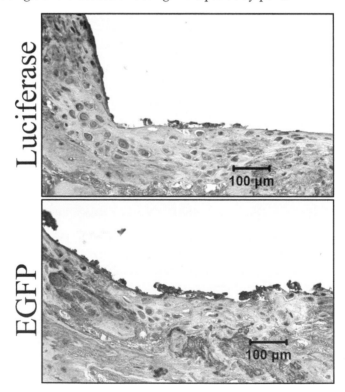

Figure 2.6 Transplanted BMSCs participated in osseointegration after implantation. Immunohistochemical staining demonstrated that, 7 d after surgery, luciferase and GFP activity could be detected in the newly formed bone area. (The figure is from Dr. Chen's laboratory, Tufts University).

Key Facts

- Bone damage and injury interfere with normal function and are often disfiguring, which may affect the patients' self-esteem and cause them to withdraw from social and public life.
- For successful bone tissue-engineered regeneration, several key elements are required, including mesenchymal stem cells (or progenitor cells), bioactive factors, and implanted materials that occupy the wound boundaries.
- In contrast, the use of BMSCs in bone tissue engineering does not bring up any ethical concerns, and BMSCs are accessible for treatments at nearly all times.
- BMSCs as adult stem cells maintain the capacity for self-renewal and are able to differentiate, when induced by appropriate chemical or mechanical signals, into chondrocytes, osteoblasts, adipocytes, myoblasts, and possibly neuron-like cells.
- BMSCs have been considered as a prerequisite for *in vitro* expansion of hematopoietic stem cells (HSCs) because HSCs may differentiate rapidly *in vitro* and lose self-renewal capacity without the presence of BMSCs.
- Transplanted BMSCs engrafted from peripheral circulation, repopulated the bone defect sites and participated in the regeneration processes.
- When BMSCs are applied in bone tissue engineering, appropriate strategies should be adopted to facilitate the therapeutic use of BMSCs, which include local transplantation of BMSCs seeded in supporting biomaterial scaffolds, local transplantation of gene-engineered BMSCs and systemic infusion of BMSCs.

Dictionary

- *BMSCs*: Also known as mesenchymal stem cells, bone marrow stromal cells are a mixed population of cells derived from the non-blood forming fraction of bone marrow. Bone marrow stromal cells are capable of growth and differentiation into a number of different cell types including bone, cartilage and fat.
- *SATB2*: Special AT-rich sequence-binding protein 2 (SATB2) also known as DNA-binding protein. SATB2 is a DNA-binding protein that specifically binds nuclear matrix attachment regions and is involved in transcriptional regulation and chromatin remodeling. SATB2 has been implicated as causative in the cleft or high palate of individuals with 2q32q33 microdeletion syndrome.

- *Runx2*: Runt-related transcription factor 2 (RUNX2) also known as *core-binding factor subunit alpha-1* (CBF-alpha-1) is a protein that in humans is encoded by the *RUNX2* gene. RUNX2 is a key transcription factor associated with osteoblast differentiation.
- *Osx*: Osterix (Osx) is a zinc-finger-containing transcription factor that is expressed in osteoblasts of all endochondral and membranous bones. Osx is essential for osteoblast differentiation and bone formation.
- *BMP*: Bone morphogenetic proteins (BMPs) are a group of growth factors also known as cytokines and as metabologens, which constitute a group of pivotal morphogenetic signals, orchestrating tissue architecture throughout the body.

Summary Points

- BMSCs can be readily isolated and manipulated. Under optimal culture conditions billions of BMSCs can be generated from a limited amount of starting material.
- BMSCs maintain the capacity for self-renewal and are able to differentiate, when induced by appropriate chemical or mechanical signals, into chondrocytes, osteoblasts, adipocytes, myoblasts, and possibly neuron-like cells.
- *Ex vivo* expanded BMSCs are a rich source of osteoprogenitor cells which can promote the regeneration of skeletal defects.
- When BMSCs are applied in bone tissue engineering, appropriate strategies include local transplantation of BMSCs seeded in supporting biomaterial scaffolds, local transplantation of gene-engineered BMSCs and systemic infusion of BMSCs.
- Normally used biomaterials in bone tissue engineering can be roughly divided into three categories: natural materials, bioactive inorganic materials and synthetic polymers.
- A second strategy for the application of BMSCs in bone engineering is to introduce osteogenic genes into the BMSCs before transplanting these cells into bone defect sites. Overexpression of these osteoinductive factors enhances and maintains a robust osteoblastic differentiation capacity in BMSCs, and therefore drastically promotes bone regeneration.
- Alternatively, BMSCs, with or without osteogenic gene modification, can be infused systemically to enhance bone regeneration. This represents an important therapeutic strategy especially for the treatment of genetic bone disorders or other bone diseases affected multiple body parts such as osteoporosis and multiple myeloma.

List of Abbreviations

ADSCs	:	Adipose tissue-derived stem cells
ALP	:	Alkaline phosphatase
BMSCs	:	Bone marrow stromal cells
BMD	:	Bone mineral densities
BMPs	:	Bone morphogenetic proteins
BMP-2	:	Bone morphogenetic protein-2
BMPRIA	:	Bone morphogenetic protein receptor type IA
BSP	:	Bone sialoprotein
CFU-f	:	Colony forming unit—fibroblasts
PLGA	:	Copolymers of PLA and PGA
DFCs	:	Dental follicle cells
DPSCs	:	Dental pulp stem cells
ESCs	:	Embryonic stem cells
EGFP	:	Enhanced green fluorescent protein
HSCs	:	Hematopoietic stem cells
iPS cells	:	Induced pluripotent stem cells
MARs	:	Matrix-attachment regions
MSCs	:	Mesenchymal stem cells
MDSCs	:	Muscle-derived stem cells
Osx	:	Osterix
PTH	:	Parathyroid hormone
PDGF	:	Platelet-derived growth factor
PGA	:	Polyglycolic acid
PLA	:	Polylactic acid
PECA	:	Poly[poly(ethylene glycol)-co-cyclic acetal]
Runx2	:	Runt-related gene 2
SATB2	:	Special AT-rich sequence-binding protein 2
SDF-1	:	Stromal-derived factor-1
TGF-β 1	:	Transforming growth factor-β 1
TGF-β	:	Transforming growth factor-β
UCB-MSCs	:	Umbilical cord blood-derived mesenchymal stem cells
VEGF	:	Vascular endothelial growth factor

References

Anklesaria, P., K. Kase, J. Glowacki, C.A. Holland, M.A. Sakakeeny, J.A. Wright, T.J. FitzGerald, C.Y. Lee and J.S. Greenberger. 1987. Engraftment of a clonal bone marrow stromal cell line *in vivo* stimulates hematopoietic recovery from total body irradiation. Proc Natl Acad Sci USA. 84(21): 7681–5.

Carcamo-Orive, I., A. Gaztelumendi, J. Delgado, N. Tejados, A. Dorronsoro, J. Fernandez-Rueda, D.J. Pennington and C. Trigueros. 2011. Regulation of human bone marrow stromal cell proliferation and differentiation capacity by glucocorticoid receptor and AP-1 crosstalk. J Bone Miner Res. 25(10): 2115–2125.

Cohnheim, J. 1867. Ueber entzundung und eiterung. Path Anat Physiol Klin Med. 1867; 40: 1–79.

Dobreva, G., M. Chahrour, M. Dautzenberg, L. Chirivella, B. Kanzler, I. Farinas, G. Karsenty and R. Grosschedl. 2006. SATB2 is a multifunctional determinant of craniofacial patterning and osteoblast differentiation. Cell. 125(5): 971–986.

Dominici, M., K. Le Blanc, I. Mueller, I. Slaper-Cortenbach, F. Marini, D. Krause, R. Deans, A. Keating, D. Prockop and E. Horwitz. 2006. Minimal criteria for defining multipotent mesenchymal stromal cells. The International Society for Cellular Therapy position statement. Cytotherapy. 8(4): 315–317.

Einhorn, T. 2003. Basic science of bone graft substitutes. Paper presented at: 2003 Annual Meeting of the Orthopaedic Trauma Association. Available at: "www.hwbf.org/ota/am/ota03/bssf/OTA03BG1.htm".

Friedenstein, A.J., J.F. Gorskaja and N.N. Kulagina. 1976. Fibroblast precursors in normal and irradiated mouse hematopoietic organs. Exp Hematol. 4(5): 267–274.

Haghani, K., S. Bakhtiyari and A.M. Nouri. 2011. *In vitro* study of the differentiation of bone marrow stromal cells into cardiomyocyte-like cells. Mol Cell Biochem.

Hofstetter, C.P., E.J. Schwarz, D. Hess, J.E.I. Widenfalk, A. Manira, D.J. Prockop and L. Olson. 2002. Marrow stromal cells form guiding strands in the injured spinal cord and promote recovery. Proc Natl Acad Sci USA. 99(4): 2199–2204.

Hou, Z., Q. Nguyen, B. Frenkel, S.K. Nilsson, M. Milne, A.J. van Wijnen, J.L. Stein, P. Quesenberry, J.B. Lian and G.S. Stein. 1999. Osteoblast-specific gene expression after transplantation of marrow cells: implications for skeletal gene therapy. Proc Natl Acad Sci USA. 96(13): 7294–7299.

Jiang, X., J. Zhao, S. Wang, X. Sun, X. Zhang, J. Chen, D.L. Kaplan and Z. Zhang. 2009. Mandibular repair in rats with premineralized silk scaffolds and BMP-2-modified bMSCs. Biomaterials. 30(27): 4522–4532.

Kaihara, S., S. Matsumura and J.P. Fisher. 2009. Cellular responses to degradable cyclic acetal modified PEG hydrogels. J Biomed Mater Res A. 90(3): 863–873.

Khadka, A., J. Li, Y. Li, Y. Gao, Y. Zuo and Y. Ma. 2011. Evaluation of hybrid porous biomimetic nano-hydroxyapatite/polyamide 6 and bone marrow-derived stem cell construct in repair of calvarial critical size defect. J Craniofac Surg. 22(5): 1852–1858.

Kojima, H. and T. Uemura. 2005. Strong and rapid induction of osteoblast differentiation by Cbfa1/Til-1 overexpression for bone regeneration. J Biol Chem. 280(4): 2944–2953.

Kuo, Y.C. and K.H. Chiu. 2011. Inverted colloidal crystal scaffolds with laminin-derived peptides for neuronal differentiation of bone marrow stromal cells. Biomaterials. 32(3): 819–831.

Leoyklang, P., K. Suphapeetiporn, P. Siriwan, T. Desudchit, P. Chaowanapanja, W.A. Gahl and V. Shotelersuk. 2007. Heterozygous nonsense mutation SATB2 associated with cleft palate, osteoporosis, and cognitive defects. Hum Mutat. 28(7): 732–738.

Li, M., M.J. Mondrinos, X. Chen, M.R. Gandhi, F.K. Ko and P.I. Lelkes. 2006. Co-electrospun poly(lactide-co-glycolide), gelatin, and elastin blends for tissue engineering scaffolds. J Biomed Mater Res A. 79(4): 963–973.

Li, S., Q. Tu, J. Zhang, G. Stein, J. Lian, P.S. Yang and J. Chen. 2008. Systemically transplanted bone marrow stromal cells contributing to bone tissue regeneration. J Cell Physiol. 215(1): 204–209.

Lien, C.Y., K. Chih-Yuan Ho, O.K. Lee, G.W. Blunn and Y. Su. 2009. Restoration of bone mass and strength in glucocorticoid-treated mice by systemic transplantation of CXCR4 and cbfa-1 co-expressing mesenchymal stem cells. J Bone Miner Res. 24(5): 837–48.

Lymperi, S., F. Ferraro and D.T. Scadden. 2010. The HSC niche concept has turned 31. Has our knowledge matured? Ann N Y Acad Sci. 1192: 12–18.

Mackay, A.M., S.C. Beck, J.M. Murphy, F.P. Barry, C.O. Chichester and M.F. Pittenger. 1998. Chondrogenic differentiation of cultured human mesenchymal stem cells from marrow. Tissue Eng. 4(4): 415–428.

Mao, J.J., W.V. Giannobile, J.A. Helms, S.J. Hollister, P.H. Krebsbach, M.T. Longaker and S. Shi. 2006. Craniofacial tissue engineering by stem cells. J Dent Res. 85(11): 966–979.

Nakashima, K., X. Zhou, G. Kunkel, Z. Zhang, J.M. Deng, R.R. Behringer and B. de Crombrugghe. 2002. The novel zinc finger-containing transcription factor osterix is required for osteoblast differentiation and bone formation. Cell. 108(1): 17–29.

Nilsson, S.K., M.S. Dooner, H.U. Weier, B. Frenkel, J.B. Lian, G.S. Stein and P.J. Quesenberry. 1999. Cells capable of bone production engraft from whole bone marrow transplants in nonablated mice. J Exp Med. 189(4): 729–734.

Otto, F., A.P. Thornell, T. Crompton, A. Denzel, K.C. Gilmour, I.R. Rosewell, G.W. Stamp, R.S. Beddington, S. Mundlos, B.R. Olsen, P.B. Selby and M.J. Owen. 1997. Cbfa1, a candidate gene for cleidocranial dysplasia syndrome, is essential for osteoblast differentiation and bone development. Cell. 89(5): 765–771.

Phinney, D.G., G. Kopen, R.L. Isaacson and D.J. Prockop. 1999. Plastic adherent stromal cells from the bone marrow of commonly used strains of inbred mice: variations in yield, growth, and differentiation. J Cell Biochem. 72(4): 570–585.

Tu, Q., P. Valverde and J. Chen. 2006. Osterix enhances proliferation and osteogenic potential of bone marrow stromal cells. Biochem Biophys Res Commun. 341(4): 1257–1265.

Tu, Q., P. Valverde, S. Li, J. Zhang, P. Yang and J. Chen. 2007. Osterix overexpression in mesenchymal stem cells stimulates healing of critical-sized defects in murine calvarial bone. Tissue Eng. 13(10): 2431–2440.

Wakitani, S., T. Goto, S.J. Pineda, R.G. Young, J.M. Mansour, A.I. Caplan and V.M. Goldberg. 1994. Mesenchymal cell-based repair of large, full-thickness defects of articular cartilage. J Bone Joint Surg Am. 76(4): 579–592.

Xu, B., J. Zhang, E. Brewer, Q. Tu, L. Yu, J. Tang, P. Krebsbach, M. Wieland and J. Chen. 2009. Osterix enhances BMSC-associated osseointegration of implants. J Dent Res. 88(11): 1003–1007.

Ye, J.H., Y.J. Xu, J.Gao, S.G. Yan, J. Zhao, Q. Tu, J. Zhang, X.J. Duan, C.A. Sommer, G. Mostoslavsky, D.L. Kaplan, Y.N. Wu, C.P. Zhang, L. Wang and J. Chen. 2011. Critical-size calvarial bone defects healing in a mouse model with silk scaffolds and SATB2-modified iPSCs. Biomaterials. 32(22): 5065–5076.

Zhang, J., Q. Tu, R. Grosschedl, M.S. Kim, T. Griffin, H. Drissi, P. Yang and J. Chen. 2011. Roles of SATB2 in osteogenic differentiation and bone regeneration. Tissue Eng Part A. 17(13–14): 1767–1776.

Zheng, H., Z. Guo, Q. Ma, H. Jia and G. Dang. 2004. Cbfa1/osf2 transduced bone marrow stromal cells facilitate bone formation *in vitro* and *in vivo*. Calcif Tissue Int. 74(2): 194–203.

CHAPTER 3

Isolation and Purification Methods for Mesenchymal Stem Cells

Kristen P. McKenzie,[a] Dana C. Mayer[b] and Jane E. Aubin[c,*]

ABSTRACT

Multipotential mesenchymal cells, now commonly known as mesenchymal stem cells (MSCs), were originally isolated from murine bone marrow as adherent colony forming units-fibroblast (CFU-F). More recently, MSCs or MSC-like cells have been isolated from a wide variety of tissues and organs, including peripheral blood, umbilical cord blood, adipose tissue, and muscle. However, the functionally heterogeneous nature of clonally-derived subpopulations isolated from MSC populations and tested *in vitro* and *in vivo*, together with the lack of a single unique MSC marker, or even a unique combination of markers amongst MSCs of different species, has made it difficult to characterize unambiguously the molecular and biological features of MSCs. Nevertheless, much progress has been made in isolation and purification strategies that include such approaches as positive and negative immunoselection, and adherence and expansion in medium

Dept. of Molecular Genetics, University of Toronto, Toronto, ON.
[a]E-mail: kristen.mckenzie@utoronto.ca
[b]E-mail: dana.mayer@utoronto.ca
[c]E-mail: jane.aubin@utoronto.ca
*Corresponding author

List of abbreviations given at the end of the text.

supplemented with a variety of growth factors and/or cytokines. In this chapter, we present an overview of MSC definitions and markers, isolation and purification techniques, and our evolving understanding of MSC characteristics as a basis for using these cells in regenerative medicine applications.

Introduction

Stem cells are the most primitive cells in the human body and are thought to be capable of both self-renewal and multilineage differentiation during development and tissue repair and regeneration. Amongst well-studied stem cell types is the adult stem cell class termed mesenchymal stem cells (MSCs). Cells designated MSCs were originally isolated from the bone marrow and are thought to give rise to mesenchymal tissue types, amongst others, and to play an important role in the repair and regeneration of these tissues (Mezey 2011). This chapter presents an overview of the current literature on MSC characteristics, including the still evolving understanding of their properties and markers, and the strategies being used to isolate and purify MSCs.

Defining MSCs

MSCs, also known as skeletal stem cells amongst other names, were originally identified by Friedenstein et al. as fibroblast colony forming units (CFU-F) found in bone marrow and adherent in culture (Friedenstein et al. 1976). Although not yet empirically proven *in vivo*, MSC characteristics include the capacity for multilineage differentiation, maintenance throughout the life of the organism and ability for self-renewal. MSCs have been shown to differentiate *in vitro* into mesenchymal cell types, including osteoblasts/osteocytes, chondrocytes, adipocytes, myocytes (Fig. 3.1), fibroblasts, tenocytes, and cardiomyocytes. They have also been shown to differentiate *in vitro* into ectodermal and endodermal lineage cells, such as neurons and hepatocytes, respectively, although their ability to differentiate into these cell types *in vivo* remains controversial (Augello et al. 2010). The effects of donor age on MSC number, proliferation, differentiation, self-renewal and responsiveness to growth factors and hormones remain uncertain (Stolzing et al. 2008), however, it is agreed that at least some MSCs are maintained with age. This suggests the possibility of MSC self-renewal, which is the ability of MSCs to divide asymmetrically into one daughter MSC and one daughter cell that further differentiates along a particular cell lineage, such as the osteoblast lineage. Self-renewal is a *theoretical* MSC characteristic, with many authors equating high replicative capacity or colony-forming

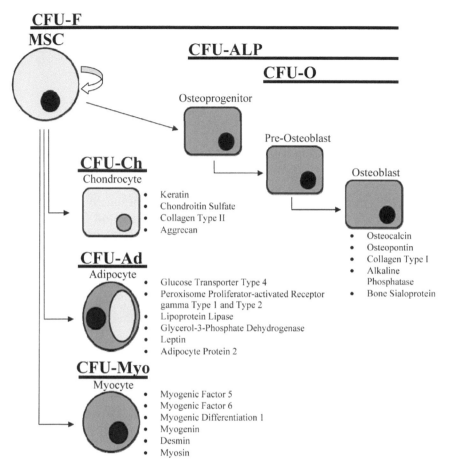

Figure 3.1 Illustration of mesenchymal stem cell (MSC) multipotency and expression of differentiation markers. Heterogeneity in size, morphology and potential for differentiation within colony forming unit-fibroblasts (CFU-Fs) is consistent with limiting dilution assays demonstrating that only a portion of CFU-Fs are colony forming unit-alkaline phosphatase (CFU-ALP) and an even smaller fraction are colony forming unit-osteoblast (CFU-O; clonogenic bone colonies or bone nodules). Osteocalcin, osteopontin, collagen type I, alkaline phosphatase and bone sialoprotein are genes expressed during osteoblast differentiation. Markers for the chondrocyte lineage or colony forming unit-chondrocyte (CFU-Ch) include the proteoglycans keratin and chondroitin sulphate, as well as collagen type II and aggrecan. When fully differentiated, adipocytes or colony forming unit-adipocyte (CFU-Ad) are characterized by expression of leptin, glucose-3-phosphate dehydrogenase, adipocyte protein 4, glucose transporter type 4, lipoprotein lipase and peroxisome proliferator-activated receptor gamma types 1 and 2. Lastly, myocytes or colony forming unit-myocyte (CFU-Myo) can be identified by the expression of myogenic factors 5 and 6, myogenic differentiation 1, myogenin, desmin and myosin.

ability of cells *in vitro* with proof of self-renewal (Phinney 2007). However, since many cell types (e.g., fibroblasts) are clonogenic, and multipotent progenitors exhibit high proliferative potential *in vitro* and *in vivo*, it is inappropriate to use proliferation or colony formation *per se* to distinguish between stem and progenitor cells, which have more limited differentiation potential. Further complicating these definitions is the fact that within the current literature, different authors use the term "MSC" differently, e.g., to designate cells with the potential to differentiate along certain cell lineages, but not others, or when describing a heterogeneous population only partly composed of cells with multilineage potential, making comparisons between studies very difficult. Nevertheless, accumulating *in vivo* and *in vitro* data support the concept of MSC self-renewal, even to the extent of identifying genes that may regulate self-renewal capacity (Bonyadi et al. 2003; Hardouin et al. 2011) (also see below), but there remains no *in vivo* assay, as available for hematopoietic stem cells (HSC) with serial transplantation, for example, to unambiguously confirm this trait.

MSC Markers

The ability to unambiguously identify MSCs by using a specific marker or combination of markers would help address the definition issues and also give insight into tissue-specific similarities and differences in MSCs. Unfortunately, the utility of current putative markers may vary between species, tissues and methods used for isolation (Table 3.1). The International Society for Cellular Therapy has developed the minimum criteria for human MSCs as cells that are plastic-adherent in culture, express CD105, CD73 and CD90, do not express CD34, CD45, CD11b, CD19 and HLA-DR and have the ability to differentiate into osteocytes, adipocytes and chondrocytes when cultured with the appropriate media (Dominici et al. 2006). The protein STRO-1 is also a well-known putative human MSC marker, since the STRO-1(-) bone marrow cell population does not contain CFU-F, and STRO-1 is not expressed by more mature cells of the osteoblast lineage, suggesting that it may be specific for immature cells. Unfortunately, the expression of STRO-1 is not exclusive to MSCs, so it is best used in combination with other markers. Additionally, the use of STRO-1 is limited to human MSCs from early passages, since there is no known mouse equivalent and its expression is lost during cell expansion (Kolf et al. 2007). Stem cell antigen (SCA-1) is a putative marker for both human and murine MSCs and, as mentioned above, is thought to be required for self-renewal because Sca-1-null mice exhibit age-dependent osteoporosis due apparently to a deficiency in MSC self-renewal (Bonyadi et al. 2003). Similar to STRO-1, SCA-1 is best used in combination with other markers, because it is known to be also expressed on HSCs (Aubin 2008). CD106 or vascular cell adhesion molecule

Table 3.1 Putative mesenchymal stem cell (MSC) markers. This table, while not exhaustive, summarizes the most frequently used putative MSC markers for various species and tissues used for isolation. + and − designate expression or not of the markers, respectively. Where the designation + or − is used in the table, there are conflicting data on marker expression. Arrows (→) delineate a gain or loss of a surface marker due to passaging. Not applicable (N/A) designates markers that have not been investigated in the cell population indicated.

Marker	Human MSCs, various tissues (Aubin 2008; Dominici et al. 2006; Kolf et al. 2007)	Human ASCs (Suga et al. 2009; Zuk et al. 2002)	Human trabecular MSCs (Sakaguchi et al. 2004)	Mouse MSCs, various tissues (Aubin 2008; Kolf et al. 2007)	Mouse bone marrow HipOPs (Itoh and Aubin 2009)	Mouse CB-MSCs (Short et al. 2009; Zhu et al. 2010)	MAPC, various tissues and species (Harting et al. 2008; Jiang et al. 2002)
7-4	−	−	N/A	−	−	−	N/A
CD5	−	−	N/A	−	−	−	N/A
CD11b	−	−	N/A	−	N/A	−	−
CD19	−	−	N/A	−	N/A	N/A	N/A
CD29	+	+	N/A	+ or −	N/A	+	+
CD31/PECAM1	−	−	−	−	+	−	−
CD34	−	+ → −	−	+ or −	+	−	−
CD44	+	+	+	+ or −	+	+	−
CD45	−	−	−	−	−	−	−
CD49(d or e)	−	+	N/A	−	N/A	N/A	+
CD71	+	+	+	+	N/A	N/A	N/A
CD73	+	+	+	+ or −	+	N/A	+
CD90	+	+	+	+ or −	+	N/A	+
CD105	+	− → +	+	+ or −	+	+	+
CD106/VCAM-1	+	+ or −	+	+	N/A	N/A	N/A
CD117/c-Kit	−	−	−	−	N/A	N/A	−
CD144/VE-Cadherin	N/A	+ or −	N/A	−	+	−	N/A
CD146/MCAM	+	+ or −	+	−	+	N/A	N/A
GR-1	−	−	N/A	−	−	−	N/A
HLA-DR	−	−	N/A	N/A	N/A	N/A	N/A
SCA-1	+	+	N/A	+	+	+	+
STRO-1	+	+	N/A	−	−	−	+ or −

1 (VCAM-1) has been successfully used in combination with STRO-1 to select a subpopulation of human bone marrow stromal cells (BMSCs) with extensive replicative capacity and retention of mulipotency (Kolf et al. 2007). CD146 was also used to select human BMSCs that showed some ability to self-renew and were able to generate bone and the hematopoietic microenvironment *in vivo* (Sacchetti et al. 2007). Despite the uncertainty as to which markers MSCs of different species express, there is a consensus that MSCs do not express such hematopoietic markers as CD117, CD45 and CD11b (immune cells) (Kolf et al. 2007). MSCs are also believed to lack expression of CD31 or platelet endothelial cell adhesion molecule 1 (PECAM1), which is expressed by endothelial and hematopoietic cells (Kolf et al. 2007). The HSC marker CD34 is also thought not to be expressed by human MSCs, but may be expressed by mouse MSCs (Aubin 2008).

Differences in marker expression may reflect the different tissues and methods used for cell isolation, the same kinds of issues that exacerbate difficulties in characterizing location of MSCs within tissues and organs, as well as the dynamics of marker expression during commitment and differentiation. For example, human adipose-derived MSCs (ASCs) share some phenotypic characteristics with human BMSCs, such as expression of CD29, CD44, CD71, CD90, STRO-1, CD105 and CD73 and lack of expression of HSC markers: CD31, CD34 and CD45, however human ASCs have been reported to be CD49d-positive and CD106-negative in contrast to BMSCs, which are thought to be CD49d-negative and CD106-positive (Zuk et al. 2002). Moreover, different laboratories have reported the presence or absence of STRO-1 on ASCs, possibly reflecting the use of fluorescence-activated cell sorting (FACS) versus immunofluorescent labelling for detection (Gimble et al. 2007). Whereas human BMSCs are thought to lack expression of CD34, human ASCs express CD34 prior to passaging (Gimble et al. 2007), although they tend to lose this surface marker after passaging, a loss that may correlate with ASC commitment towards a specific lineage and a decrease in proliferative capacity (Suga et al. 2009).

Sources of MSCs

Comparisons between studies are further confounded by differences in species and tissue types from which the MSCs are isolated. MSCs were originally isolated from murine yellow bone marrow (Friedenstein et al. 1976), which is found in the medullary cavity of long bones. They are now posited to migrate through peripheral circulation to different sites in the body in order to participate in tissue repair, although this remains without unequivocal evidence (He et al. 2007). The possibility is supported by the fact that cells with the same or similar properties to human BMSCs have been isolated from many other species and tissues of various ages, including amniotic fluid, umbilical cord blood, peripheral blood, cortical and trabecular bone, periosteum, adipose tissue, cartilage, muscle, skin, brain, liver, pancreas, among others. Some of these tissues, such as peripheral blood and adipose tissue, were initially investigated due to their relative ease of access compared to bone marrow. It has been reported that some tissues may have a higher yield and/or frequency of MSCs than bone marrow. For example, some report higher yields of MSC-like cells from adipose tissue (i.e., ASCs) than from bone marrow due to the relative abundance and accessibility of adipose tissue. CFU-Fs have been successfully isolated from the peripheral blood of many different species, however their frequency in peripheral blood is lower than that in the bone marrow (He et al. 2007).

It is currently unclear whether MSCs from different tissues are inherently different or whether they have mobilized from the bone marrow via peripheral blood and have taken up residence in other tissues, with changes in response to their different microenvironments. Although MSCs derived from different tissues share many similarities, they also exhibit some tissue-specific differences in marker expression, as noted above, and differentiation potential. For example, ASCs may lack osteogenic potential and MSCs from umbilical cord blood may lack adipogenic potential when compared to their bone marrow counterparts (Feng et al. 2010). These differences may be deciding factors when choosing a source of MSCs for clinical use. The caveats on some of the apparent differences described in MSC-containing populations from different tissues are that often comparisons are made amongst populations that are not homogeneous and that significantly different protocols have been used for their isolation and expansion from different sources. However, when multipotent adult progenitor cells (MAPCs) were isolated from murine bone marrow, muscle and brain tissue with the same isolation protocol, they were quite similar in terms of growth rate, onset of senescence, surface phenotype, gene expression profiles and differentiation potential *in vitro* (Jiang et al. 2002). More studies directly comparing tissue sources by using identical protocols are required.

In vivo MSC niche

Regardless of the species or tissue used for isolation of MSCs, the microenvironment(s) within which MSCs reside *in vivo* remains to be established, although there is evidence to support their association with endothelial, pericyte, hematopoietic and/or osteoblast niches, including the periosteum and endosteum, sharing these niches with HSCs (Bianco 2011) (Fig. 3.2). A stem cell niche is defined as a specific microenvironment in which the behaviour, phenotype, self-renewal, proliferation and differentiation of stem cells are regulated by direct interactions with other

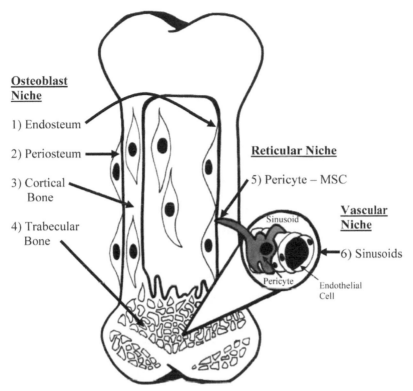

Figure 3.2 Putative mesenchymal stem cell (MSC) niches illustrated in long bone. There are three main niches, which are thought to be shared by hematopoietic stem cells (HSCs) (Bianco 2011). The Osteoblast Niche is composed of the endosteum (1), periosteum (2), cortical bone (3) and trabecular bone (4), where MSCs are in close contact with osteoblasts. The recently-described Reticular Niche is thought to contain MSCs associated with the processes of adventitial reticular cells/pericytes/chemokine ligand-12 (CXCL12)-abundant reticular (CAR) cells (5), which may extend into the endosteal niche (Bianco 2011). The Vascular Niche encompasses sinusoids (6), which are found in the bone marrow, trabecular bone and cortical bone. Within the sinusoids, MSCs are found perivascularly associated with endothelial cells and smooth muscle cells. Pericytes have been suggested to be the *in vivo* counterpart to the cells termed MSCs (da Silva Meirelles et al. 2008).

cells and paracrine and autocrine factors (Bianco 2011). With the discovery of chemokine ligand-12 (CXCL12)-abundant reticular (CAR) cells, a reticular/perivascular niche has recently been postulated for both MSCs and HSCs (Bianco 2011). These adventitial reticular cells, or pericytes, are "adventitial" because they reside on the abluminal surface of blood vessels and sinusoids and "reticular" because they have cellular processes that project into adjacent hematopoietic tissue. HSCs and MSCs are thought to associate with these processes, which may extend toward endosteal surfaces, partially accounting for the proposed endosteal stem cell niche (Bianco

2011). Furthermore, human CAR cells are themselves MSCs that recreate the hematopoietic microenvironment *in vivo* (Sacchetti et al. 2007), and mouse CAR cells can differentiate along osteoblast and adipocyte lineages (Bianco 2011). Consistent with these findings is the possibility that pericytes could be the *in vivo* counterpart to *in vitro* MSCs (da Silva Meirelles et al. 2008). Populations containing relatively high frequencies of MSCs have been isolated from pericyte-containing populations from various tissues (Corselli et al. 2010). Nevertheless, these notions are still challenged by the fact that cells with properties of MSCs have been isolated from avascular tissue, such as articular cartilage (da Silva Meirelles et al. 2008). It therefore seems likely that CAR cells/pericytes are a subset of MSCs, and that different tissue-specific MSCs exist.

Identification of a MSC Signature

The search for a combination of markers or a signature unique to MSCs isolated from various species and tissues has been augmented by microarray analyses. Global gene expression profiles of MSC populations isolated from human bone marrow and adipose tissue are very similar, and those derived from umbilical cord are surprisingly more similar to those from the bone marrow than embryonic stem cells despite their fetal origin (Roche et al. 2009). However, whereas umbilical cord-derived MSCs express more genes involved in neurogenesis, consistent with their *in vitro* differentiation capacity, bone marrow-derived MSCs express more genes involved in mesodermal and endodermal developmental processes, and adipose-derived MSCs are enriched in expression of immune-related genes (Jansen et al. 2010). Peripheral blood-derived MSCs (PB-MSCs) appear to express a wider array of genes than their bone marrow counterparts, which has been interpreted as the former being more primitive (He et al. 2007). Knockdowns of human transcription factors preferentially expressed in undifferentiated MSC populations as compared to osteocytes, chondrocytes and adipocytes have identified nine transcription factors—Ets variant genes 1 and 5 (ETV1 and ETV5), forkhead box P1 (FOXP1), GATA6, high-mobility group AT-hook 2 (HMGA2), Krueppel-like factor 12 (KLF12), PR domain containing 16 (PRDM16), single-minded homolog 2 (SIM2) and sex-determining region Y-box 11 (SOX11)—thought to be involved in MSC self-renewal and retention of multipotency, proxied by *in vitro* proliferation and differentiation assays, respectively, and thus comprising an essential molecular signature of MSCs (Kubo et al. 2009). From a meta-analysis of microarray studies comparing mouse MSCs isolated from various tissues with other cell types, Pedemonte and colleagues found that MSCs highly express transcription factors and components involved in the Wnt signalling pathway, HSC niche maintenance and the modulation of T cells

(Pedemonte et al. 2007). Virtually all of the analyses are complicated by the usually phenotypically and functionally heterogeneous nature of the MSC-containing populations compared, underscoring the need for additional analyses of MSC isolation strategies and marker analyses.

Isolation and Purification Methods

Serial passaging and adherence techniques

As previously mentioned, Friedenstein and colleagues initially identified multipotent stromal cells or CFU-F from bone marrow by their adherence in culture (Friedenstein et al. 1976) and this plastic adherence is now considered part of the definition of MSCs (Dominici et al. 2006). However, many studies have documented heterogeneity amongst the adherent cells from bone marrow and amongst individual colonies or clones in stromal cell populations isolated by adherence: heterogeneity of lineages present, heterogeneity in marker expression, and heterogeneity in proliferative and differentiation capacity (reviewed in (Aubin 2008)). Serial passaging of cells in culture has been used to remove nonadherent hematopoietic cells. For example, the expression of hematopoietic markers in rat BMSC populations decreases with serial passaging, while the expression of putative MSC markers increases, although the latter tend to decrease after more than 10 passages (Harting et al. 2008). The success of passaging may depend on the medium used; for example, Harting and colleagues showed that a stromal cell population devoid of hematopoietic cells could only be obtained when cells were cultured in medium supplemented with certain growth factors, but not in basal medium (Harting et al. 2008). Additionally, the trypsinization time used for passaging may affect the resulting cell population. Nadri et al. observed that decreasing trypsinization time to 2 min. results in a passaged population with a lower frequency of HSC antigens and an adherent population incapable of differentiating down the adipogenic and osteogenic lineages *in vitro* and thus devoid of MSCs (Nadri et al. 2007). Passaging and adherence techniques have also been employed to isolate ASCs from adipose tissue (Zuk et al. 2002).

Although passaging techniques are simple to use, they generally fail to result in homogeneous MSCs. For example, subpopulations of such cells as mast cells, lymphocytes, macrophages, endothelial cells, and/or smooth muscle cells have been identified in ASC-containing (Sengenes et al. 2005) and BMSC populations (Javazon et al. 2004). These cell types, especially macrophages, have been particularly problematic in mouse cell cultures, with studies showing that mouse macrophages are not effectively eliminated until after passage 10 (Schrepfer et al. 2007), at which point proliferative multipotent MSCs may also have been lost (Javazon et al. 2004).

These observations underscore the difficulty of isolating and maintaining putative MSC/ASC populations in culture, and the need for monitoring and quantifying phenotype and functionality amongst subpopulations of adherent mesenchymal cell fractions.

Collagenase and bone explant techniques

Cortical, trabecular and calvarial bone, typically used to isolate osteoblast populations, have vascular channels lined with endothelial cells on the inner surface as well as reticular, adipocytic, pericytic and smooth muscle cells lining the abluminal side, and may also be a source of MSCs. Collagenase release can be combined with bone explant techniques, in which MSCs are thought to migrate out of the bone fragments and adhere to the culture plate. Sakaguchi et al. harvested both collagenase-released trabecular and bone marrow aspirates from donors to isolate and compare MSC phenotype and functionality between the two sources (Sakaguchi et al. 2004). The collagenase-released cells from trabecular bone were phenotypically similar to BMSCs, expressing CD44, CD90, CD105, CD147, CD166 and CD54, while devoid of hematopoietic markers CD31, CD34, CD45 and CD117, and rivalled BMSCs in osteogenic, adipogenic and chondrogenic differentiation capacity. Zhu et al. flushed the bone marrow from mouse long bones and used collagenase, explanting and passaging to isolate a putative MSC population from the cortical bone (Zhu et al. 2010).

Cell selection techniques

To overcome limitations of the above isolation techniques, such as contaminating adherent non-MSC cells, two main cell selection techniques have been employed to purify MSCs via FACS or magnetic-activated cell sorting (MACS). Positive selection is based on selection of MSCs with antibodies directed to putative MSC surface markers, such as CD90, SCA-1, CD44, SH2 (endoglin; CD105), SH3-SH4 (CD73) among others. In many previous studies where positive selection techniques have been used to enrich MSCs, problems have been noted. For example, when FACS and anti-SCA-1-antibody were used to positively select osteogenic cells from mouse bone marrow or fetal calvaria, the ability of the resulting cell population to mineralize was lost upon subcultivation *in vitro* (Van Vlasselaer et al. 1994). Similarly, when immunomagnetic (MACS) cell sorting was used to select SCA-1-expressing cells, the SCA-1(+) population did not retain its adipogenic potential after *in vitro* expansion (Steenhuis et al. 2008). On the other hand, negative selection is based on removing contaminating hematopoietic or other cells with antibodies directed to their surface markers. For example, immunodepletion with anti-CD11b, CD34 and CD45

(HSC markers) resulted in MSCs with slow growth and poor proliferation (Baddoo et al. 2003). Choosing markers to use in selection may depend on desired functional outcomes. For example, since CD271 has recently been shown to be a useful marker for isolating human MSCs with more potent allosuppressive properties than unfractionated plastic-adherent MSCs (Kuci et al. 2011), it may be suited for selection of MSCs for use in treatment of graft-versus-host disease (GvHD) or of other inflammatory disorders, such as rheumatoid arthritis.

The following are more in-depth examples of how cell selection has resulted in purer or enriched populations of mesenchymal progenitors from various tissues.

Highly Purified Osteoprogenitors (HipOPs) from Mouse Bone Marrow

Using negative selection with anti-CD5, CD45, CD11b, GR-1, 7-4, TER-119 and CD45R conjugated magnetic micro-beads, our laboratory recently showed that osteoprogenitor cells can be significantly enriched (Itoh and Aubin 2009), as indicated by a high frequency of colony forming units-osteoblast (CFU-O, Fig. 3.3) and high expression levels of osteoblast differentiation markers *in vitro*. In addition, the highly purified osteoprogenitor population (HipOPs) was also enriched for cells with

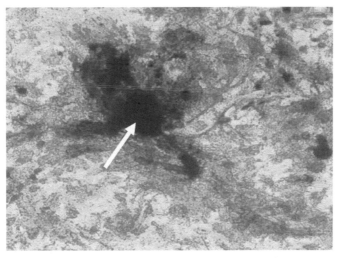

Figure 3.3 The highly purified osteoprogenitor (HipOP) population contains a high proportion of colony forming unit-osteoblast (CFU-O). A CFU-O double-stained with alkaline phosphatase and von Kossa to detect osteogenic cells and phosphate mineral deposition, respectively. Arrow indicates CFU-O. Please refer to (Ochotny et al. 2011), Figure 2F, to view a CFU-O in colour.

capacity to differentiate into chondrocytes and adipocytes, suggesting significant enrichment for MSCs (Fig. 3.4). Notably, when transplanted within collagen sponges into immunodeficient mice, donor HipOPs had far greater capacity than unfractionated stromal cells to produce a bone-like organ with differentiated osteoblasts, osteocytes, sinusoidal cells and bone marrow cells of donor origin. HipOPs were able to reconstitute a bone niche for HSCs, which was supported by high expression of HSC niche markers *in vitro*. The *in vivo* data also suggested the possibility of the presence of an angiogenic cell population since YFP-labelled donor HipOPs were observed around sinusoids. Interestingly, we found that further selection with CD73, an accepted human MSC marker (Dominici et al. 2006), did not prove useful in our murine model (McKenzie and Aubin, unpublished observation).

$$\text{CFU-O} \quad \frac{\text{HipOPs}}{\text{BMSCs}} > 100 \qquad \text{CFU-Ad} \quad \frac{\text{HipOPs}}{\text{BMSCs}} > 2$$

$$\text{CFU-ALP} \quad \frac{\text{HipOPs}}{\text{BMSCs}} > 10 \qquad \text{CFU-Ch} \quad \frac{\text{HipOPs}}{\text{BMSCs}} > 5$$

$$\text{CFU-F} \quad \frac{\text{HipOPs}}{\text{BMSCs}} > 5$$

Figure 3.4 Fractionated highly purified osteoprogenitors (HipOPs) are enriched in mesenchymal progenitors compared to unfractionated bone marrow stromal cells (BMSCs). HipOPs contain ~100-fold more colony forming unit-osteoblast (CFU-O), 10-fold more colony forming unit-alkaline phosphatase (CFU-ALP), 5-fold more colony forming unit-fibroblast (CFU-F), 2-fold more colony forming unit-adipocyte (CFU-Ad) and 5-fold more colony forming unit-chondrocyte (CFU-Ch) than unfractionated BMSCs (Itoh and Aubin 2009).

Cortical Bone Sorted Cells

Lack of understanding of the precise location of MSCs within the bone marrow, combined with their low frequency, has prompted studies to look for MSCs residing within other bone compartments such as trabecular and cortical bone (also see above). Indeed, some authors have speculated that cortical bone may contain a lower frequency of hematopoietic cells and a higher frequency of MSCs than bone marrow (Short et al. 2009; Zhu et al. 2010), although the lack of definitive markers to localize MSCs *in vivo* means that compact bone, endosteum and/or surrounding blood vessels have all been proposed to house MSCs (Corselli et al. 2010; Zhu et al. 2010). Additionally, hematopoietic cells are present in, and may travel to the blood vessels that are embedded within the Haversian canals of cortical bone. In fact, depleting hematopoietic cells results in the removal of 95–98 percent of total cortical bone-derived cells (Short et al. 2009) and 90 percent of adherent bone marrow stromal cells (Itoh and Aubin 2009), suggesting that there may not be a significant difference in number of contaminating hematopoietic cells between the cortical bone and bone marrow as previously thought.

In contrast to the Zhu et al. method using collagenase-released cells and passaging discussed above, Short and colleagues first digested, minced and then washed (to remove marrow) mouse bone fragments. They then used MACS to deplete mature hematopoietic cells with antibodies against CD3, CD4, CD5, CD8, CD11b (MAC-1), GR-1, B220, and TER-119 (Lin-) and subsequent positive (SCA-1) and negative (CD45, and CD31) selection by FACS to obtain a Lin-CD45-CD31-SCA-1+ population that was highly enriched in highly clonogenic multipotent MSCs (Short et al. 2009) (Fig. 3.5).

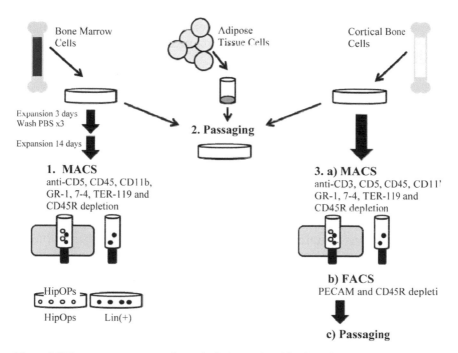

Figure 3.5 Three common approaches to isolation and purification of mesenchymal stem cells (MSCs) from mouse femur, tibia and adipose tissue. Bone marrow is harvested, plated and either passaged (2) or expanded and then sorted by magnetic-activated cell sorting (MACS) (1) (Itoh and Aubin 2009). Adipose tissue is harvested, digested with collagenase, centrifuged and passaged (2) (Zuk et al. 2002). Cortical bone is harvested and either plated and passaged (2) or immediately sorted via MACS and then fluorescence-activated cell sorting (FACS), followed by passaging (3) (Short et al. 2009).

Conclusion

Several issues need to be investigated to improve the biological understanding and clinical utility of MSCs. Currently, there is no single marker that is able to identify an MSC, since none of the putative markers, including those discussed above, is expressed exclusively on MSCs nor has any one marker been shown to be expressed on all MSC-like cells. Further complicating the search for a marker is the notion that MSC populations are heterogeneous, and markers may differ between species, strains, tissues and isolation procedures. In all cases, unambiguous characterization of MSCs and comparisons with their more differentiated progeny at a biological and molecular level remain difficult. Ability to isolate a purer population of MSCs that is not contaminated by other cell types would be helpful to acquire a better understanding of their biology and to better estimate their frequency in different tissues and changes in frequency due to aging, disease or particular kinds of interventions. Further studies comparing tissue sources and the subsequent differentiation capacity of the cells isolated will also affect the choice of cell source for basic and translational medicine studies.

Key Facts of MSC Isolation and Purification Methods

- MSCs are adult stem cells that can form different types of cells, such as bone-producing osteoblasts, cartilage-producing chondrocytes and fat cells or adipocytes, amongst other cell types.
- Human MSCs express certain cell surface markers such as CD105, CD73 and CD90, and lack expression of the hematopoietic cell markers CD34, CD45, CD11b, CD19 and HLA-DR, however, this phenotype varies between species, tissues and methods used for MSC isolation.
- Populations containing cells that are functionally and phenotypically similar to BMSCs have been isolated from various species and tissues.
- MSCs are thought to reside in bone and vascular niches, where they can interact with bone and vascular cells, as well as with HSCs, which share their niches.
- There are many different techniques for the isolation and purification of MSCs:
    - Passaging (washing and re-plating) of cells in culture to remove nonadherent contaminating cell populations.
    - Collagenase-treating and explanting bone fragments to allow MSCs to migrate out from bone matrix and adhere to plastic culture dishes.

- o Fluorescence-Activated Cell Sorting (FACS) or Magnetic-Activated Cell Sorting (MACS).
  - Positive selection to select for cells expressing putative MSC markers.
  - Negative depletion of contaminating cells.
- Improvement in methods for isolating and purifying MSCs is required to further characterize and assess MSCs for clinical applications.

Dictionary of Key Terms

- *Adipogenic*: Having capacity to differentiate along the adipocyte lineage.
- *Allosuppressive*: Donor cell ability to suppress the recipient's immune response.
- *Bone Marrow Stromal Cells*: Plastic-adherent cells from bone marrow. A subset of these is thought to comprise MSCs.
- *Chondrogenic*: Having capacity to differentiate along the chondrogenic lineage.
- *Endosteum/endosteal*: Thin layer of cells lining the inner surface of cortical/compact bone, surfaces of trabecular/cancellous/spongy bone.
- *FACS*: A technique involving antibodies that recognize antigens on cell surfaces and used to separate cells using fluorescence and electrostatic charge.
- *MACS*: A technique using antibodies that recognize antigens on cell surfaces; the antibodies are conjugated to magnetic beads and cells are passed through a magnetic column for fractionation.
- *Multipotent*: Having the ability to differentiate into more than one but a limited number of distinct cell lineages.
- *Niche*: A specific microenvironment in which the behaviour, phenotype, self-renewal, proliferation and differentiation of stem cells are regulated by direct interactions with other cells and paracrine and autocrine factors.
- *Osteogenic*: Having capacity to differentiate along the osteoblast lineage.
- *Periosteum/periosteal*: Soft tissue lining the outer surfaces of bone.
- *Pericyte*: A cell that surrounds blood vessels, also known as adventitial reticular cell or CAR cell. It is thought that MSCs reside near pericytes or that pericytes are themselves MSCs.

Summary Points

- MSCs, like other stem cell populations, are defined by their capacity for multilineage differentiation, maintenance throughout the life of the organism and ability for self-renewal.
- The utility of current putative markers may vary between species, tissues and methods used for MSC isolation, as well as the dynamics of marker expression during commitment and differentiation.
- MSCs isolated from various sources may differ in phenotype, differentiation capacity, frequency and gene expression, and therefore choosing a source may depend on the desired application.
- Further studies are required to determine whether MSCs isolated from various sources are inherently different or if they have migrated from a single, original location, such as bone marrow, and acquired different molecular and functional properties in response to micrenvironmental effects.
- Three main niches, which are thought to be shared by HSCs, have been described for MSCs:
  - Osteoblast niche: MSCs are found on periosteal and endosteal surfaces, as well as within compact and trabecular bone.
  - Vascular niche: MSCs are found perivascularly associated with endothelial cells and smooth muscle cells. The perivascular cells, pericytes, have been suggested to be the *in vivo* counterpart to the cells termed MSCs based on *in vitro* properties.
  - Reticular niche: It was recently proposed that MSCs are associated with the processes of adventitial reticular cells/pericytes/CAR cells, which may extend into the endosteal niche.
- Various MSC isolation and purification techniques have been described:
  - Isolation of plastic-adherent stromal cells and their serial passaging are simple to use, but usually do not result in homogenous MSCs, and repeated passaging may result in loss of proliferative or multipotential capacity.
  - Collagenase-release and explant techniques are useful for MSCs from bone sources, but are subject to similar limitations as those based on adherence and serial passaging.
  - Positive cell selection with putative MSC markers and/or negative selection with known markers of contaminating cell types, such as hematopoietic cells, results in populations enriched in MSC-like cells.
- Further studies are required to investigate the utility of putative MSC markers, how these differ across species and tissues, and how to best apply them in isolation and purification techniques.

List of Abbreviations

ASC	:	Adipose-derived Stem Cell
BMSC	:	Bone Marrow Stromal Cell
CAR	:	CXCL12-Abundant Reticular
CB	:	Cortical Bone
CFU-Ad	:	Colony Forming Unit—Adipocyte
CFU-ALP	:	Colony Forming Unit—Alkaline Phosphatase
CFU-Ch	:	Colony Forming Unit—Chondrocyte
CFU-F	:	Colony Forming Unit—Fibroblast
CFU-O	:	Colony Forming Unit—Osteoblast
CXCL12	:	Chemokine Ligand-12
FACS	:	Fluorescence-Activated Cell Sorting
HipOP	:	Highly Purified Osteoprogenitor Population
HSC	:	Hematopoietic Stem Cell
MACS	:	Magnetic-Activated Cell Sorting
MAPC	:	Multipotent Adult Progenitor Cells
MSC	:	Mesenchymal Stem Cell
PB	:	Peripheral Blood
VEC	:	Vascular Endothelial Cell
YFP	:	Yellow Fluorescent Protein

References

Aubin, J.E. Mesenchymal Stem Cells and Osteoblast Differentiation. pp. 83–106. *In:* J.P. Bilezikian, L.G. Raisz and T.J. Martin [eds.]. 2008.Principles of Bone Biology Academic Press, Inc., San Diego, California.

Augello, A., T.B. Kurth and C. De Bari. 2010. Mesenchymal stem cells: a perspective from *in vitro* cultures to *in vivo* migration and niches. Eur Cell Mater. 20: 121–133.

Baddoo, M., K. Hill, R. Wilkinson, D. Gaupp, C. Hughes, G.C. Kopen and D.G. Phinney. 2003. Characterization of mesenchymal stem cells isolated from murine bone marrow by negative selection. J Cell Biochem. 89: 1235–1249.

Bianco, P. 2011. Minireview: The stem cell next door: skeletal and hematopoietic stem cell "niches" in bone. Endocrinology. 152: 2957–2962.

Bonyadi, M., S.D. Waldman, D. Liu, J.E. Aubin, M.D. Grynpas and W.L. Stanford. 2003. Mesenchymal progenitor self-renewal deficiency leads to age-dependent osteoporosis in Sca-1/Ly-6A null mice. Proc Natl Acad Sci USA. 100: 5840–5845.

Corselli, M., C.W. Chen, M. Crisan, L. Lazzari and B. Peault. 2010. Perivascular ancestors of adult multipotent stem cells. Arterioscler Thromb Vasc Biol. 30: 1104–1109.

da Silva Meirelles, L., A.I. Caplan and N.B. Nardi. 2008. In search of the *in vivo* identity of mesenchymal stem cells. Stem Cells. 26: 2287–2299.

Dominici, M., K. Le Blanc, I. Mueller, I. Slaper-Cortenbach, F. Marini, D. Krause, R. Deans, A. Keating, D. Prockop and E. Horwitz. 2006. Minimal criteria for defining multipotent mesenchymal stromal cells. The International Society for Cellular Therapy position statement. Cytotherapy. 8: 315–317.

Feng, J., A. Mantesso and P.T. Sharpe. 2010. Perivascular cells as mesenchymal stem cells. Expert Opin Biol Ther. 10: 1441–1451.

Friedenstein, A.J., J.F. Gorskaja and N.N. Kulagina. 1976. Fibroblast precursors in normal and irradiated mouse hematopoietic organs. Exp Hematol. 4: 267–274.

Gimble, J.M., A.J. Katz and B.A. Bunnell. 2007. Adipose-derived stem cells for regenerative medicine. Circ Res. 100: 1249–1260.

Hardouin, S.N., R. Guo, P.H. Romeo, A. Nagy and J.E. Aubin. 2011. Impaired mesenchymal stem cell differentiation and osteoclastogenesis in mice deficient for Igf2-P2 transcripts. Development. 138: 203–213.

Harting, M., F. Jimenez, S. Pati, J. Baumgartner and C. Cox Jr. 2008. Immunophenotype characterization of rat mesenchymal stromal cells. Cytotherapy. 10: 243–253.

He, Q., C. Wan and G. Li. 2007. Concise review: multipotent mesenchymal stromal cells in blood. Stem Cells. 25: 69–77.

Itoh, S. and J.E. Aubin. 2009. A novel purification method for multipotential skeletal stem cells. J Cell Biochem. 108: 368–377.

Jansen, B.J., C. Gilissen, H. Roelofs, A. Schaap-Oziemlak, J.A. Veltman, R.A. Raymakers, J.H. Jansen, G. Kogler, C.G. Figdor, R. Torensma and G.J. Adema. 2010. Functional differences between mesenchymal stem cell populations are reflected by their transcriptome. Stem Cells Dev. 19: 481–490.

Javazon, E.H., K.J. Beggs and A.W. Flake. 2004. Mesenchymal stem cells: paradoxes of passaging. Exp Hematol. 32: 414–425.

Jiang, Y., B. Vaessen, T. Lenvik, M. Blackstad, M. Reyes and C.M. Verfaillie. 2002. Multipotent progenitor cells can be isolated from postnatal murine bone marrow, muscle, and brain. Exp Hematol. 30: 896–904.

Kolf, C.M., E. Cho and R.S. Tuan. 2007. Mesenchymal stromal cells. Biology of adult mesenchymal stem cells: regulation of niche, self-renewal and differentiation. Arthritis Res Ther. 9: 204.

Kubo, H., M. Shimizu, Y. Taya, T. Kawamoto, M. Michida, E. Kaneko, A. Igarashi, M. Nishimura, K. Segoshi, Y. Shimazu et al. 2009. Identification of mesenchymal stem cell (MSC)-transcription factors by microarray and knockdown analyses, and signature molecule-marked MSC in bone marrow by immunohistochemistry. Genes Cells. 14: 407–424.

Kuci, Z., S. Kuci, S. Zircher, S. Koller, R. Schubert, H. Bonig, R. Henschler, R. Lieberz, T. Klingebiel and P. Bader. 2011. Mesenchymal stromal cells derived from CD271(+) bone marrow mononuclear cells exert potent allosuppressive properties. Cytotherapy. 13: 1193–11204.

Mezey, E. 2011. The therapeutic potential of bone marrow-derived stromal cells. J Cell Biochem. 112: 2683–2687.

Nadri, S., M. Soleimani, R.H. Hosseni, M. Massumi, A. Atashi and R. Izadpanah. 2007. An efficient method for isolation of murine bone marrow mesenchymal stem cells. Int J Dev Biol. 51: 723–729.

Ochotny, N., A.M. Flenniken, C. Owen, I. Voronov, R.A. Zirngibl, L.R. Osborne, J.E. Henderson, S.L. Adamson, J. Rossant, M.F. Manolson and J.E. Aubin. 2011. The V-ATPase a3 subunit mutation R740S is dominant negative and results in osteopetrosis in mice. J Bone Miner Res. 26: 1484–1493.

Pedemonte, E., F. Benvenuto, S. Casazza, G. Mancardi, J.R. Oksenberg A. Uccelli and S.E. Baranzini. 2007. The molecular signature of therapeutic mesenchymal stem cells exposes the architecture of the hematopoietic stem cell niche synapse. BMC Genomics. 8: 65.

Phinney, D.G. 2007. Biochemical heterogeneity of mesenchymal stem cell populations: clues to their therapeutic efficacy. Cell Cycle. 6: 2884–2889.

Roche, S., B. Delorme, R.A. Oostendorp, R. Barbet, D. Caton, D. Noel, K. Boumediene, H.A. Papadaki, B. Cousin, C. Crozet et al. 2009. Comparative proteomic analysis of human mesenchymal and embryonic stem cells: towards the definition of a mesenchymal stem cell proteomic signature. Proteomics. 9: 223–232.

Sacchetti, B., A. Funari, S. Michienzi, S. Di Cesare, S. Piersanti, I. Saggio, E. Tagliafico, S. Ferrari, P.G. Robey, M. Riminucci and P. Bianco. 2007. Self-renewing osteoprogenitors in bone marrow sinusoids can organize a hematopoietic microenvironment. Cell. 131: 324–336.

Sakaguchi, Y., I. Sekiya, K. Yagishita, S. Ichinose, K. Shinomiya and T. Muneta. 2004. Suspended cells from trabecular bone by collagenase digestion become virtually identical to mesenchymal stem cells obtained from marrow aspirates. Blood. 104: 2728–2735.

Schrepfer, S., T. Deuse, C. Lange, R. Katzenberg, H. Reichenspurner, R.C. Robbins and M.P. Pelletier. 2007. Simplified protocol to isolate, purify, and culture expand mesenchymal stem cells. Stem Cells Dev. 16: 105–107.

Sengenes, C., K. Lolmede, A. Zakaroff-Girard, R. Busse and A. Bouloumie. 2005. Preadipocytes in the human subcutaneous adipose tissue display distinct features from the adult mesenchymal and hematopoietic stem cells. J Cell Physiol. 205: 114–122.

Short, B.J., N. Brouard and P.J. Simmons. 2009. Prospective isolation of mesenchymal stem cells from mouse compact bone. Methods Mol Biol. 482: 259–268.

Steenhuis, P., G.J. Pettway and M.A. Ignelzi Jr. 2008. Cell surface expression of stem cell antigen-1 (Sca-1) distinguishes osteo-, chondro-, and adipoprogenitors in fetal mouse calvaria. Calcif Tissue Int. 82: 44–56.

Stolzing, A., E. Jones, D. McGonagle and A. Scutt. 2008. Age-related changes in human bone marrow-derived mesenchymal stem cells: consequences for cell therapies. Mech Ageing Dev. 129: 163–173.

Suga, H., D. Matsumoto, H. Eto, K. Inoue, N. Aoi, H. Kato, J. Araki and K. Yoshimura. 2009. Functional implications of CD34 expression in human adipose-derived stem/progenitor cells. Stem Cells Dev. 18: 1201–1210.

Van Vlasselaer, P., N. Falla, H. Snoeck and E. Mathieu. 1994. Characterization and purification of osteogenic cells from murine bone marrow by two-color cell sorting using anti-Sca-1 monoclonal antibody and wheat germ agglutinin. Blood. 84: 753–763.

Zhu, H., Z.K. Guo, X.X. Jiang, H. Li, X.Y. Wang, H.Y. Yao, Y. Zhang and N. Mao. 2010. A protocol for isolation and culture of mesenchymal stem cells from mouse compact bone. Nat Protoc. 5: 550–560.

Zuk, P.A., M. Zhu, P. Ashjian, D.A. De Ugarte, J.I. Huang, H. Mizuno, Z.C. Alfonso, J.K. Fraser, P. Benhaim and M.H. Hedrick. 2002. Human adipose tissue is a source of multipotent stem cells. Mol Biol Cell. 13: 4279–4295.

CHAPTER 4

The Human Nose Offers a New Stem Cell Source for Bone Injuries

Pierre Layrolle,[1] Bruno Delorme,[2]
Francois Feron[3,4,a,]* and Emmanuel Nivet[3,b,5,]*

ABSTRACT

Bone defects, due to pathological and traumatic reasons, face great clinical challenges. When the bone self-healing process is insufficient or compromised, autologous bone graft is considered the best option but has the limitation of donor sites. Nowadays, the delivery of cells able to participate to the bone regeneration process either by growth factor secretion or differentiation into osseous cells is an appealing approach for the regenerative medicine. Among the different cell-based strategies

[1]Inserm U957, Laboratory of physiology of bone resorption, Faculty of Medicine, University of Nantes, 1 rue Gaston Veil, 44035 Nantes, France.
E-mail: pierre.layrolle@inserm.fr
[2]MacoPharma, Tourcoing.
E-mail: bruno.delorme@macopharma.com
[3]Aix Marseille Univ, NICN, CNRS UMR 6184, Bd P Dramard, 13015 Marseille, France.
[a]E-mail: francois.feron@univmed.fr
[b]E-mail: enivet@salk.edu
[4]Aix Marseille Univ, Centre d'Investigations Cliniques en Biothérapie CIC-BT 510, AP-HM—Institut Paoli Calmettes—Inserm, 232 Bd de Sainte Marguerite-13273 Marseille, France.
[5]Gene Expression Laboratory, The Salk Institute for Biological Studies, 10010 North Torrey Pines Road, La Jolla, CA 92037, USA.
*Corresponding authors

List of abbreviations given at the end of the text.

envisaged, tissue engineering endeavours to repair bone losses using cell filled three-dimensional scaffolds inserted at the defective site. However, cell-based applications are limited by a shortage of "highly qualified cells". Much evidence has highlighted mesenchymal stem cells as good cell candidates because of their capacity to enhance/ support bone formation after transplantation in animal models as well as in humans. Although multiple adult tissues harbour a niche of mesenchymal stem cells displaying osteogenic properties, differences in bone regenerative capacity have been pointed out. In this chapter, we present a new stem cell candidate for cell-based bone regeneration studies that has been recently identified in the nasal cavity, namely olfactory ectomesenchymal stem cells. This cell type shares several features with mesenchymal stem cells. In addition, they display a high mitogenic activity and a strong inclination to differentiate into osseous cells, *in vitro* and *in vivo*. Considering their properties, olfactory ecto-mesenchymal stem cells and their bone-associated derivatives offer new perspectives for cell-based therapy targeting bone injuries.

Introduction

Cell-based therapy, and more particularly the use of stem cells, represents one of the most promising areas of research for the future of regenerative medicine. Bones provide the scaffolding upon which the body is built and the development of methods to cure bone damages or anomalies is of importance for the clinic. Albeit bones have the capacity to self-repair in response to traumatic injury, numerous bone-associated diseases/ traumas, in which the regenerative process is compromised or insufficient for self-healing, remain incurable. The bone regeneration process is made of a succession of physiological events well orchestrated, spatially and temporally, by a partially overlapping interplay of various signals and cell types. The different lineages and substances involved in this process display osteoinductive, osteoconductive and/or osteogenic properties (Giannoudis et al. 2007). Several strategies aiming to reproduce or boost this natural process are envisioned to enhance bone repair and restore skeletal function after critical damages. Among the methods under investigation, the use of molecules known to regulate the process of bone regeneration such as Bone Morphogenic Proteins, growth hormone, parathyroid hormone and so forth, represents an active field of research (Nauth et al. 2010). Paralleling the use of bioactive molecules, the development of biomaterials used as scaffolds promoting the migration, proliferation and differentiation of bone cells is under constant progress (Ohtsuki et al. 2009). Beside these two strategies, the emergence of cell therapy as a highly promising alternative for regenerative medicine raises new hopes for the field of bone regeneration. Such an approach relies on the delivery of cells, displaying osteogenic properties, to the injury sites. Several studies have shown the potential value of stem cells in repairing major injuries involving the loss of bone structure (El Tamer and Reis 2009). Mesenchymal stem cells have been identified as one of the most attractive stem cell type (Jones and Yang 2011). Of importance is that mesenchymal stem cell (MSC) can give rise to osteoprogenitors that are able to differentiate into osteoblasts/osteocytes, when placed in appropriate conditions either *in vitro* (Mostafa et al. 2011) or *in vivo* (Petite et al. 2000; Quarto et al. 2001). For example, it has been demonstrated in different animal models (Arinzeh et al. 2003; Kon et al. 2000; Petite et al. 2000) but also in humans (Horwitz et al. 1999; Quarto et al. 2001) that bone marrow-derived mesenchymal stem cells (BM-MSCs) are an effective cellular product to support bone regeneration. Alternatively, other cell types such as pluripotent stem cells, umbilical cord blood-derived mesenchymal stem cells, adipose tissue-derived mesenchymal stem cells, muscle-derived stem cells and periosteal stem cells (for review see El Tamer and Reis 2009), as well as dental pulp stem cells (Mori et al. 2011) or peripheral blood stem cells (Matsumoto et al. 2006) have the capability to differentiate into osteogenic lineages. This extending list can now be updated with a new candidate stem cell subtype, the olfactory ecto-mesenchymal stem cell (OE-MSC) (Delorme et al. 2010), which is a new cellular product to consider for cell-based bone regeneration.

The olfactory mucosa, located in the nasal cavity, is the peripheral organ in charge of detecting odorants. The mature sensory neurons extend their dendrites to the top end of the epithelium and emerge in the lumen of the nose, leaving the olfactory knobs and their olfactory receptor-bearing cilia at risk of being regularly insulted by xenobiotics or infectious agents. As a result, the mean lifespan of an olfactory sensory neuron is around 30 d. Fortunately, a sustained and permanent neurogenesis allows every individual to avoid acute or chronic anosmia.

In addition, the olfactory mucosa is readily accessible in every living individual and can be biopsied safely without any loss of sense of smell (Girard et al. 2011). The procedure lasts 10–15 min and is carried out by an Ear Nose and Throat (ENT) surgeon, in accordance with the relevant local ethical committee(s), after having collected a signed informed consent form from every outpatient. Consequently, brain diseases can be studied with live olfactory tissue from cohorts of patients and autologous transplantations can be envisioned.

The olfactory mucosa consists of two tissues—olfactory neuro-epithelium and lamina propria—separated by a thin basement membrane. Within the epithelium, new neurons arise from germinal layers that include two types of basal cells: horizontal and globose basal cells, able to generate neurons and supporting cells. Nonetheless, these neuro-epithelial stem cells exhibit a poor proliferative aptitude and a restricted differential potential.

For these reasons, we studied another stem cell subtype, located in the lamina propria.

Isolation of human nasal olfactory lamina propria-derived stem cells

This procedure should be carried out by an ENT surgeon to avoid complications such as cerebrospinal fluid leak and to handle the potential complication of a potential severe life threatening nosebleed, as previously described (Girard et al. 2011). The biopsy is excised on the nasal septum, approximately halfway back along the length of the middle turbinate and about a centimetre from the roof of the nasal cavity, the cribriform plate. The nasal cavity is prepared with a local anaesthetic solution which could be Lidocaine with adrenalin (epinephrin) topical solution or 10 percent cocaine solution. The surgeon then uses a number 15 scalpel blade to make a small vertical cut 6–8 mm in the nasal septum in the front of the area raised as a bleb by the local anaesthetic injection. A semi sharp dissector such as a Freer's knife raises a small subperiosteal flap back for 2–3 mm. A small through-cut ethmoid forceps or similar small ENT grabbing forceps is put into the nose and a small piece of the raised mucosa with underlying periosteum is snipped off. This piece of mucosa should be at least 2 mm x 1 mm. Once collected, the olfactory mucosa is then transferred, using a sterile needle, into a sterile 2 ml tube filled with 1 ml of cold Dulbecco's Modified Eagle Medium/Ham F-12 (DMEM/F12) containing 10 percent fetal bovine serum (FBS) and 1 percent Streptomycin/Penicillin.

In order to isolate the underlying lamina propria from the epithelium, the mucosa is incubated in a Petri dish filled with 1 ml of Dispase II solution (2.4 IU/ml), for 1 hr at 37°C. Next, under a dissecting microscope with a diffracted inverted light, the olfactory epithelium is removed using a micro spatula (Girard et al. 2011). The lamina propria is then sliced into three to four pieces with a thickness ranging from 200 to 500 µm and each strip is isolated, inserted into its own 2 cm diameter culture dish, covered with sterile 1.3 cm diameter glass coverslips and cultivated with the culture medium mentioned above. Five to seven days after, stem cells invade the culture dish and, 2 wk later, the culture is confluent. When confluency is reached, stem cells are passaged and transferred to culture flasks.

Once expanded by passage, stem cells can be banked down in aliquots after harvest by storage in liquid nitrogen with 90 percent FBS and 10 percent dimethyl sulfoxide. Frozen aliquots are then thawed and grown under standard conditions on tissue culture plastic in DMEM/F12/10% FBS at 37°C and 5% CO_2.

A promising candidate for cell therapy

We first showed that human olfactory lamina propria-derived stem cells can be grown in large numbers and differentiated into neural and non-neural cell types *in vitro* and *in vivo* (Murrell et al. 2005), the latter feature being not observed with neural stem cells. Then, using animal models of spinal cord trauma (Xiao et al. 2005, 2007), cochlear damage (Pandit et al. 2011), Parkinson's disease (Murrell et al. 2008) or amnesia (Nivet et al. 2011), we and others demonstrated that human nasal olfactory stem cells are promising candidates for cell therapy.

In parallel, we attempted to thoroughly characterize this new stem cell subtype. Using immuno-markers and flow cytometry, we first observed that human olfactory lamina propria-derived stem cells had little in common with neural stem cells and hematopoietic stem cells. Then, after having shown that these cells can give rise to mesodermal cell types (Delorme et al. 2010), we postulated that, as resident of a connective tissue originating from a mesenchyme, this cell type could be part of the family of mesenchymal stem cells (MSC). This hypothesis led us to comprehensively compare human olfactory lamina propria stem cells with human bone marrow mesenchymal stem cells, at the cellular and molecular levels. More specifically, we also assessed their respective osteogenic potential.

A member of the mesenchymal stem cell superfamily

For this study, nasal biopsies were collected from four individuals, two males and two females aged 23 to 52 yr, at North University Hospital (Marseille, France); bone marrow cells were collected from four age- and gender-matched individuals, two males and two females aged 25 to 56 yr, at Tours University Hospital (Tours, France). Both cell types were cultivated as previously described (Delorme and Charbord 2007).

We first compared the two cell types using flow cytometry and a panel of 34 membrane markers (Table 4.1). Out of 21 markers known to be expressed by BM-MSCs, 21 were also expressed by OE-MSCs. Only the protein CD200 was not found in OE-MSCs. In addition, two other membrane proteins were differentially regulated in the two populations: CD9 (TSPAN29) was over-expressed while CD146 (MCAM) was under-expressed in OE-MSCs, when compared to BM-MSCs.

We then performed a transcriptome study. Total RNA from passage 3 (P3) stem cells was isolated and OE-MSC samples were compared to BM-MSC samples, prepared by the same operator under the same conditions. Hybridization on Affymetrix HG-U133 Plus 2.0 microarrays was performed according to the standards supplied by the manufacturer. Group comparison and gene retrieval was performed using the SiPaGene database. Hierarchical

Table 4.1 Expression of stem cell markers in OE-MSCs and BM-MSCs.

CD	Symbol	OE-MSC rMFI (Mean ± SEM)	BM-MSC rMFI (Mean ± SEM)
CD 9	TSPAN29	11,5 ± 7,4 *	6,2 ± 0,5
CD 13	ANPEP	312,0 ± 158,4	425 ± 50,4
CD 15	CD15	1,0 ± 0,1	1,0 ± 0,1
CD 18	ITGB2	1,0 ± 0,1	1,0 ± 0,2
CD 26	DPP4	5,2 ± 3,1	3,4 ± 0,4
CD 29	ITGB1	109,3 ± 47,1	115,7 ± 17,3
CD 31	PECAM	1,0 ± 0,1	1,1 ± 0,2
CD 34	CD34	1,0 ± 0,1	1,0 ± 0,1
CD 38	CD38	1,0 ± 0,1	1,1 + 0,2
CD 44	PGP1	175.2 ± 17.35	138.7 ± 16.37
CD 45	PTPRC	1,0 ± 0,2	1,0 ± 0,1
CD 47	IAP	21,3 ± 7,8	32,3 ± 4,2
CD 49 a	ITGA1	4,2 ± 1,0	3,4 ± 0,4
CD 49 e	ITGA5	72,0 ± 9,8	60,9 ± 7,4
CD 51	ITGAV	42,6 ± 3,192	59.46 ± 1,179
CD 54	ICAM1	12,7± 8,7	7,2 ± 0,4
CD 56	NCAM	1,4 ± 0,4	1,2 ± 0,2
CD 61	ITGB3	2,4 ± 0,4	3,9 ± 0,3
CD 63	TSPAN30	32,9 ± 16,3	25,5 ± 6,5
CD 73	NT5E	172,7 ± 36,4	169,5 ± 23,2
CD 81	TSPAN28	183,3 ± 14,4	156,3 ± 11,4
CD 90	THY1	334,8 ± 82,1	542,3 ± 71,2
CD 105	ENG	54,9 ± 21,5	133,8 ± 10,1
CD 106	VCAM1	3,5 ± 2	6,5 ± 2,9
CD 109	CD109	1,0 ± 0,1	1,0 ± 0,2
CD 117	KIT	1,3 ± 0,1	1,1 ± 0,3
CD 126	IL6R	1,0 ± 0,1	1,2 ± 0,4
CD 133	PROM1	1,0 ± 0,1	1,0 ± 0,2
CD 146	MCAM	2,3 ± 0,6 *	19,1 ± 3,4
CD 151	TSPAN24	70,3 ± 26,7	138, ± 17,6
CD 166	ALCAM	14,0 ± 2,7	19,6 ± 1,7
CD 200	CD200	1,4 ± 0,3 *	4,9 ± 0,7
VEGFR-1	FLT1	1,1 ± 0,1	1,0 ± 0,1
VEGFR-2	KDR	1,1 ± 0,0	1,1 ± 0,1

Expression of membrane markers on undifferentiated OE-MSCs in comparison to undifferentiated BM-MSCs. The number, the symbol and the fold change for each protein are indicated. Over-expressed proteins on OE-MSCs are highlighted in grey; under-expressed proteins on OE-MSCs are highlighted in black. *Indicates that P < 0.05, according to the Mann Whitney test, n = 4 for each group. OE-MSC: olfactory ecto-mesenchymal stem cell; BM-MSC: bone marrow mesenchymal stem cell.
With permission from Stem Cells and Development.

clustering and principal component analyses (PCA) were performed using Genesis software.

For comparison purposes, we used a panel of 147 CDs that were previously described as specific for BM-MSCs. Figure 4.1 shows that the OE-MSC population is homogenous and, within the mesenchymal lineage, closely related to BM-MSCs. In contrast, OE-MSCs and BM-MSCs segregated apart from the other stem cell populations studied—synovial fibroblasts (SFb), periosteal cells (POC)—and from three hematopoietic cell populations (CD11+, CD45+ and GlyA+ (CD235a) cells).

A gene expression profiling supporting a propensity to differentiate into osseous cells

Bone marrow mesenchymal stem cells are known to be able to differentiate into adipocytes, chondrocytes and osteocytes. Osteogenesis is a strictly controlled process requiring the activation of osteoblast-specific signalling proteins and transcription factors for osteoblast differentiation. That being said, we analyzed the genetic profile of OE-MSCs and assessed whether this cell type presents a genetic signature that could support osteogenic properties. Interestingly, the list of differentially regulated transcripts in the OE-MSC group (Table 4.2) indicates that inhibitors of chondrogenesis and adipogenesis while activators of neurogenesis and osteogenesis were over-expressed. It was therefore predicted that these stem cells will exhibit i) a poor aptitude to produce chondrocytes and adipocytes and ii) an enhanced ability to give rise to osteoblasts. This prediction was supported by an RT-PCR experiment showing that OE-MSCs express PTHR1 when they are undifferentiated while AGC and COL2A1 are undetectable in OE-MSCs before differentiation into chondrocytes. In addition, COL2A1 was not found after attempted chondrogenesis induction (Fig. 4.2).

OE-MSCs can be efficiently driven to differentiate in vitro into osteoblasts

The assumptions derived from the gene expression analysis were further evaluated by using standard *in vitro* differentiation assays. The osteogenic differentiation capacity of OE-MSCs was assessed after 21 d of incubation in a defined media containing Dexamethasone (0.1 µM), L-ascorbic acid (0.15 mM), NaH_2PO_4 (1 mM) and fetal bovine serum (10 percent). In such a media it has been previously demonstrated that BM-MSCs, the gold standard cell type for osseous reparation, differentiate towards an osteogenic phenotype. When placed in the same culture conditions, OE-MSCs have demonstrated a high capacity to generate osteoblasts (Delorme et al. 2010). Albeit OE-MSCs derived from different patient samples have

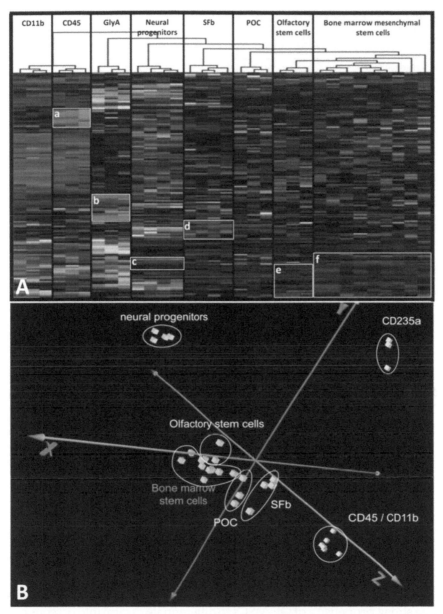

Figure 4.1 Transcriptomic profile of olfactory ecto-mesenchymal stem cells (OE-MSC). (A) Hierarchical clustering was performed using a panel of 147 CD transcripts on 6 independent samples of Bone marrow mesenchymal stem cells (BM-MSC), 3 samples of OE-MSCs, 3 samples of CD45+ (CD45) cells, CD11b+ (CD11b) cells and CD235a+ (GlyA) hematopoietic cells, 3 samples of periosteal cells (POC) and 4 samples of synovial fibroblasts (SFb). (B) Principal component analysis (PCA) was carried out using the same CD transcripts and the same samples described above. Each plotted data point represents a single profile. Unpublished.

Table 4.2 Selected dysregulated transcripts involved in adipogenesis, osteogenesis and chondrogenesis.

Symbol	Gene name	Fold Δ
Osteogenesis		
OSR2	Protein odd-skipped-related 2	+ 8,33
CYP27B1	25-hydroxyvitamin D-1 alpha hydroxylase	+ 2,64
PTN	Pleiotrophin (Osteoblast-specific factor)	+ 2,35
AEBP1	Osteoclast-like cell cDNA	+ 2,28
VDR	Vitamin D3 receptor	–3,16
Chondrogenesis		
ASPN	Asporin	+ 9,98
SOX9	Transcription factor SOX-9	–3,64
XYLT1	Xylosyltransferase 1	–3,83
CHI3L1	Chitinase-3-like protein 1	–58,22
Adipogenesis		
GATA2	Endothelial transcription factor GATA-2	+ 9,77
C10orf116	Adipose most abundant gene transcript 2 protein	+ 7,11
LPIN1	Lipin 1	+ 4,44
Stem cell markers		
NES	Nestin	+ 7,09
CD9	CD9 antigen	+ 5,56
CD44	CD44 antigen	+ 4,57
CD200	OX-2 membrane glycoprotein	–14,85

Selected dysregulated transcripts in undifferentiated OE-MSCs in comparison to undifferentiated BM-MSCs. Genes are clustered by main biological functions and ranked by decreasing fold change. The symbol, the full name and the fold change of each gene are indicated. Duplicate or triplicate transcripts are reported only once with the appropriate mean value. Under expressed transcripts are highlighted in grey. Alternative names are mentioned in bracket. OE-MSC: olfactory ecto-mesenchymal stem cell; BM-MSC: bone marrow mesenchymal stem cell. Unpublished.

shown some variability, all tested lines have exhibited a strong inclination to generate osteoblasts. The osteoblast phenotype was corroborated by Alizarin Red and Von Kossa staining for matrix mineralization (Fig. 4.3). Paralleling these observations, we have demonstrated that once placed in culture conditions supporting adipocyte or chondrocyte differentiation, a limited number of OE-MSCs can differentiate into adipocytes but none into chondrocytes (Delorme et al. 2010). All together, these data have confirmed the strong propensity of OE-MSCs to differentiate *in vitro* into osseous cells and brought evidence that this cell type could be an interesting alternative to BM-MSCs.

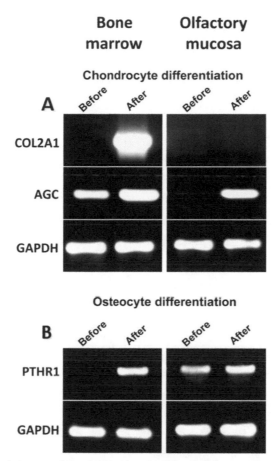

Figure 4.2 Transcript expression of genes involved in chondrogenesis and osteogenesis. BM-MSCs (left lane) and OE-MSCs (right lane) were cultivated either in a proliferating medium (before) or in a differentiating medium for 21 d (after). RT-PCR shows that Agc and Col2a1 are undetectable in OE-MSCs, before and after a differentiation into chondrocytes. Conversely, Pthr1 is expressed in undifferentiated OE-MSCs and not in undifferentiated BM-MSCs. OE-MSC: olfactory ecto-mesenchymal stem cell; BM-MSC: bone marrow mesenchymal stem cell. Unpublished.

OE-MSCs differentiate in vivo into osteoblasts

The *in vivo* osteogenic potential of OE-MSCs was studied by implantation in ectopic subcutis sites of immune compromised mice (Swiss NU/NU, Charles River Laboratories, France). As previously shown (Arinzeh et al. 2005), macro- and microporous biphasic calcium phosphate ceramic discs

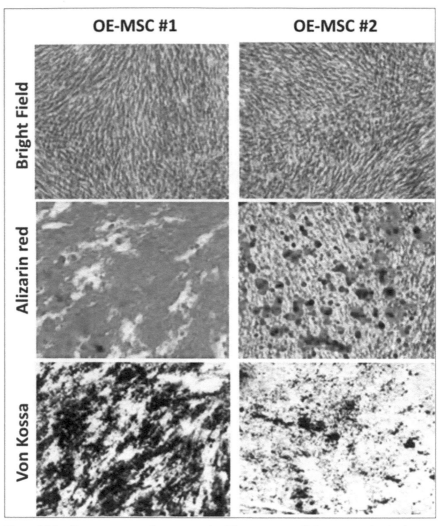

Figure 4.3 *In vitro* assessment of osteogenesis. Olfactory stem cells from two individuals were cultivated in a differentiating medium for 21 d. In both cases, a high proportion of olfactory stem cells turned into osteoblasts, stained according to Alizarin Red or von Kossa procedures. OE-MSC: olfactory ecto-mesenchymal stem cell. Unpublished.

(MBCP, Biomatlante, Vigneux de Bretagne, France) were used as scaffolds for olfactory stem cells. The porous ceramic discs, 8 mm in diameter and 3 mm in thickness, were made by compaction of calcium phosphate powder and pore makers, burning and sintering at 1050°C. The optimal composition of MBCP ceramic discs for ectopic bone formation was 20 percent Hydroxyapatite and 80 percent beta-tricalcium phosphate. The total porosity was approximately 70 percent with pore size in the range

300–600 µm and micropores of 0.3 µm favourable for entrapment of proteins. The MBCP ceramic discs were double packaged and sterilized by gamma irradiation at 25 kGy. Prior to cell seeding, the discs were soaked for 48 hr in 5 ml of DMEM/HAM's F12 culture medium supplemented with 10 percent FBS, 100 U/mL penicillin, and 100 µg/mL streptomycin in 6-well plates. After discarding excess medium, twice passaged olfactory stem cells were then seeded on ceramics at a density of 10^5 cells/disc. The cells were allowed to attach on the ceramic surface for 3 hr. Then, osteogenic differentiation media (3 mL) composed of DMEM high glucose supplemented with 10 percent serum, 1 percent penicillin/streptomycin, dexamethasone (0.1 µM), l-ascorbic acid (0.15 mM), and $NaH2PO4$ (2 mM) was added in each well. The cell/material constructs were then cultured for 12 d in osteogenic medium at 37°C, 5 percent CO_2, with refreshment every 2 d. Two discs per animal consisting of 10 cell/ceramic constructs and four controls were implanted in subcutaneous pockets in the back of seven nude mice under general anesthesia with isoflurane. After 7 wk of implantation, animals were euthanized by inhalation of an overdose of carbon dioxide gas. Explants were dissected and immediately fixed in neutral buffered formaldehyde (4 percent) solution and then processed for non-decalcified histology. Briefly, samples were dehydrated in ascending series of ethanol, acetone and embedded in polyglycol methylmethacrylate resin. The blocks were sawed (Leica SP1600), polished using silicon carbide paper, and metalized with Au-Pd prior to scanning electron microscopy (Zeiss Leo VP1450). Back-scattered electron microscopy allowed the discrimination of ceramic material, mineralized bone, and non mineralized tissue based on their respective grey levels. Thin (7 µm) sections were cut with a microtome (Leica SM2500), stained by Goldner's trichrome technique and observed under light micrcropy (Zeiss Axioplan 2).

As shown in Fig. 4.4, mineralized bone tissue has formed in contact to the porous MBCP ceramic after 7 wk in subcutis of nude mice. All samples on which olfactory stem cells were cultured, exhibited bone tissue *in vivo* (n=10) while ceramic discs implanted without cells were only encapsulated by a fibro vascularized tissue (n=4, data not shown). Mineralized bone tissue with osteocyte lacunae was clearly visible in the pores of the MBCP discs (Fig. 4.4A). Histology corroborated ectopic bone formation with mature bone with osteocytes lining on the ceramic surface. Osteoid tissue and adipose rich bone marrow were also observed. Alike bone marrow derived mesenchymal stem cells, olfactory stem cells cultured on MBCP ceramics in osteogenic conditions formed mature bone tissue in ectopic sites. This study demonstrated that the lamina propria is a potent source of stem cells for bone regeneration.

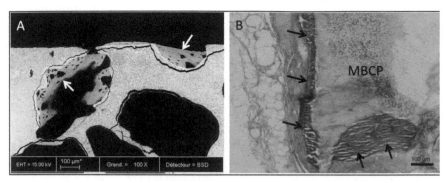

Figure 4.4 *In vivo* **ectopic bone formation by olfactory stem cells cultured on MBCP ceramic, 7 wk after subcutis implantation in nude mice.** (A) Back-scattered electron microscopy showing mineralized bone (arrows) with osteocytes *lacunae* in the pores of the MBCP ceramic. (B) Non decalcified histology (Goldner's trichrome staining) showing mature bone (arrows) lining in contact to the MBCP ceramic encapsulated by fibrous vascularized tissue. MBCP: Microporous biphasic calcium phosphate. Unpublished.

Clinical advantages of OE-MSCs

A series of evidence demonstrates that the olfactory mucosa is a new source of cells displaying osteogenic properties. It is worth mentioning in this regard that this cell type represents an alternative to other MSCs (e.g., bone marrow, adipose tissue), currently being tested and used in clinical applications aiming to heal bone injuries. Because of its straightforward accessibility in every individual, the olfactory mucosa is a tissue from which OE-MSCs can be easily isolated, expanded and transplanted back to the donor. Similar to other MSCs, the immunogenicity of OE-MSCs is not an obstacle hampering the translation of this cell type to the clinic because of the possibility to perform autologous transplantations. More important, studies have shown that the quality of BM-MSCs may vary significantly, depending on the individual and especially in older people. Efficacy studies have shown that the number of progenitors in the graft, and the number of progenitors available in bone marrow aspirated from the iliac crest appears to be critical for successful bone regeneration after BM-MSC transplantation (Hernigou et al. 2005). In this context, our data indicate that OE-MSCs possess strong osteogenic properties, suggesting the generation of high number of progenitors upon differentiation, and proliferate faster than BM-MSCs. Consequently, the number of cells available, especially the number of osteoprogenitors for further transplantation, is not really an issue when considering the possibility to expand cells *in vitro*, prior to transplantation.

Perspectives

With about one million procedures annually in Europe, bone is the most transplanted tissue in humans. Although regarded as the current gold standard in clinical practice, autologous bone graft requires surgical sites and is limited in quantity. An alternative may be to use mesenchymal stem cells in association with a biomaterial for bone regeneration. Adult stem cells are present in very low numbers in many different tissues. These cells are easily harvested under local anaesthesia, isolated and amplified in culture up to several hundred million within 2 to 3 wk. The lamina propria should be regarded as a potent source of osteoprogenitor cells for bone regeneration, particularly for maxillofacial reconstruction of large bone defects following cancers.

Key Facts of Olfactory Stemness

- The olfactory mucosa, located in the nasal cavity, is the only nervous tissue which is easily accessible in every individual. It is also the sole nervous organ whose neuron endings lie outside the body. As a consequence, the mortality rate of olfactory sensory neurons is high. Fortunately, a permanent neurogenesis is at play and newly formed neurons allow us to avoid anosmia.
- The olfactory mucosa is composed of two adjacent tissues. At the top, facing the outside world, the neuroepithelium contains uni/bipotent stem cells giving rise to olfactory neurons; at the bottom, lying on the nasal cartilegeous septum and turbinates, the lamina propria harbours a niche of multipotent stem cells displaying mesenchymal stem cells-associated properties.
- Stem cells located in the lamina propria can give rise to various cell types, including muscle, cardiac, neural and skeletal cells. They are multipotent stem cells.
- Nasal olfactory stem cells have been used in animal models of brain diseases and trauma. It is established that they are able to migrate across biological barriers, home into inflamed areas and differentiate into various neural cell types.
- Stem cells located in the olfactory lamina propria share many similarities with bone marrow mesenchymal stem cells. We compared their respective potential to differentiate into adipocytes, osteocytes and chondrocytes. We found that, under appropriate conditions, olfactory stem cells give rise to high numbers of osteocytes. Conversely, they poorly differentiate into adipocytes and chondrocytes.

Key Terms

- Neurogenesis is the process by which neurons are engendered from stem or progenitor cells.
- Anosmia relates to a non-functioning sense of smell. The anosmic individual is unable to perceive any odour.
- A neuroepithelium is a tissue combining the characteristics of two basic types of animal tissues, an epithelium and a nervous tissue.
- The lamina propria is a layer of connective tissue which lies beneath the epithelium. It contains axons and their surrounding ensheathing cells, blood vessels, glands that secrete mucus and stem cells.
- Multipotent stem cells have the potential to give rise to many but not all cell types.
- Mesenchymal stem cells are multipotent stromal cells displaying fibroblastoid morphology and adherent properties, when placed in appropriate culture conditions.
- Homing is the ability of a cell to navigate toward an area through various barriers and surrounding tissues.
- Adipocytes are the cells that mostly compose the adipose tissue.
- Osteocytes are star-shaped cells found in abundance in compact bones.
- Chondrocytes are the only cells found in cartilage.

Summary Points

- The human olfactory lamina propria contains a subpopulation of adult stem cells that are multipotent.
- Olfactory lamina propria-derived stem cells live in the mesenchymal compartment of a nervous tissue and display similarities with bone marrow mesenchymal stem cells. For these reasons, we have named them olfactory ecto-mesenchymal stem cells.
- Gene expression profiling indicates that human olfactory ecto-mesenchymal stem cells are prone to differentiate into osseous cells.
- *In vitro*, when compared with human bone marrow mesenchymal stem cells, human olfactory ecto-mesenchymal stem cells display a high propensity to give rise to osteoblasts.
- *In vivo*, when plated onto ceramics and subcutaneously transplanted into nude mice, human olfactory ecto-mesenchymal stem cells differentiate into osteoblasts.
- When compared with bone marrow mesenchymal stem cells, olfactory ecto-mesenchymal stem cells are highly mitogenic.
- Being easily collected in human individuals, they can be used for autologous transplantation.

Acknowledgements

This work was financially supported by ANR (Agence nationale de la Recherche), AFM (Association Française contre les Myopathies), FEDER in PACA and IRME (Institut de Recherche sur la Moelle épinière et l'Encéphale).

List of Abbreviations

AGC	:	Aggrecan
BM-MSC	:	Bone marrow mesenchymal stem cells
COL2A1	:	Collagen type II alpha 1
DMEM/F12	:	Dulbecco's Modified Eagle Medium/Ham F-12
ENT	:	Ear, nose and throat
FBS	:	Fetal bovine serum
MBCP	:	Microporous biphasic calcium phosphate
MSC	:	Mesenchymal stem cells
OE-MSC	:	Olfactory ecto-mesenchymal stem cells
POC	:	Periosteal cells
PTHR1	:	Parathyroid hormone receptor 1
SFb	:	Synovial fibroblasts

References

Arinzeh, T.L., T. Tran, J. Mcalary and G. Daculsi. 2005. A comparative study of biphasic calcium phosphate ceramics for human mesenchymal stem-cell-induced bone formation. Biomaterials. 26(17): 3631–8. doi:10.1016/j.biomaterials.2004.09.035.

Arinzeh, T.L., S.J. Peter, M.P. Archambault, C. van den Bos, S. Gordon, K. Kraus, A. Smith et al. 2003. Allogeneic mesenchymal stem cells regenerate bone in a critical-sized canine segmental defect. The Journal of Bone and Joint Surgery. American volume. 85-A(10): 1927–35.

Delorme, B. and P. Charbord. 2007. Culture and characterization of human bone marrow mesenchymal stem cells. Methods in Molecular Medicine. 140: 67–81.

Delorme, B., E. Nivet, J. Gaillard, T. Häupl, J. Ringe, A. Devèze, J. Magnan et al. 2010. The human nose harbors a niche of olfactory ectomesenchymal stem cells displaying neurogenic and osteogenic properties. Stem Cells and Development. 19(6): 853–66. doi:10.1089/scd.2009.0267.

El Tamer, M.K. and R.L. Reis. 2009. Progenitor and stem cells for bone and cartilage regeneration. Journal of Tissue Engineering and Regenerative Medicine. 3(5): 327–37. doi:10.1002/term.173.

Giannoudis, P.V., T.A. Einhorn and D. Marsh. 2007. Fracture healing: the diamond concept. Injury. 38 Suppl. 4: S3–6.

Girard, S.D., A. Devéze, E. Nivet, B. Gepner, F.S. Roman and F. Féron. 2011. Isolating nasal olfactory stem cells from rodents or humans. Journal of Visualized Experiments: JoVE, (54). doi:10.3791/2762.

Hernigou, P., A. Poignard, F. Beaujean and H. Rouard. 2005. Percutaneous autologous bone-marrow grafting for nonunions. Influence of the number and concentration of progenitor cells. The Journal of Bone and Joint Surgery. American volume. 87(7): 1430–7. doi:10.2106/ JBJS.D.02215.

Horwitz, E.M., D.J. Prockop, L.A. Fitzpatrick, W.W. Koo, P.L. Gordon, M. Neel, M. Sussman et al. 1999. Transplantability and therapeutic effects of bone marrow-derived mesenchymal cells in children with osteogenesis imperfecta. Nature Medicine. 5(3): 309–13. doi:10.1038/6529.

Jones, E. and X. Yang. 2011. Mesenchymal stem cells and bone regeneration: Current status. Injury. 42(6): 562–8. doi:10.1016/j.injury.2011.03.030.

Kon, E., A. Muraglia, A. Corsi, P. Bianco, M. Marcacci, I. Martin, A. Boyde et al. 2000. Autologous bone marrow stromal cells loaded onto porous hydroxyapatite ceramic accelerate bone repair in critical-size defects of sheep long bones. Journal of Biomedical Materials Research. 49(3): 328–37.

Matsumoto, T., A. Kawamoto, R. Kuroda, M. Ishikawa, Y. Mifune, H. Iwasaki, M. Miwa et al. 2006. Therapeutic potential of vasculogenesis and osteogenesis promoted by peripheral blood CD34-positive cells for functional bone healing. The American Journal of Pathology. 169(4): 1440–57. doi:10.2353/ajpath.2006.060064.

Mori, G., G. Brunetti, A. Oranger, C. Carbone, A. Ballini, L.L. Muzio, S. Colucci et al. 2011. Dental pulp stem cells: osteogenic differentiation and gene expression. Annals of the New York Academy of Sciences. 1237(1): 47–52. doi:10.1111/j.1749-6632.2011.06234.x.

Mostafa, N.Z., R. Fitzsimmons, P.W. Major, A. Adesida, N. Jomha, H. Jiang and H. Uludağ. 2011. Osteogenic Differentiation of Human Mesenchymal Stem Cells Cultured with Dexamethasone, Vitamin D3, Basic Fibroblast Growth Factor, and Bone Morphogenetic Protein-2. Connective Tissue Research. doi:10.3109/03008207.2011.611601.

Murrell, W., F. Féron, A. Wetzig, N. Cameron, K. Splatt, B. Bellette, J. Bianco et al. 2005. Multipotent stem cells from adult olfactory mucosa. Developmental dynamics : an official publication of the American Association of Anatomists. 233(2): 496–515. doi:10.1002/ dvdy.20360.

Murrell, W., A. Wetzig, M. Donnellan, F. Féron, T. Burne, A. Meedeniya, J. Kesby et al. 2008. Olfactory mucosa is a potential source for autologous stem cell therapy for Parkinson's disease. Stem Cells (Dayton, Ohio). 26(8): 2183–92. doi:10.1634/stemcells.2008-0074.

Nauth, A., P.V. Giannoudis, T.A. Einhorn, K.D. Hankenson, G.E. Friedlaender, R. Li and E.H. Schemitsch. 2010. Growth factors: beyond bone morphogenetic proteins. Journal of Orthopaedic Trauma. 24(9): 543–6. doi:10.1097/BOT.0b013e3181ec4833.

Nivet, E., M. Vignes, S.D. Girard, C. Pierrisnard, N. Baril, A. Devèze, J. Magnan et al. 2011. Engraftment of human nasal olfactory stem cells restores neuroplasticity in mice with hippocampal lesions. The Journal of Clinical Investigation. 121(7): 2808–20. doi:10.1172/ JCI44489.

Ohtsuki, C., M. Kamitakahara and T. Miyazaki. 2009. Bioactive ceramic-based materials with designed reactivity for bone tissue regeneration. Journal of the Royal Society, Interface / the Royal Society, 6 Suppl. 3: S349–60. doi:10.1098/rsif.2008.0419.focus.

Pandit, S.R., J.M. Sullivan, V. Egger, A.A. Borecki and S. Oleskevich. 2011. Functional effects of adult human olfactory stem cells on early-onset sensorineural hearing loss. Stem Cells (Dayton, Ohio). 29(4): 670–7. doi:10.1002/stem.609.

Petite, H., V. Viateau, W. Bensaïd, A. Meunier, C. de Pollak, M. Bourguignon, K. Oudina et al. 2000. Tissue-engineered bone regeneration. Nature Biotechnology. 18(9): 959–63. doi:10.1038/79449.

Quarto, R., M. Mastrogiacomo, R. Cancedda, S.M. Kutepov, V. Mukhachev, A. Lavroukov, E. Kon et al. 2001. Repair of large bone defects with the use of autologous bone marrow stromal cells. The New England Journal of Medicine. 344(5): 385–6. doi:10.1056/ NEJM200102013440516.

Xiao, M., K.M. Klueber, C. Lu, Z. Guo, C.T. Marshall, H. Wang and F.J. Roisen. 2005. Human adult olfactory neural progenitors rescue axotomized rodent rubrospinal neurons and promote functional recovery. Experimental Neurology. 194(1): 12–30. doi:10.1016/j. expneurol.2005.01.021.

Xiao, M., K.M. Klueber, J. Zhou, Z. Guo, C. Lu, H. Wang and F.J. Roisen. 2007. Human adult olfactory neural progenitors promote axotomized rubrospinal tract axonal reinnervation and locomotor recovery. Neurobiology of Disease. 26(2): 363–74. doi:10.1016/j. nbd.2007.01.012.

CHAPTER 5

Multipotential Mesenchymal Stromal/Stem Cells in Skeletal Tissue Repair

Agnieszka Arthur,[1,a] Andrew Zannettino[2] and Stan Gronthos[1,b,*]

ABSTRACT

Mesenchymal stromal/stem cells (MSC) are increasingly becoming a prime candidate for tissue engineering and regenerative medicine due to the unique combination of their accessibility, multipotency, paracrine effects, and immunomodulatory properties. MSC are a subpopulation of clonogenic marrow stromal cells originally identified in the 1960s, and are now thought to reside in a quiescent state within a perivascular niche within the bone marrow microenvironment. The function of MSC during normal bone homeostasis/remodeling is to instigate bone formation through their differentiation into osteoblasts, which has also led to the investigation of how MSC may be utilized specifically in skeletal tissue repair. Skeletal tissues include not only

[1]Mesenchymal Stem Cell Group, Department of Haematology, SA Pathology, Frome Road, Adelaide 5000, SA, Australia
[a]E-mail: agnes.arthur@health.sa.gov.au
[b]E-mail: stan.gronthos@health.sa.gov.au
[2]Myeloma Research Laboratory, Department of Haematology, SA Pathology, Adelaide 5000, SA, Australia.
E-mail: andrew.zannettino@health.sa.gov.au
*Corresponding author

List of abbreviations given at the end of the text.

bone and cartilage, but also tendon and ligament tissue. A multitude of approaches using MSC in different skeletal tissue engineering applications have been performed in small and larger animal pre-clinical models, where some of this work has been extended to early human clinical trials with varying success. One strategy utilizes endogenous MSC through the application of soluble cytokines, growth factors or hormones to stimulate local MSC proliferation and differentiation. Alternatively, the isolation of exogenous MSC has been studied more extensively, ranging from freshly isolated or culture expanded autologous, allogeneic or genetically modified MSC. These populations have then been delivered either by systemic infusion or directly into the site of injury in combination with a variety of biocompatible scaffold materials only or additionally supplemented with cytokines, growth factors or hormones. While a number of approaches have proven to be clinically efficacious in skeletal repair, a standardized approach for the repair of specific skeletal tissues is yet to be elucidated. The present chapter summarizes the isolation, characterization and application of MSC for skeletal tissue regeneration.

BONE MARROW STROMAL CELLS

The Origin

Adult bone marrow microenvironment is comprised of various stromal elements including, adipocytes, smooth muscle cells, reticular cells and osteoblasts (Weiss 1976). These cell populations, not only provide environmental cues to support the survival, proliferation and differentiation of hematopoietic stem cells (HSC), but also play a pivotal role in regulating skeletal tissue homeostasis. Friedenstein and colleagues were the first to report the isolation of a population of stem/progenitor stromal cells located within rodent bone marrow (Friedenstein 1976). These stem/progenitor stromal cells were identified by their capacity to form clonogenic adherent fibroblast clusters, containing 50 or greater cells in number when cultured *in vitro*, denoted as colony forming unit-fibroblasts (CFU-F) (Friedenstein 1976). Since these initial seminal findings, human studies have also confirmed that multipotential MSC or bone marrow stromal cells (BMSC) reside within the bone marrow spaces (Gronthos et al. 2003b; Pittenger et al. 1999). The developmental potential of MSC encompasses many of the stromal populations found in skeletal tissues such as osteoblasts, chondrocytes, adipocytes, myoblasts and tendon/ligament fibroblasts (Fig. 5.1).

More recently it has been shown that various tissues have also been found to contain MSC-like populations using the same criteria as described for bone marrow derived MSC. These include dental pulp tissue, periodontal ligament, cartilage, tendon, muscle, periosteum, synovium, synovial fluid,

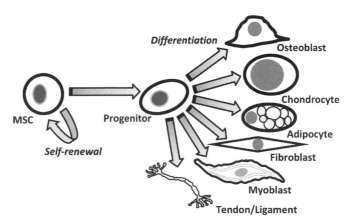

Figure 5.1 The skeletal differentiation capacity of MSC. MSC have the capacity to self-renew, they also proliferate and give rise to committed progenitors and subsequently differentiate into skeletal derivatives.

adipose, umbilical cord blood, placenta, liver, spleen and thymus, where some of these tissues have a different developmental origin to bone marrow (Huang et al. 2009). The different types of MSC-like populations are also thought to reside in a common microenvironment at perivascular sites within their respective tissues (Covas et al. 2008; Gronthos et al. 2003b; Sacchetti et al. 2007; Shi and Gronthos 2003). While it is tempting to classify different stem/progenitor cell populations under the banner of "MSC", more research is needed to assess the origins, properties and functional of different MSC-like populations in skeletal tissue regeneration applications.

Stem Cell Niche

Stem cells reside in a microenvironment, commonly referred to as a niche, within the tissue where they can remain quiescent until required (Schofield 1983). The perivascular niche has been identified as a potential location for MSC in different tissues using the perivascular marker, CD146 (MUC-18) which is located in close proximity within the vasculature wall (Covas et al. 2008; Sacchetti et al. 2007). CD146 in addition to stromal precursor antigen-1 (STRO-1) identify a subset of bone marrow, adipose tissue and dental pulp tissue-derived MSC (Gronthos et al. 2003b; Shi and Gronthos 2003). *In situ*, these markers also co-localize with the pericyte markers 3G5 and α-smooth muscle actin (Gronthos et al. 2003b; Shi and Gronthos 2003). It is postulated that MSC identified in the different tissues may in fact be STRO-1[+] pericytes (Doherty et al. 1998; Farrington-Rock et al. 2004). Furthermore, it is suggested that the vasculature in different organs may associate with stromal elements to support the maintenance and regulation of local parenchyma and assist in local tissue regeneration. It is thought that a number of factors assist in maintaining the MSC within their niche until they are mobilized for normal tissue homeostasis following insult due to disease or injury. However, to properly investigate the function and properties of the MSC, better protocols are needed for the purification of homogeneous populations of multi-potential MSC.

Identification and Isolation

The Mesenchymal and Tissue Stem Cell Committee of the International Society for Cellular Therapy have proposed three criteria to define MSC; including "(1) the plastic adherence of the isolated cells in culture, (2) the expression of CD105, CD73 and CD90 in greater than 95% of the culture, and their lack of expression of markers including CD34, CD45, CD14 or CD11b, CD79α or CD19 and HLA-DR in greater than 95% of the culture, (3) the differentiation of the MSC into osteoblasts, adipocytes and chondrocytes *in vitro*" (Dominici et al. 2006). Other key markers also used to identify MSC include, STRO-1, STRO-3, STRO-4; CD106, CD146, CD49a/CD29, EGF-R, IGF-R, PDGF-R, or NGF-R, where antibodies reactive with these molecules have been used to isolate populations of human MSC with multi-lineage differentiation potential *in vivo* (Arthur et al. 2008) (Fig. 5.2).

The conventional methodology by which MSC populations are isolated is predominantly carried out by centrifugation using cell separation density gradients (ficoll, percoll, dextran and sucrose), followed by plastic adherence (Pittenger et al. 1999). Alternatively, a more sophisticated method is to utilize fluorescence or magnetic activated cell sorting based on the expression of one or more specific stromal cell surface markers described above, to prospectively isolate MSC from a background of hematopoietic cells and mature stromal and vascular elements (Fig. 5.2). Negative immunodepletion strategies using antibodies to haematopoietic cell populations have also been used to partially purify MSC populations from bone marrow. The stem cell potential of these MSC preparations is assessed by their ability to generate primary CFU-F and to differentiate into bone, fat and cartilage. It is important to note that a number of cell surface molecules expressed by MSC isolated from bone marrow aspirates are down regulated following *ex vivo* expansion, and that cultured MSC also exhibit a differential gene expression profile *in vitro* (Gronthos et al. 2003b). It must also be stressed that these molecules are not indicators of a homogeneous population of MSC following *ex vivo* expansion, as illustrated by the differential growth and developmental potentials exhibited by individually expanded MSC clones (Gronthos et al. 2003b; Kuznetsov et al. 1997; Pittenger et al. 1999). However, *ex vivo* expanded MSC that continue to express high levels of

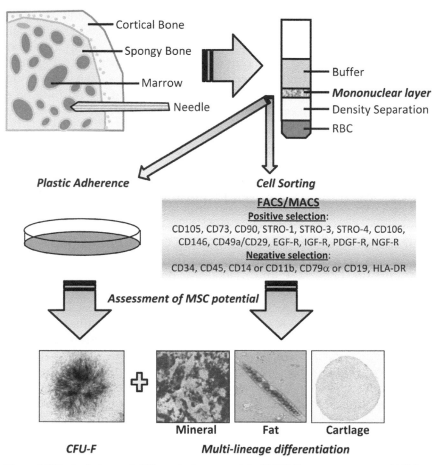

Figure 5.2 The isolation and differentiation potential of MSC. Bone marrow stromal MSC are aspirated from the marrow of the iliac crest, separation techniques isolate the mononuclear cells using density gradients. These mononuclear cells undergo plastic adherence directly or purified further using FACS or MACS with either single or multiple positive or negative selection markers. The stem cell capacity is measured by CFU-F and their capacity to differentiate into mineral producing osteoblasts, adipocytes and chondrocytes.

STRO-1 appear to maintain an immature stem cell-like phenotype (Psaltis et al. 2010). The maintenance of immature MSC populations during the *ex vivo* expansion process, whether by secondary immunoselection or by the use of specialized media formulations, is currently an area of intense research for the development of therapeutic cell preparations that maintain multi-lineage potential, high expression of paracrine and immunomodulatory factors, in order to optimize the potency and minimize the risk of immune rejection following transplantation.

Skeletal Tissue Regeneration Using MSC

Skeletal integrity and mechanical strength of the bone requires the continuous remodeling of bone throughout life. This is known as bone homeostasis, bone metabolism, remodeling or coupling. It is a complex process that involves the coordinated function of bone resorbing osteoclasts and bone forming osteoblasts to replace old bone or to repair skeletal damage, respectively. The normal bone remodeling process occurs in a cyclical manner which is referred to as the activation-resorption-formation (ARF) cycle and takes place in bone multicellular units (BMU). Osteoclasts, which initiate the remodeling process, originating from the hematopoietic lineage, are multinucleated cells that function by resorbing damaged bone. In contrast, the bone formation stage recruits osteoblasts arising from bone marrow MSC, to form new osteoid within the resorption pit that eventually becomes mineralized to remodel the bone surface (Henriksen et al. 2009) (Fig. 5.3). The coupling of osteoclast and osteoblast function is tightly regulated by hormones in addition to local molecular signals. However,

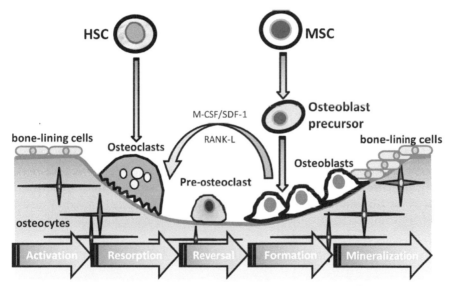

Figure 5.3 The process of bone homeostasis/remodeling. The bone remodeling process begins with the activation of osteoclast maturation, instigated by signals sent from osteoblasts, the main signaling molecules include macrophage-colony stimulating factor (M-CSF), stromal derived factor 1 (SDF-1) and receptor activator of nuclear factor kappa-β ligand (RANKL) (Vandyke et al. 2011). Osteoclasts originate from HSC; mature osteoclasts resorb the old bone and form a pit. Mononuclear cells are present following the resorption stage at an intermediate stage of the remodeling process, called reversal. Osteoblast precursors derived from MSC are then recruited to the resorbed pit, where they proliferate and differentiate into mature osteoblasts (formation stage). These mature osteoblasts produce new bone matrix (mineralization stage) and fill the resorption pit.

in certain situations the normal repair and remodeling processes are often impaired; this includes non-union fractures and diseases that affect the long bones, cranial bones, cartilage, ligament or tendon such as osteoporosis, osteopenia, osteoarthritis, cleft plate or craniofacial disorders, including mandibular deformities, cancer and infection. Therefore, MSC from different skeletal sources and preparations have been assessed as novel cell-based therapies to facilitate the developmental/remodeling processes required for the repair of damaged skeletal tissues. However, the regeneration of any three-dimensional tissue is a complex process, where it has been proposed that a number of key elements are required to coordinate the generation of functional skeletal repair including: the isolation of the most suitable stem/progenitor cell populations; whether to use an autologous or allogeneic tissue source; its efficient administration including cell concentration, biomaterial or scaffold used and location of administration; and the use of MSC in combination with appropriate factors required for skeletal repair.

Processes of Skeletal Repair

It is important to note that different clinical applications may require alternate cell therapy strategies to optimize the therapeutic benefits of MSC. The site of transplantation may also contribute to the efficacy of the transplanted cells, alternatively pre-incubating the cells with factors that induce osteogeneic or chondrogenic differentiation may also help facilitate the repair response. Furthermore, there is evidence to suggest that the immaturity of the MSC may influence the survival of human MSC and their capacity to stimulate angiogenesis and endogenous progenitor populations (Gronthos et al. 2003a; Psaltis et al. 2010). Whilst the survival and differentiation of transplanted MSC is essential for skeletal tissue engineering, it is equally important that the regenerated structure is mechanically stable. The cell-seeding density and tensile strain may also be contributing factors to the repair process (Sumanasinghe et al. 2008).

The use of Autologous or Allogeneic MSC

Generally, two types of MSC have been assessed in different tissue engineering approaches for skeletal repair, MSC isolated from an individual's own body (autologous) and MSC isolated from an unrelated donor from the same species (allogeneic). Furthermore, it remains to be proven whether it is better to sequester endogenous MSC to the site of injury or whether it is more efficacious to isolate and culture expand autologous MSC for transplantation back into the site of injury. Alternatively, *in vitro* and pre-clinical studies have demonstrated the efficacy of delivering allogeneic MSC preparations that display immunotolerance and immunmodulatory properties (Patel et al. 2008; Sotiropoulou et al. 2006), supporting the concept of the "off-the-self" use of allogeneic MSC therapies prepared from normal healthy donors.

Long bone fractures

Long bone fracture studies using canine pre-clinical models have assessed the therapeutic efficacy of either autologous or allogeneic culture expanded MSC in combination with biocompatible scaffolds transplanted into non-union femur fractures. In these studies, both allogeneic and autologous MSC were shown to facilitate new bone formation at similar rates without initiating an immunological response (Arinzeh et al. 2003; Bruder et al. 1998). Similarly, pre-clinical ovine studies have reported comparable levels of bone regeneration in segmental long bone defects for both autologous and allogeneic *ex vivo* expanded MSC preparations seeded onto HA and HA/TCP based ceramic scaffolds (Field et al. 2011; Giannoni et al. 2008). To date, there are only a small number of pilot clinical studies demonstrating the efficacy of autologous MSC preparations in long bone fracture repair (Marcacci et al. 2007; Quarto et al. 2001).

Autoimmune skeletal disorders

Rheumatoid arthritis (RA) is a skeletal disease which is also an autoimmune disorder associated with inflammation of the joints predominantly due to self-reactive T cell that cause the deterioration of cartilage matrix between the joints. There are many studies demonstrating that MSC have the capacity to suppress T cell responses that induce the chronic inflammation seen in RA and, therefore, are likely candidates as potential cellular therapies for RA (Zheng et al. 2008). Studies using animal disease models closely mimicking human RA, demonstrated that infusion of either allogeneic MSC, administration of *ex vivo* expanded autologous MSC, or freshly isolated autologous MSC and platelets, helped prevent severe and irreversible damage to the bone and cartilage by causing hyporesponsiveness of T lymphocytes (Augello et al. 2007; Centeno et al. 2008). Collectively, these pre-clinical studies showed that MSC based cellular therapies may be a viable treatment option for RA in order to reduce the onset of osteoarthritis and facilitate some functional improvement in the affected joints over time. While the precise mechanisms of action are unknown various factors produced by MSC following interactions with activated immune cells are able to inhibit NK and T-cell proliferation while stimulating Treg responses to dampen inflammatory reactions (Augello et al. 2007; Centeno et al. 2008).

Collagenous skeletal tissue disorders

Tendon and ligament are essentially elastic collagenous tissues which are important for motion by providing the attachment of bone/cartilage to muscle or joint with joint respectively. While conventional treatments for tendon and ligament related injuries are limited to surgery and physically therapy, MSC preparations have been recently used for tendon repair of race horses. Undifferentiated autologous MSC were administered to horses with partial tendon damage, where the findings indicated that nearly 90 percent of the horses showed varying levels of functional recovery (Pacini et al. 2007). Parallel studies of different tendon injury equine model involving the administration of autologous and allogeneic MSC showed a lack of any host immune response to the transplanted cells (Guest et al. 2008). Moreover, the delivery of either culture expanded MSC or bone marrow mononuclear cells also reduced scar tissue formation in the tendons of treated animals when compared with the untreated control animals (Crovace et al. 2007).

While studies investigating the use of MSC treatment for ligament repair have been scant, published reports suggest that both autologous and allogeneic derived MSC display phenotypic characteristics of the endogenous surrounding collagen type I-rich tissue following transplantation. These small animal studies showed that the transplantation of MSC reduced the size of the injury site due to the infiltration of the transplanted cells, suggesting that MSC administration at the site of injury may be a feasible treatment for ligament repair (Li et al. 2007).

Transplanted vs. Endogenous Use of MSC

Transplantation of Freshly Isolated vs. Culture expanded MSC

The use of freshly isolated bone marrow appears to be a common first step in large animal models and human trials of skeletal repair. Fresh marrow aspirates consist of a low incidence of MSC (0.0001 percent) diluted in a heterogeneous mix of hematopoietic cells (Gronthos et al. 2003b). Alternatively, immunoselection of MSC from fresh marrow aspirates may be more potent for bone repair and may mediate a higher rate of functional improvement. Whilst, direct comparative studies investigating freshly isolated MSC compared to culture expanded MSC have not been performed, one study proposed that freshly isolated marrow cells grown in a 3D matrix formed the same amount of bone using 10-fold fewer cells than 2D culture expanded MSC (Scaglione et al. 2006). However, the use of enriched preparations of freshly isolated MSC without culture expansion is generally viewed as a non-practical alternative cellular preparation for the majority of clinical settings. Therefore, researchers need to develop strategies to improve the efficacy of using culture expanded MSC for skeletal repair. One way of overcoming this issue is to utilize the variety of scaffolds that are available for the treatment of specific skeletal tissue engineering applications.

Scaffolds

Skeletal tissue requires structural and mechanical support, where many studies have focused on identifying the optimal supportive structure or scaffold. The ideal scaffold should provide appropriate support or mechanical stability, including the appropriate architecture to produce the correct elasticity or tensile strength. Other properties of scaffolds include support of biological activity via the infiltration of MSC and incorporation of signaling molecules, which is influenced by the porosity of the material and the biocompatibility of materials when transplanted with the living tissue to avoid rejection by the host immune system. Many biomaterials provide a conductive environment for appropriate guided skeletal tissue formation, such as bone, cartilage, or connective tendon or ligament tissues, which can also be designed to biodegrade at different rates during tissue repair. The scaffolds that have been primarily used in skeletal tissue regeneration, thus far, include synthetic and natural biodegradable materials composed of specific polymers or ceramic composites. These scaffolds can range from biodegradable natural materials including collagen, gelatin, silk, fibrin, and matrigel to synthetic polymers such as poly glycolic acid (PGA), poly lactic acid (PLA), and poly $_{DL}$-lactic-glycolic acid (PLGA), silicon or ceramic materials HA and TCP that can be manipulated to form diverse structures such as hydrogels or porous sponges, depending on the application (reviewed by (Arthur et al. 2008; Sundelacruz and Kaplan 2009).

Alternatives to the traditional biomaterials include the TheriForm 3D-printing of scaffolds, or the use of cultured hyperconfluent MSC sheets attached onto bone grafts. These alternatives provide the versatility to form the correct shape and size of the scaffold compatible with the defect site. Recently, two other unique scaffolds have been suggested for mandibular repair, including coral and injectable tissue-engineered bone. Firstly, coral, a porous natural material very similar in composition to bone is composed of 1 percent organic material (amino acids) and 99 percent calcium carbonate (form of aragonite) (Hou et al. 2007). While the injectable tissue-engineered bone was composed of insoluble plasma rich gel that was combined with culture expanded pre-differentiated MSC and injected directly into the defect in combination with titanium plates for stability (Yamada et al. 2004).

A multitude of scaffolds are being developed for tissue engineering purposes as specific materials are more advantageous to use when trying to repair particular skeletal tissues. For example, the repair of long or cranial bones requires the material to provide strength and stability, particularly for load bearing bony tissue, while repair of cartilaginous tissue requires the scaffold to provide elasticity and flexibility.

Scaffolds for long bones and cranial skeletal tissue

Porous ceramic scaffolds composed of either HA, TCP or HA/TCP composites are ideal biocompatible materials for long bone or cranial skeletal repair, due to the osteo-conductive environment provided to the transplanted cells, and because of the provision of the initial structural integrity required at the defect site. However, the extent of biodegradability is a limiting factor depending on the percentage of HA, which may compromise the biomechanics of the newly formed bone. Despite this limitation, the HA/TCP ceramic composite scaffolds in combination with autologous MSC have significantly enhanced bone formation by facilitating cellular integration, strength and repair of critical sized bony defects in large pre-clinical canine, caprine and ovine models (Field et al. 2011; Giannoni et al. 2008; Kuznetsov et al. 2008).

Subsequent studies in rodents and sheep, utilizing ceramic scaffolds consisting of silicon and TCP rather than HA, found that the scaffolds reduced in size from 3 mon post-transplantation and were completely resorbed within 2 yr, coincided with new bone formation (Giannoni et al. 2008; Mastrogiacomo et al. 2006). However, species differences in the capacity of MSC to form bone *in vivo* have been previously reported for different biomaterial carriers (Zannettino et al. 2011). It appears that human MSC, but not ovine or mouse MSC, require a biomaterial composite containing approximately 15 percent HA to stimulate appropriate bone formation *in vivo* (Quarto et al. 2001; Zannettino et al. 2011).

Scaffolds for cartilaginous tissues

The repair of osteo-chondral defects require the generation of hyaline-like cartilage that can integrate with the surrounding cartilage and repair the defect. Therefore, there has been a trend to use collagen based scaffolds for cartilage repair in pre-clinical animal and human clinical studies, where these materials lack osteo-conductive properties and are more flexible and resorbable, unlike the HA/TCP ceramic scaffolds preferred for long bone fracture and craniofacial defect repair. Human case studies have reported that all the patients who received autologous MSC therapy in combination with a collagen-based scaffold demonstrated evidence of cartilage repair, albeit most likely in the form of fibrocartilage, and were able to resume their daily and sporting activities, with no evidence of adverse effects 5 yr post MSC therapy (Kuroda et al. 2007; Wakitani et al. 2007).

Furthermore, other scaffolds have been reported to mediate osteochondral repair when combined with MSC include poly (-caprollactone) (PCL)-TCP, polylactic acid (PLA) matrix, fibrin gel, hyaluronan gel and chitosan, which is a linear polysaccharide consisting of $_D$-glucosamine reviewed by Arthur et al. (2008). An interesting alternative to the traditional solid scaffolds is an atelocollagen gel, a liquid scaffold that hardens at body temperature, with flexible properties to allow for the correct shaping of the graft size. This type of scaffold has been assessed in a rabbit intervertebral disc injury model and demonstrated improved bone formation and repair following transplantation in combination with autologous rabbit MSC (Sakai et al. 2003). Other carrier vehicles in combination with MSC have been used for the repair of osteochondral defects of the knee and intervertebral disc degeneration in pre-clinical animal models, include hyaluronan and pentosan polysulfate, where these compounds have been shown to stimulate cartilage matrix development (Ghosh et al. 2011; Goldschlager et al. 2011).

Scaffolds for connective tissues

The use of MSC seeded scaffolds has not been studied extensively in tendon or ligament repair. The limited studies to date have focused on *in vitro* studies and animal models using horse, rabbit, pig or rat. The scaffold materials that have demonstrated promising findings at the initial stages of tendon repair include collagen, laminin or fibrin based compounds in combination with MSC (Hairfield-Stein et al. 2007; Kajikawa et al. 2007). However, it is important to note that over extended periods of time, minimal differences could be detected in the mechanical properties between MSC and control implants (Awad et al. 2003). This observation implies that either the scaffold alone is sufficient for tendon repair; or potentially other mechanisms may also be acting to assist in the repair process, such as paracrine mechanisms that may sequester endogenous stem cell populations, stimulate angiogenesis and inhibit inflammatory processes. One study supporting this notion showed that when cultured MSC sheets were attached to frozen tendon grafts, the co-cultured MSC sheets were able to respond to environmental cues produced by the tendon tissue. The mechanism of action was based on the ability of MSC to integrate within the tendon tissue which displayed characteristics of tenocyte-like cells at the site of injury (Ouyang et al. 2006). Interestingly, enforced expression of Scleraxis in cultured MSC, which is known to be a critical transcription factor in tendon formation during embryogenesis, improved rotator cuff tendon healing and reduced the incidence of re-tears following transplantation into a damaged rotator cuff rat injury model (Gulotta et al. 2011).

Manipulation of endogenous MSC for skeletal repair

While the focus thus far has been on manipulating /enhancing the function of exogenous MSC, other studies have focused on augmenting the function of endogenous MSC. The rationale for this approach is that the release of specific growth factors or cytokines important for proliferation and more importantly osteogenesis, may also stimulate the therapeutic potential of endogenous MSC. These factors predominantly include growth factors from the members of the transforming growth factor β (TGF β) family, the BMPs (BMP-2, BMP-4, BMP-7), the fibroblast growth factor (bFGF) family (Hou et al. 2007) and parathyroid hormone (PTH) (Takahata et al. 2011). These factors have all shown that they are able to improve the capacity of endogenous MSC to form bone. Specifically BMPs have been used in pre-clinical animal and human clinical studies of non-union fracture, craniofacial repair and lumbar spinal fusion (Delawi et al. 1976; White et al. 2007). For example the administration of soluble BMP-7 alone or in combination with autologous cancellous bone resulted in healing of non-union fracture in 87 percent of patients within 5–6 mon of treatment (Giannoudis et al. 2009; Kanakaris et al. 2009). It is also important to note that caution should be taken for the unrestricted use of BMP therapy to avoid severe adverse effects in specific applications such as those that have arisen in cervical spinal fusion trials using BMP-2 (Carragee et al. 2011; Lindley et al. 2011).

An alternative to growth factors is the use of the bone stimulating hormone, PTH, which induces not only osteogenesis and chondrogenesis but also stimulates MSC proliferation and is important particularly during the callus formation stage of fracture repair. Large animal models and case reports have demonstrated successful results in fracture repair. However, the latest clinical trial failed to demonstrate any significant difference in repair between the experimental and placebo groups (reviewed by (Takahata et al. 2011)).

Similarly other potential chemokines, mitogens and osteo-inductive factors have also been considered as candidates for inducing endogenous MSC therapeutic potential including, CXCL12/SDF-1, bFGF, VEGF, EGF and TGFβ3. Collectively, these observations suggest that while certain growth factors, chemokines and hormones may hold great therapeutic potential, further investigations are clearly required in order to optimize the efficacy of biological factors as single therapeutic agents for skeletal tissue regeneration.

Key Facts

- During the normal bone remodeling process, MSC give rise to osteoblasts, which are bone forming cells that also mediate signals that activate bone resorbing osteoclasts to remove damaged regions of bone.
- MSC are an ideal therapy for skeletal repair as they are easy to access, isolate, expand and manipulate.
- MSC have immune-modulatory properties, which is protective and reduces the chances of rejection following transplantation. Therefore there is a potential dual role for MSC to mediate skeletal repair and inhibit inflammatory/autoimmune responses.
- Skeletal repair includes repair of long bones, cranial bones, cartilaginous tissue, ligaments and tendon. Therefore, MSC need to be manipulated accordingly to facilitate the specific repair.
- MSC combined with the appropriate scaffolds may be an advantageous approach for skeletal tissue repair.
- Appropriate scaffold design is dependent on: **mechanical properties**, providing strength or flexibility, **architecture**, influencing structural integrity, **porosity**, important for the incorporation, survival, differentiation and vascularization, **biodegradability**, to allow for appropriate remodeling of newly formed skeletal tissue, **biocompatability**, to avoid rejection by the host, and **biological activity**, which includes MSC function and also refers to the incorporation of molecular factors that may assist in the proliferation and/or differentiation of MSC.
- Manipulation of endogenous MSC with factors essential during normal osteogenesis is an alternative therapeutic approach used for skeletal repair; it has been studied in large animal models and expanded to clinical trials. Factors predominantly investigated include members of the BMPs and PTH.

Dictionary

- *Bone remodeling/homeostasis:* the removal of damaged bone and replacement with newly formed bone utilizing bone forming osteoblasts, derived from MSC and bone resorbing osteoclasts originating from HSC (Fig. 5.3).
- *Mesenchymal Stem Cells (MSC):* are a population of specialized cells that reside within a specific location within the body (a niche) where they remain dormant until required. These cells have the capacity to self-renew, proliferate and differentiate into multiple lineages (skeletal, myogenic and neurogenic).

- *Perivascular niche:* a microenvironment located within the vascular wall of many tissues where MSC are thought to reside.
- *Skeletal tissue:* encompasses bones, cartilaginous tissue, and connective tissues including ligaments and tendons.
- *Autologous MSC:* isolated from an individual's own body.
- *Allogeneic MSC:* isolated from an unrelated donor from the same species.

Summary

- MSC can be isolated from many tissues and are thought to reside in the perivascular niche until they are sequestered to mediate a response.
- MSC, are able to differentiate into bone, fat, cartilage, tendon and ligament tissues both *in vitro* and *in vivo*.
- As a therapeutic agent, MSC are becoming increasingly more popular and viable, due to their differentiation capacity, immune-modulatory properties, accessibility and easy manipulation.
- Important factors to consider for appropriate skeletal repair using MSC include the immaturity of the MSC, the seeding density, mechanical stability and tensile strain.
- Researchers have employed a number of approaches of utilizing MSC to achieve the greatest skeletal repair for the specific skeletal tissue (Fig. 5.4). These include stimulation with cytokines, chemokines, growth factors and hormones, isolation of autologous or allogeneic MSC (freshly isolated, culture expanded or genetically manipulated) and varying transplantation process (direct injection or utilizing scaffold material).
- MSC mediated therapy shows varying potential to improve skeletal regeneration; however, challenges remain to overcome a number of technical facets, including the precise mechanism of action of MSC, the source and dose, the survival rate, the appropriate scaffold/carrier material, time of transplantation, and whether supplementation of cytokine/growth factors are necessary.

List of Abbreviations

ARF	:	activation-resorption-formation
bFGF	:	fibroblast growth factor
BMSC	:	bone marrow stromal cells
BMSSC	:	bone marrow stromal stem cells
BMU	:	bone multicellular units
BrdU	:	Bromodeoxyuridine

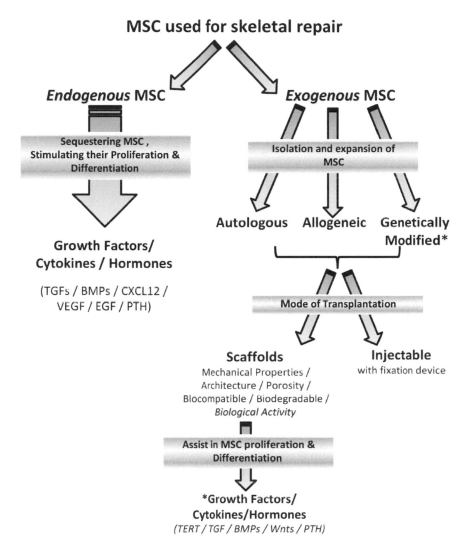

Figure 5.4 A summary of MSC based therapeutics for skeletal tissue repair. MSC utilized for skeletal tissue engineering have focused on stimulating endogenous MSC to proliferate and differentiate into skeletal derivatives. Alternatively, MSC have been isolated, culture expand and transplanted back into the injury site, directly or in combination with a scaffold. These exogenous MSC are autologous, allogeneic or genetically modified with specific factors that may assist in MSC proliferation or differentiation. Numerous scaffold materials have been investigated for their efficacy and functional repair. Correct skeletal repair is dependent on appropriate scaffold design and potentially the use of specific molecules important for MSC proliferation and differentiation.

CFU-F	:	colony forming unit-fibroblasts
CXCL12/ SDF-1	:	stromal derived factor 1
EGF	:	epidermal growth factor
EGF-R	:	epidermal growth factor receptor
FACS	:	fluorescence activated cell sorting
HA	:	hydroxyapatite
HLA	:	human leukocyte antigen
HSC	:	hematopoietic stem cells
IGF-R	:	insulin-like growth factor 1 receptor
MACS	:	magnetic activated cell sorting
MAPC	:	multipotent adult progenitor cells
MASCS	:	mesenchymal adult stem cells
M-CSF	:	macrophage-colony stimulating factor
MIAMI	:	marrow-isolated adult multipotent inducible cells
MSC	:	Mesenchymal/Stromal Stem Cells
NGF-R	:	nerve growth factor receptor
PCL-TCP	:	poly-caprollactone
PDGF-R	:	platelet-derived growth factor receptor
PGA	:	poly glycolic acid
PLA	:	poly lactic acid
PLA	:	polylactic acid
PLGA	:	poly $_{DL}$-lactic-glycolic acid
PTH	:	parathyroid hormone
RA	:	Rheumatoid arthritis
RANKL	:	receptor activator of nuclear factor kappa-β ligand
STRO-1	:	stromal precursor antigen-1
STRO-3	:	stromal precursor antigen-3
STRO-4	:	stromal precursor antigen-4
TCP	:	β-tricalcium phosphate
TGF β	:	transforming growth factor β
VEGF	:	Vascular endothelial growth factor

References

Arinzeh, T.L., S.J. Peter, M.P. Archambault, C. van den Bos, S. Gordon, K. Kraus, A. Smith and S. Kadiyala. 2003. Allogeneic mesenchymal stem cells regenerate bone in a critical-sized canine segmental defect. J Bone Joint Surg Am. 85-A: 1927–1935.

Arthur, A., A. Zannettino and S. Gronthos. 2008. The therapeutic applications of multipotential mesenchymal/stromal stem cells in skeletal tissue repair. J Cell Physiol. 218: 237–245.

Augello, A., R. Tasso, S.M. Negrini, R. Cancedda and G. Pennesi. 2007. Cell therapy using allogeneic bone marrow mesenchymal stem cells prevents tissue damage in collagen-induced arthritis. Arthritis Rheum. 56: 1175–1186.

Awad, H.A., G.P. Boivin, M.R. Dressler, F.N. Smith, R.G. Young and D.L. Butler. 2003. Repair of patellar tendon injuries using a cell-collagen composite. J Orthop Res. 21: 420–431.

Bruder, S.P., K.H. Kraus, V.M. Goldberg and S. Kadiyala. 1998. The effect of implants loaded with autologous mesenchymal stem cells on the healing of canine segmental bone defects. J Bone Joint Surg Am. 80: 985–996.

Carragee, E.J., E.L. Hurwitz and B.K. Weiner. 2011. A critical review of recombinant human bone morphogenetic protein-2 trials in spinal surgery: emerging safety concerns and lessons learned. Spine J. 11: 471–491.

Centeno, C.J., D. Busse, J. Kisiday, C. Keohan, M. Freeman and D. Karli. 2008. Increased knee cartilage volume in degenerative joint disease using percutaneously implanted, autologous mesenchymal stem cells. Pain Physician. 11: 343–353.

Covas, D.T., R.A. Panepucci, A.M. Fontes, W.A. Silva, Jr., M.D. Orellana, M.C. Freitas, L. Neder, A.R. Santos, L.C. Peres, M.C. Jamur and M.A. Zago. 2008. Multipotent mesenchymal stromal cells obtained from diverse human tissues share functional properties and gene-expression profile with CD146+ perivascular cells and fibroblasts. Exp Hematol. 36: 642–654.

Crovace, A., L. Lacitignola, R. De Siena, G. Rossi and E. Francioso. 2007. Cell therapy for tendon repair in horses: an experimental study. Vet Res Commun. 31 Suppl. 1: 281–283.

Delawi, D., W.J. Dhert, L. Rillardon, E. Gay, D. Prestamburgo, C. Garcia-Fernandez, E. Guerado, N. Specchia, J.L. Van Susante, N. Verschoor, H.M. van Ufford and F.C. Oner. 1976. A prospective, randomized, controlled, multicenter study of osteogenic protein-1 in instrumented posterolateral fusions: report on safety and feasibility. Spine (Phila Pa). 35: 1185–1191.

Doherty, M.J., B.A. Ashton, S. Walsh, J.N. Beresford, M.E. Grant and A.E. Canfield. 1998. Vascular pericytes express osteogenic potential *in vitro* and *in vivo*. J Bone Miner Res. 13: 828–838.

Dominici, M., K. Le Blanc, I. Mueller, I. Slaper-Cortenbach, F. Marini, D. Krause, R. Deans, A. Keating, D. Prockop and E. Horwitz. 2006. Minimal criteria for defining multipotent mesenchymal stromal cells. The International Society for Cellular Therapy position statement. Cytotherapy. 8: 315–317.

Farrington-Rock, C., N.J. Crofts, M.J. Doherty, B.A. Ashton, C. Griffin-Jones and A.E. Canfield. 2004. Chondrogenic and adipogenic potential of microvascular pericytes. Circulation. 110: 2226–2232.

Field, J.R., M. McGee, R. Stanley, G. Ruthenbeck, T. Papadimitrakis, A. Zannettino, S. Gronthos and S. Itescu. 2011. The efficacy of allogeneic mesenchymal precursor cells for the repair of an ovine tibial segmental defect. Vet Comp Orthop Traumatol. 24: 113–121.

Friedenstein, A.J. 1976. Precursor cells of mechanocytes. Int Rev Cytol. 47: 327–359.

Ghosh, P., J. Wu, S. Shimmon, A.C. Zannettino, S. Gronthos and S. Itescu. 2011. Pentosan polysulfate promotes proliferation and chondrogenic differentiation of adult human bone marrow-derived mesenchymal precursor cells. Arthritis Res Ther. 12: R28.

Giannoni, P., M. Mastrogiacomo, M. Alini, S.G. Pearce, A. Corsi, F. Santolini, A. Muraglia, P. Bianco and R. Cancedda. 2008. Regeneration of large bone defects in sheep using bone marrow stromal cells. J Tissue Eng Regen Med. 2: 253–262.

Giannoudis, P.V., N.K. Kanakaris, R. Dimitriou, I. Gill, V. Kolimarala and R.J. Montgomery. 2009. The synergistic effect of autograft and BMP-7 in the treatment of atrophic nonunions. Clin Orthop Relat Res. 467: 3239–3248.

Goldschlager, T., P. Ghosh, A. Zannettino, S. Gronthos, J.V. Rosenfeld, S. Itescu and G. Jenkin. 2011. Cervical motion preservation using mesenchymal progenitor cells and pentosan polysulfate, a novel chondrogenic agent: preliminary study in an ovine model. Neurosurg Focus. 28: E4.

Gronthos, S., S. Chen, C.Y. Wang, P.G. Robey and S. Shi. 2003a. Telomerase accelerates osteogenesis of bone marrow stromal stem cells by upregulation of CBFA1, osterix, and osteocalcin. J Bone Miner Res. 18: 716–722.

Gronthos, S., A.C. Zannettino, S.J. Hay, S. Shi, S.E. Graves, A. Kortesidis and P.J. Simmons. 2003b. Molecular and cellular characterisation of highly purified stromal stem cells derived from human bone marrow. J Cell Sci. 116: 1827–1835.

Guest, D.J., M.R. Smith and W.R. Allen. 2008. Monitoring the fate of autologous and allogeneic mesenchymal progenitor cells injected into the superficial digital flexor tendon of horses: preliminary study. Equine Vet J. 40: 178–181.

Gulotta, L.V., D. Kovacevic, J.D. Packer, X.H. Deng and S.A. Rodeo. Bone marrow-derived mesenchymal stem cells transduced with scleraxis improve rotator cuff healing in a rat model. Am J Sports Med. 39: 1282–1289.

Hairfield-Stein, M., C. England, H.J. Paek, K.B. Gilbraith, R. Dennis, E. Boland and P. Kosnik. 2007. Development of self-assembled, tissue-engineered ligament from bone marrow stromal cells. Tissue Eng. 13: 703–710.

Henriksen, K., A.V. Neutzsky-Wulff, L.F. Bonewald and M.A. Karsdal. 2009. Local communication on and within bone controls bone remodeling. Bone. 44: 1026–1033.

Hou, R., F. Chen, Y. Yang, X. Cheng, Z. Gao, H.O. Yang, W. Wu and T. Mao. 2007. Comparative study between coral-mesenchymal stem cells-rhBMP-2 composite and auto-bone-graft in rabbit critical-sized cranial defect model. J Biomed Mater Res A. 80: 85–93.

Huang, G.T., S. Gronthos and S. Shi. 2009. Mesenchymal stem cells derived from dental tissues vs. those from other sources: their biology and role in regenerative medicine. J Dent Res. 88: 792–806.

Kajikawa, Y., T. Morihara, N. Watanabe, H. Sakamoto, K. Matsuda, M. Kobayashi, Y. Oshima, A. Yoshida, M. Kawata and T. Kubo. 2007. GFP chimeric models exhibited a biphasic pattern of mesenchymal cell invasion in tendon healing. J Cell Physiol. 210: 684–691.

Kanakaris, N.K., N. Lasanianos, G.M. Calori, R. Verdonk, T.J. Blokhuis, P. Cherubino, P. De Biase and P.V. Giannoudis. 2009. Application of bone morphogenetic proteins to femoral non-unions: a 4-year multicentre experience. Injury. 40 Suppl. 3: S54–61.

Kuroda, R., K. Ishida, T. Matsumoto, T. Akisue, H. Fujioka, K. Mizuno, H. Ohgushi, S. Wakitani and M. Kurosaka. 2007. Treatment of a full-thickness articular cartilage defect in the femoral condyle of an athlete with autologous bone-marrow stromal cells. Osteoarthritis Cartilage. 15: 226–231.

Kuznetsov, S.A., P.H. Krebsbach, K. Satomura, J. Kerr, M. Riminucci, D. Benayahu and P.G. Robey. 1997. Single-colony derived strains of human marrow stromal fibroblasts form bone after transplantation *in vivo*. J Bone Miner Res. 12: 1335–1347.

Kuznetsov, S.A., K.E. Huang, G.W. Marshall, P.G. Robey and M.H. Mankani. 2008. Long-term stable canine mandibular augmentation using autologous bone marrow stromal cells and hydroxyapatite/tricalcium phosphate. Biomaterials. 29: 4211–4216.

Li, F., H. Jia and C. Yu. 2007. ACL reconstruction in a rabbit model using irradiated Achilles allograft seeded with mesenchymal stem cells or PDGF-B gene-transfected mesenchymal stem cells. Knee Surg Sports Traumatol Arthrosc. 15: 1219–1227.

Lindley, T.E., N.S. Dahdaleh, A.H. Menezes and K.O. Abode-Iyamah. 2011. Complications associated with recombinant human bone morphogenetic protein use in pediatric craniocervical arthrodesis. J Neurosurg Pediatr. 7: 468–474.

Marcacci, M., E. Kon, V. Moukhachev, A. Lavroukov, S. Kutepov, R. Quarto, M. Mastrogiacomo and R. Cancedda. 2007. Stem cells associated with macroporous bioceramics for long bone repair: 6- to 7-year outcome of a pilot clinical study. Tissue Eng. 13: 947–955.

Mastrogiacomo, M., A. Corsi, E. Francioso, M. Di Comite, F. Monetti, S. Scaglione, A. Favia, A. Crovace, P. Bianco and R. Cancedda. 2006. Reconstruction of extensive long bone defects in sheep using resorbable bioceramics based on silicon stabilized tricalcium phosphate. Tissue Eng. 12: 1261–1273.

Ouyang, H.W., T. Cao, X.H. Zou, B.C. Heng, L.L. Wang, X.H. Song and H.F. Huang. 2006. Mesenchymal stem cell sheets revitalize nonviable dense grafts: implications for repair of large-bone and tendon defects. Transplantation. 82: 170–174.

Pacini, S., S. Spinabella, L. Trombi, R. Fazzi, S. Galimberti, F. Dini, F. Carlucci and M. Petrini. 2007. Suspension of bone marrow-derived undifferentiated mesenchymal stromal cells for repair of superficial digital flexor tendon in race horses. Tissue Eng. 13: 2949–2955.

Patel, S.A., L. Sherman, J. Munoz and P. Rameshwar. 2008. Immunological properties of mesenchymal stem cells and clinical implications. Arch Immunol Ther Exp (Warsz). 56: 1–8.

Pittenger, M.F., A.M. Mackay, S.C. Beck, R.K. Jaiswal, R. Douglas, J.D. Mosca, M.A. Moorman, D.W. Simonetti, S. Craig and D.R. Marshak. 1999. Multilineage potential of adult human mesenchymal stem cells. Science. 284: 143–147.

Psaltis, P.J., S. Paton, F. See, A. Arthur, S. Martin, S. Itescu, S.G. Worthley, S. Gronthos and A.C. Zannettino. 2010. Enrichment for STRO-1 expression enhances the cardiovascular paracrine activity of human bone marrow-derived mesenchymal cell populations. J Cell Physiol. 223: 530–540.

Quarto, R., M. Mastrogiacomo, R. Cancedda, S.M. Kutepov, V. Mukhachev, A. Lavroukov, E. Kon and M. Marcacci. 2001. Repair of large bone defects with the use of autologous bone marrow stromal cells. N Engl J Med. 344: 385–386.

Sacchetti, B., A. Funari, S. Michienzi, S. Di Cesare, S. Piersanti, I. Saggio, E. Tagliafico, S. Ferrari, P.G. Robey, M. Riminucci and P. Bianco. 2007. Self-renewing osteoprogenitors in bone marrow sinusoids can organize a hematopoietic microenvironment. Cell. 131: 324–336.

Sakai, D., J. Mochida, Y. Yamamoto, T. Nomura, M. Okuma, K. Nishimura, T. Nakai, K. Ando and T. Hotta. 2003. Transplantation of mesenchymal stem cells embedded in Atelocollagen gel to the intervertebral disc: a potential therapeutic model for disc degeneration. Biomaterials. 24: 3531–3541.

Scaglione, S., A. Braccini, D. Wendt, C. Jaquiery, F. Beltrame, R. Quarto and I. Martin. 2006. Engineering of osteoinductive grafts by isolation and expansion of ovine bone marrow stromal cells directly on 3D ceramic scaffolds. Biotechnol Bioeng. 93: 181–187.

Schofield, R. 1983. The stem cell system. Biomed Pharmacother. 37: 375–380.

Shi, S. and S. Gronthos. 2003. Perivascular niche of postnatal mesenchymal stem cells in human bone marrow and dental pulp. J Bone Miner Res. 18: 696–704.

Sotiropoulou, P.A., S.A. Perez, M. Salagianni, C.N. Baxevanis and M. Papamichail. 2006. Characterization of the optimal culture conditions for clinical scale production of human mesenchymal stem cells. Stem Cells. 24: 462–471.

Sumanasinghe, R.D., J.A. Osborne and E.G. Loboa. 2008. Mesenchymal stem cell-seeded collagen matrices for bone repair: Effects of cyclic tensile strain, cell density, and media conditions on matrix contraction *in vitro*. J Biomed Mater Res A.

Sundelacruz, S. and D.L. Kaplan. 2009. Stem cell- and scaffold-based tissue engineering approaches to osteochondral regenerative medicine. Semin Cell Dev Biol. 20: 646–655.

Takahata, M., H.A. Awad, R.J. O'Keefe, S.V. Bukata and E.M. Schwarz. 2011. Endogenous tissue engineering: PTH therapy for skeletal repair. Cell Tissue Res.

Vandyke, K., S. Fitter, A.L. Dewar, T.P. Hughes and A.C. Zannettino. 2011. Dysregulation of bone remodeling by imatinib mesylate. Blood. 115: 766–774.

Wakitani, S., M. Nawata, K. Tensho, T. Okabe, H. Machida and H. Ohgushi. 2007. Repair of articular cartilage defects in the patello-femoral joint with autologous bone marrow mesenchymal cell transplantation: three case reports involving nine defects in five knees. J Tissue Eng Regen Med. 1: 74–79.

Weiss, L. 1976. The hematopoietic microenvironment of the bone marrow: an ultrastructural study of the stroma in rats. Anat Rec. 186: 161–184.

White, A.P., A.R. Vaccaro, J.A. Hall, P.G. Whang, B.C. Friel and M.D. McKee. 2007. Clinical applications of BMP-7/OP-1 in fractures, nonunions and spinal fusion. Int Orthop. 31: 735–741.

Yamada, Y., M. Ueda, H. Hibi and T. Nagasaka. 2004. Translational research for injectable tissue-engineered bone regeneration using mesenchymal stem cells and platelet-rich plasma: from basic research to clinical case study. Cell Transplant. 13: 343–355.

Zannettino, A.C., S. Paton, S. Itescu and S. Gronthos. Comparative assessment of the osteoconductive properties of different biomaterials *in vivo* seeded with human or ovine mesenchymal stem/stromal cells. Tissue Eng Part A. 16: 3579–3587.

Zheng, Z.H., X.Y. Li, J. Ding, J.F. Jia and P. Zhu. 2008. Allogeneic mesenchymal stem cell and mesenchymal stem cell-differentiated chondrocyte suppress the responses of type II collagen-reactive T cells in rheumatoid arthritis. Rheumatology (Oxford). 47: 22–30.

Adipose-derived Stem Cells (ASCs) and Bone Repair: An Overview in Regenerative Medicine

Natalina Quarto[a],* and Michael T. Longaker[b]

ABSTRACT

Research in regenerative medicine is developing at a significantly rapid pace. Regenerative medicine and tissue engineering both harness the potency of human cells to repair, regenerate, and even recreate tissues and organs with the goal of restoring their architecture and functionality. The emerging field of regenerative medicine will require a reliable source of stem cells in addition to biomaterial scaffolds and cytokine growth factors. Adipose tissue represents an abundant and accessible source of adult stem cells with the ability to differentiate along multiple lineage pathways. Adipose derived stem cells (ASCs) have emerged as a new and promising type of stem cells with two clear advantages over previously used types (e.g., bone marrow, blood, or skeletal muscle): first, the easy and repeatable access that makes it

Hagey Laboratory for Pediatric Regenerative Medicine, Department of Surgery, Stanford University, School of Medicine, Stanford, CA, USA.
[a]E-mail: quarto@unina.it
[b]E-mail: longaker@stanford.edu
*Corresponding author

List of abbreviations given at the end of the text.

possible to harvest large amounts of adipose tissue by a minimally invasive method and, secondly, their increased proliferative potential in culture, either because of the properties of the cells themselves or because of the greater frequency of stem cells within the population used to initiate the culture. The aim of this chapter is to review some aspects of ASCs biology, their osteoinductive ability and potential use in skeletal tissue repair.

Introduction

The repair of large bone defects due to trauma, inflammation, osteoporosis or tumor represents a major challenge in orthopaedics. Regenerative medicine is a rapidly expanding field that has emerged as a potential solution in addressing the problems associated with current approaches to skeletal defects. The central focus of regenerative medicine is human cells. These may be adult, embryo-derived or somatic cells and now there are versions of the latter cells that have been reprogrammed from adult cells so that both can be conveniently collected under the heading of "pluripotent cells" (Takahashi and Yamanaka 2006). In the past decade, the field of stem cell biology has undergone a remarkable evolution sparked by reports demonstrating that adult stem cells possess greater plasticity than dictated by established paradigms of embryonic development.

Tissue engineering is becoming a promising and appealing alternative for reconstructive surgery procedures. The main goal behind tissue engineering is to mimic the natural wound healing cascade by providing suitable biochemical and mechanical cellular microenvironments which augment the proliferation and differentiation of recruited host cells or implanted cells at the defect site. Cell-based modalities incorporating the use of multipotent mesenchymal cells, particularly those derived from adipose tissue, have generated much attention both within the research and clinical realms. Stem cells are ideal candidates for the use in regenerative medicine because of their ability to self-renew and to commit to multiple cell lineages (Pittenger et al. 1999). Stem cells for regenerative medicine should meet the following criteria: they can be found in abundant numbers (millions to billions of cells); can be harvested by a minimally invasive procedure with little morbidity; can differentiate along multiple lineage pathways in a controlled and reproducible manner; and can be safely and effectively transplanted to either an autologous or allogenic host.

Current surgical strategies for the repair of skeletal defects employ either autogenous grafts or alloplastic materials. Although these approaches are in large part successful, they both have inherent disadvantages. For instance, alloplastic materials have problems with rejection, infection and eventual breakdown, while autologous tissue such as bone and bone marrow grafts are often limited in availability and require a substantive operation with potential morbidity (Sawin et al. 1998; Younger and Chapman 1989). Therefore, there is still a strong clinical need for a suitable alternative to currently available techniques for bone tissue repair.

Historically, adipose tissue has been considered a metabolic reservoir for packaging, storing and releasing high-energy substrates, and for years lipoaspirate has been discarded as surgical waste. Today, adipose tissue is considered a "valuable" compartment for its abundant population of mesenchymal stem cells, namely adipose derived stem cells (ASCs). With the increased occurrence of obesity, subcutaneous adipose tissue is readily accessible and thus ASCs can be harvested in large quantities with minimal risk. Because of the existence of multipotent stem cells within the adipose tissue, this tissue has acquired increased importance for regenerative medicine. Several laboratories included ours, focus on harnessing the osteogenic capability of ASCs for eventual repair of non-healing skeletal defects.

Herein, we describe recent data pertaining to the plasticity and therapeutic potential of stem/progenitor cells derived from adult adipose tissues. Our focus is to provide an overview on current knowledge of ASCs considering the tissue origin, their phenotype, their optimal niche, and describe how the unique biology of ASCs provides a basis for their therapeutic efficacy in the context of bone regeneration.

Biology of ASCs

Nomenclature

The nomenclature is part of the history of ASCs. As in many rapidly developing fields, a variety of names have been initially used to describe the plastic adherent cell population present in the stromal vascular fraction (SVF) isolated from collagenase digests of adipose tissue. The following terms have been used to identify the same adipose tissue cell population: adipose derived adult stem (ADAS) cells, adipose derived stromal cells (ADSC), adipose stromal cells (ASC), adipose mesenchymal stem cells (AdMSC), human multipotent adipose-derived stem cells (hMADS), processed lipoaspirate (PLA) cells and pre-adipocytes. The disparate use of this nomenclature has lead to significant confusion in the literature until when in 2004, at the Second International Fat Applied Technology Society (IFATS) meeting scientists reached a consensus to adopt the term "adipose-derived stem cells" (ASCs) to identify the isolated, plastic-adherent, multipotent cell population. Thus, these days it is widely accepted by investigators to use the acronym ASCs when referring to adipose-derived stem cells.

ASCs and BMSCs: *vis a vis*

Mesenchymal stem cells (MSCs) represent a promising tool for new clinical concepts in supporting cellular therapy. MSCs are found in many adult tissues and are an attractive stem cell source for the regeneration of damaged tissues in clinical applications because they are characterized as undifferentiated cells, able to self-renew with a high proliferative capacity, and possess a mesodermal differentiation potential. Bone marrow (BM) was the first source reported to contain MSCs (Caplan 1991; Pittenger et al. 1999). Although bone marrow (BM) has been the main source for the isolation of multipotent MSCs, the harvest of BM is an invasive procedure and the number, differentiation potential, and maximal life span of MSCs from BM decline with increasing age (D'Ippolito et al. 1999; Huibregtse et al. 2000). Therefore, alternative sources from which to isolate MSCs are subject to intensive investigation. With the ASCs, 40 yr after the identification of bone marrow stem cells, a new era of active stem cell therapy has opened. Zuk and collegues were the first to report the presence of stem cells in the adipose tissue, publishing two important papers, the first demonstrating the multilineage differentiation capacity of ASCs (Fig. 6.1) into cells of mesenchymal origin such as adipocytes, osteocytes, chondrocytes and myocytes (Zuk et al. 2001), and the second one showing the presence of all the proteins recognized as markers for the stem cells (Zuk et al. 2002). Isolated ASCs can be cryopreserved and expanded easily *in vitro*. Under the conditions commonly used, these cells develop a fibroblast-like morphology.

Comparative studies indicated that ASCs and BMSCs are remarkably similar with respect to multilineage differentiation capability, growth and morphology, telomerase activity and gene expression (Izadpanah et al. 2006). Other common characteristics of ASCs and BMSCs can be found in the transcriptional and cell surface profile. ASCs and BMSCs both express embryonic stem cell markers, Oct-4, Rex-1 and Sox-2 for at least 10 passages (Izadpanah et al. 2006). They also display similar immune-phenotypic profiles with no expression of the hematopoietic markers (CD14, CD34, CD45), or the stem cell marker CD133 or the endothelial cell marker CD144. In contrast, they have high expression of the typical MSC marker proteins CD44, CD73, CD29, CD90 CD105 (Kern et al. 2006). However, ASCs have unique features making them an appealing and convenient tool for tissue engineering, perhaps more than BMSCs. For example, contrary to bone marrow, adipose tissue is probably the most abundant and accessible source of adult stem cells (Fraser et al. 2006). Therefore, ASCs hold great promise for use in tissue repair and regeneration representing an alternative source of autologous adult stem cells that can be obtained repeatedly in large quantities under local anesthesia by liposuction procedures having minimal

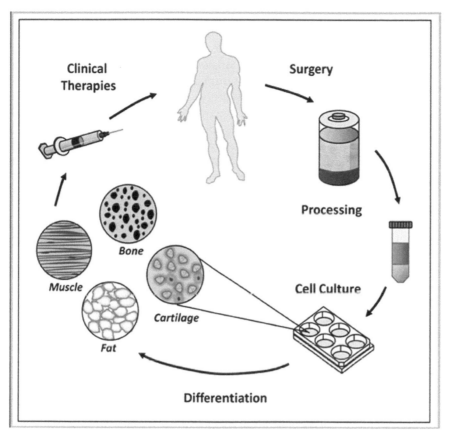

Figure 6.1 Schematic outline of differentiation potential of ASCs isolated from lipoaspirate tissues. Human lipoaspirate is an abundant source of ASCs that under appropriate culture conditions can differentiate along different lineages.

Color image of this figure appears in the color plate section at the end of the book.

donor side morbidity and patient discomfort, whereas MSCs harvest requires aspiration from the iliac crest or from bone marrow biopsies, both of which can be painful and yield low numbers (Pittenger et al. 1999; Prockop 1997). Additionally, experience with large volume bone marrow has also demonstrated that scaling from small to large harvest volumes compromises drastically the yield of BMSCs (100 times less) due to a substantial increase of whole blood contaminants (Bacigalupo et al. 1992; Batinic et al. 1990). In contrast, adipose tissue harvested by liposuction is not subject to this "dilution effect". Therefore, for clinical use, BMSCs may be a less attractive cell source due to the invasive donation procedure and the decline in MSC number in large harvest volume. Moreover, the abundance of ASCs harvested from adipose tissue enables one to instantly apply primary cells without culture expansion. On the contrary, BMSCs are present at low frequencies and require extensive *in vitro* expansion before sufficient numbers are obtained for matrix loading, while the ASCs are abundant in adipose tissue, easy to harvest, and have rapid *in vitro* expansion (De Ugarte et al. 2003; Zuk et al. 2002; Zuk et al. 2001). One gram of human adipose tissue could yield over 70,000 ASCs within 24 hr of culture (Gimble and Guilak 2003). Characteristics, such as the colony frequency and the maintenance of proliferating ability in culture, are also superior in ASCs compared to BMSCs (Lee et al. 2007). Further, ASCs are shown to survive in a low oxygen environment (Follmar et al. 2006), making them a good candidates for cell-based therapies by secreting angiogenic cytokines such as VEGF and HGF (Gimble et al. 2007; Moon et al. 2006; Rehman et al. 2004) perhaps to a higher degree than for instance BMSCs (Kilroy et al. 2007).

However, the number of clinical trials performed using BMSCs is much higher than those performed with ASCs. More than 500 BMSCs trials can be found that are underway or completed, while searching for clinical trials using either SVF or ASCs yields only 18 trials, of which two are so far completed. This discrepancy is due to two reasons: first, ASCs are relatively "young" therefore, represent a new generation that needs further investigations prior clinical trials, second, the United States FDA has not yet approved their use in clinical trials.

ASCs and Bone Regeneration

The demand for new therapeutic approaches to treat bone defects and fractures is increasing in trauma surgery and orthopaedics because the number of patients with degenerative diseases is continuously growing. "Tissue Engineering" offers promising new technologies that combine the three components—cells, growth factors and matrix. Efforts are targeted at improving and accelerating recovery, especially for long bone fractures, and reducing the risk of delayed or non bone healing or pseudoarthrosis. Bone regeneration is a coordinated cascade of events regulated by various bioactive molecules such as growth factors whose concentrations, temporal gradients, spatial gradients, sequences of their release are precisely controlled (Kanczler and Oreffo 2008). ASCs can differentiate into osteoblasts in an osteogenic environment (Cowan et al. 2004; Izadpanah et al. 2006; Quarto et al. 2008; Quarto et al. 2006; Wan et al. 2006; Zuk et al. 2001). Bone morphogenetic protein-2 (BMP-2) accelerates and initiates this differentiation. ASCs have the capacity to differentiate into a multitude of cell types and participate in tissue regeneration directly or by helping to recruit additional cell types. This regenerative potential suggests that adult progenitor cells may be used to fulfill the mounting needs of patients with genetic disorders, degenerative diseases, and traumatic or post-surgical tissue deficits of the adult skeleton. Indeed, several *in vivo* studies using animal models support the potential translation for ASCs use in the treatment of human skeletal defects (Fig. 6.2).

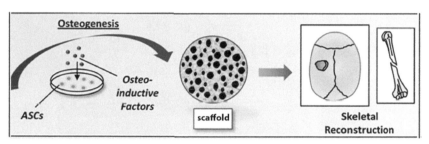

Figure 6.2 Schematic outline of ASCs isolation from human liposaspirate and their use for bone regeneration. Human lipoaspirate in a sterile container is processed for isolation of ASCs. Plated ASCs are grown and subsequentially treated with osteoinductive factors (e.g., BMPs, FGF-2, VEGF), seeded on scaffolds and implanted on injured bones to promote skeletal reconstruction.

Color image of this figure appears in the color plate section at the end of the book.

In vivo ASCs osteogenesis: animal models

To adequately translate *in vitro* findings to the clinical realm, compelling *in vivo* data must be obtained to demonstrate the osteogenic capacity of ASCs. The ability of ASCs to regenerate bony tissue has been widely documented in various animal models. A calvarial defect model is largely employed for the evaluation of ASCs to heal critical sized (or non-healing) defects. Several studies demonstrated the ability of ASCs to heal calvarial defect (Aalami et al. 2004; Behr et al. 2011; Cowan et al. 2004). This *in vivo* model is currently also used in our laboratory. A standard procedure for this model consists in creating a full-thickness calvarial defect (4 mm in diameter) in the non-suture associated right parietal bone of adult either wild type CD-1 or CD-1 nude mice using diamond-coated trephine bits and saline irrigation. Importantly, great care is devoted to avoiding injury to the underlying dura mater (Fig. 6.3). For implantation, scaffolds are seeded with autologous or heterologous ASCs with or without treatment with different growth

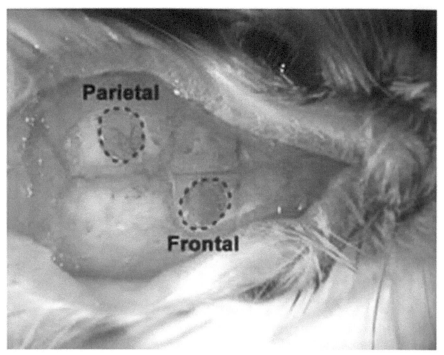

Figure 6.3 Model of murine calvarial bone defect employed to assess the ability of ASCs to regenerate bone. An example of murine calvarial defects created in the right frontal and left parietal bones of adult post-natal day 60 mice with meticulous care to avoid injury to the dura mater. Dotted circles mark the defects. Note the continuous blood vessels, indicating intact dura mater.

Color image of this figure appears in the color plate section at the end of the book.

factors (e.g., BMP-2, VEGF, FGF-2). Micro-computed tomography (μCT) is performed at 1, 2, 4, and 8 wk following osteotomy and grafting of ASC/ scaffold constructs to assess healing of the calvarial defects.

Recently it has been reported that freshly isolated hASCs loaded either onto a porous ceramic scaffold HA/β-TCP, apatite coated PLGA scaffold or coral scaffold successfully heal critical sized defect without the need for pre-differentiation (Figs. 6.4 and 6.5) (Behr et al. 2011; Levi et al. 2010; Lin et al. 2008). The ability to use hASCs immediately without any pre-differentiation and pretreatment with growth factors is of clinically relevance, as the less time hASCs need to expand *in vitro*, the less likelihood of *in vitro* contamination. From translational perspective the ability to use the harvested hASCs immediately will allow the surgeon to perform a reconstruction in one step. Recombinant differentiation factors such as BMP-2 VEGF and FGF-2 may be used to supplement hASCs mediated bone repair (Behr et al. 2011; Kang et al. 2011; Kwan et al. 2011; Levi et al. 2010).

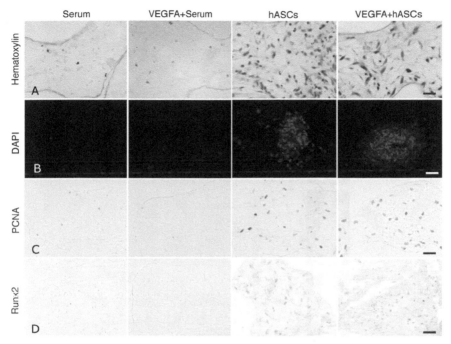

Figure 6.4 hASCs attachment to coral-apatite scaffolds and initiation of *in vivo* **osteogenic differentiation at 48 hr post implantation on calvarial defects.** (A) staining for Hematoxylin and (B) DAPI revealed hASCs attachment *in vivo* after 48 hr. (C) immunohistochemistry for the proliferation cell nuclear antigen PCNA to monitor cell proliferation of untreated and VEGFA treated hASCs. (D) immunohistochemistry for Runx2 detecting the presence of osteoprogenitors in both VEGFA treated and untreated hASCs. In contrast, no staining is observed in the control groups. Scale bars: A, C, D 50 µm, B, 100 µm (Modified from Behr et al. 2011 Stem Cells).

Color image of this figure appears in the color plate section at the end of the book.

The ability of ASCs to repair cranial bones has been also investigated in canine, rabbit and rat calvarial models in an autologous or heterologous fashion (Bohnenblust et al. 2009; Cui et al. 2007; Dudas et al. 2006; Pieri et al. 2010; Yoon et al. 2007).

To assess the osteogenic potential and utility of using ASCs to regenerate bone different models have been employed, such as repair of long bones, and critical sized femoral defects. A study by de Girolamo and co-workers used expanded autologous ASCs for the regeneration of full-thickness bone defects created in the proximal epiphysis of rabbit tibia (de Girolamo et al. 2011). They demonstrated that autologous ASCs–HA constructs are a potential treatment for the regeneration of bone defects. In addition, a heterologous collagen-ceramic scaffold loaded with BMP-2 and human ASCs successfully healed a critical sized femoral defect in nude rats (Peterson et al. 2005). Interestingly, a vertebral bone defect in rat model

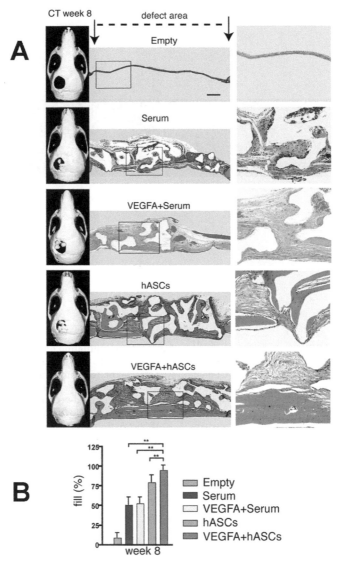

Figure 6.5 Repair of critical sized calvarial defects with VEGFA treated hASCs. (A) Micro-computed tomography (µCT) and corresponding H&E staining of critical-sized calvarial defects in parietal bones of nude mice at postoperative week 8, to assess bone healing. Defects were treated with coral apatite scaffolds loaded with either 2 mg VEGFA-treated or VEGFA-untreated hASCs. Treatment of defects loaded with serum or VEGFA + serum were controls, as well empty defect. Boxed areas are enlarged in the right column. Scale bar: 400 µm. (B) Quantification of defect healing according to µCT results revealed significantly increased healing of defects with VEGFA-treated and VEGFA-untreated ASCs loaded coral scaffolds. (Modified from Behr et al. 2011 Stem Cells).

Color image of this figure appears in the color plate section at the end of the book.

revealed that ASCs overexpressing BMP-6 were capable of regenerating the lost bone tissue, therefore laying the foundation for a novel treatment of vertebral compression fractures (VCFs) (Sheyn et al. 2011). Moreover, a periodontal tissue defect model generated in the rats has shown the ability of autologous ASCs to promote alveolar bone regeneration as well (Tobita et al. 2008).

Bioactive Agents for ASC Bone Tissue Regeneration

Scaffolds

A critical strategy for tissue engineering is to provide the signals necessary for tissue regeneration by mimicking the tissue microenvironment. Therefore, multifunctional scaffolds with sophisticated delivery strategies are needed to induce and complete bone regeneration in bone defects. The choice of scaffold is crucial for successful bone tissue engineering. These scaffolds serve the dual purpose of both cell support and delivery of bioactive agents. A suitable scaffold should have specific characteristics such as: (1) three-dimensional and highly porous for cell growth; (2) biocompatible and bioresorbable at rate to match cell/tissue growth *in vivo*; (3) a chemistry surface permissive for cell attachment, proliferation and differentiation. Indeed, extensive research showed that the success of ASCs to regenerate a functional bone tissue is achieved by loading them on biodegradable scaffold with or without bioactive molecules (i.e., growth factors and cytokines). One of the most popular scaffolds employed for the bone tissue repair by ASCs is the apatite-coated polylactic-coglycolic acid (PLGA) scaffold. First, Cowan demonstrated that osteo-induced ASCs along with the apatite-coated polylactic-coglycolic acid (PLGA) scaffold could repair a critical-sized calvarial defect of a mouse (Cowan et al. 2004). Later, Dudas and co-workers showed that ASCs in combination with gelatin gel could repair non-critical sized defect in a rabbit model (Dudas et al. 2006). Repair of cranial bone defects with ASCs and coral scaffolds in a canine model has also been described (Behr et al. 2011; Cui et al. 2007). One major advantage of using coral scaffold for bone engineering is that the degradation rate of coral suitably matches the kinetics of new bone formation *in vivo* (Petite et al. 2000).

The use of a biodegradable biphasic calcium phosphate nanocomposite (NanoBCP) comprising beta-tricalcium phosphate matrix and hydroxyl apatite nanofibers also demonstrated the ability of ASCs to promote both ectopic and *in situ* bone formation (Lin et al. 2007).

The ability of ASCs to regenerate bone in rat critical-size cranial defects has been determined also by filling the defect with alginate gel combined with BMP-2 transfected ASCs (Lin et al. 2008). The alginate gel is one of the most extensively applied biomaterials in bone tissue engineering. In the presence of calcium ions, the semisolid gel can be formed with cross-linking of alginate chains under mild conditions (Cai et al. 2007).

A porous ceramic scaffold loaded with hydroxyapatite (HA) tricalcium phosphate (TCP) (HA/β-TCP), represents an additional scaffold utilized to promote bone repair using ASCs. More recently a nanocomposite layer of polycaprolactone (PLC) and hydroxyapatite nanoparticles (nHA) (BCP/ PCL-nHA) has also been used as a scaffold to drive osteogenic differentiation of ASCs. That study reported that BCP/PCL-nHA scaffolds induced early osteogenic differentiation of ASCs through-integrin-α2 and an extracellular signal-regulated kinase (ERK) signaling pathway (Lu et al. 2011).

Growth factors

Indeed, the aim of tissue engineering is to achieve implantable bioengineering constructs, made out of cells, together with biocompatible materials. Many growth factors such as BMP-2, BMP-7, TGF-β1 PDGF, FGF and VEGF are known for their important roles in bone formation. Delivery systems for growth factors (bioactive agents) have attracted the attention of researchers including those working with ASCs. Growth factors can be incorporated into the biomaterials either during or after the fabrication process. Their delivery to the target tissue can be accomplished in the form of microparticles, nanoparticles (Basmanav et al. 2008; Jeon et al. 2008; Zhang et al. 2009) or these particles/growth factors incorporated into scaffolds (Kempen et al. 2008; Lan Levengood et al. 2010). A flourishing number of studies indicated that proosteogenic factors such as BMP-2 may be used to accelerate and improve ASCs mediated bone repair, either when loaded onto the scaffold together with the cells or when overexpressed in gene-transduced ASCs (Cowan et al. 2004; Dragoo et al. 2003; Dragoo et al. 2005; Levi et al. 2011; Levi et al. 2010). Sheyn and coworkers reported that gene-modified ASCs overexpressing BMP-6 and suspended in fibrin gel (FG) regenerated vertebral bone defects in a rat model (Sheyn et al. 2011). Peterson tested the ability of BMP-2-producing human ASCs to heal a critically sized femoral defect in a nude rat model (Peterson et al. 2005).

An alternative approach to improve the ability of ASCs for bone repair is through the BMP mediated signaling to potentially reduce antagonists of the BMP pathway. In a recent study Levi and co-workers demonstrated that knocking down the BMP antagonist *noggin*, using either lentiviral or non-integrating minicircle techniques, increased the osteogenic capability of human ASCs. Furthermore, addition of ASCs with noggin suppression to BMP-2 loaded biomimetic scaffolds significantly accelerated bone formation *in vivo* (Levi et al. 2011). This "sophisticated"customized-scaffold-design enhancing the potential of ASCs in skeletal repair might be beneficial to reduce the dose of BMP-2 which is expensive and has a significant risk profile.

A fundamental step in bone tissue regeneration is a sufficient vascular supply, as implantation of large bone grafts without adequate vascularity results in apoptosis and cartilage formation (Muschler et al. 2004). Accordingly, bone repair has been shown to benefit from increased angiogenesis (Santos and Reis 2010). In addition to the well-known angiogenic activity of VEGF, its role in osteogenesis has been established during the last years (Geiger et al. 2005; Gerber et al. 1999; Street et al. 2002). Recently, it has been also reported that hASCs have the capacity to differentiate toward endothelial cells in addition to their capability to differentiate to osteoblasts (Behr et al. 2011; Miranville et al. 2004; Planat-Benard et al. 2004; Scherberich et al. 2007). Locally applied VEGFA has been proved to be a valuable growth factor that can mediate both osteogenesis and angiogenesis of hASCs in the context of calvarial bone regeneration. Behr and co-workers demonstrated that mouse critical sized calvarial defects treated with coral scaffolds loaded with ASCs and VEGF healed faster and better than defects treated with ASCs alone (Figs. 6.4 and 6.5), with bone repair accompanied by a striking enhancement of angiogenesis (Behr et al. 2011).

Other important bioactive factors able to potentiate the osteogenic differentiation of ASCs are members of FGF family such as FGF-2. It has been described that FGF-2 is critical for the ASCs self-renewal and maintains their proliferative and osteogenic potential state to sustain their osteogenic potential *in vitro* (Quarto and Longaker 2006; Zaragosi et al. 2006). The ability of locally applied FGF-2 to further promote ASCs bone repair has been also reported *in vivo*. For instance, an elegant animal model employing a tunable chemical control of FGF-2 release (Fig. 6.6) targeting ASCs loaded on a biomimetic scaffold demonstrated significant increased bone regeneration in mouse calvarial critical-sized defects compared to untreated defects (Kwan et al. 2011).

ASCs in Translational Medicine

The use of ASCs has not yet been approved by the FDA for clinical applications in the USA; however, studies with human ASCs are currently being conducted with individual cases in various countries. Thus far, only few clinical case studies have been reported where the capacity of ASCs in bone tissue repair has been investigated (Lendeckel et al. 2004; Mesimaki et al. 2009) In a first case the patient, a 7-yr-old girl with a severe head injury had her calvarial bone defect treated with autologous SVF applied in combination with milled autologous bone from the iliac crest and autologous fibrin. After 3 mon, new bone tissue formation and a near

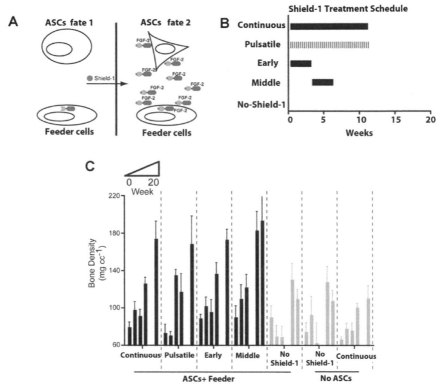

Figure 6.6 Chemical tunable control of FGF-2 release for calvarial bone defect regeneration with ASCs. (A) schematic of feedercells (DD-FGF-2 cells) engineered tosecrete a FGF-2 fusion protein under a regulation system driven by the presence of the synthetic ligand (Shield-1). These DD-FGF-2 cells induce paracrine/osteogenic inductive signals on ASCs loaded on scaffold placed into mouse calvarial defects. (B) schematic of *in vivo* treatment groups. Treatment groups (*n* = 4–6 mice/group) consisted of CD-1 nude mice with calvarial defects implanted with scaffolds on which were seeded DD-FGF-2 cells and ASCs or scaffolds loaded only with DD-FGF-2 cells. Each group was administered with intraperitoneal injections of Shield-1 (6 mg/kg) every other day for first 12 wk (**Continuous**), every 4th day for the first 12 wk (**Pulsatile**), every other day for the first 3 wk (**Early**), or every other day during wk 4–6 (**Middle**). (C) Quantification of defects healing according to µCT results obtained from treatment and control groups at wk 0, 4, 8, and 20. (Modified from Kwan et al. 2011, 286 J Biol Chem).

Color image of this figure appears in the color plate section at the end of the book.

complete calvarial reconstruction were observed (Lendeckel et al. 2004). In a second clinical case, autologous ASCs combined with hrBMP-2 and loaded onto a βTCP scaffold were employed for a maxillary reconstruction. Mature bone tissue developed within the construct after 4 mon (Mesimaki et al. 2009). Recently cranioplasty procedures have been reported , using autologous ASCs seeded in βTCP granules in four patients who had large calvarial defects (Thesleff et al. 2011).

Key Facts

- Bone regeneration can be successfully achieved employing ASCs.
- Optimizations of "ASCs-growth factors-scaffold" bioengineered complex(s) accelerate bone regeneration.

Key Terms

- *Regenerative medicine*: process replacing or regenerating tissues or organs to re-establish normal functions.
- *Osteoinductive*: factor stimulating osteogenesis.
- *Dura mater*: the outermost membrane covering the brain.
- *Cranioplasty*: surgical repair of a skull defect or deformity.
- *Calvaria critical-sized defect*: a defect that will not spontaneously regenerate bone.
- *Micro Computed Tomography*: 3-D image analysis reconstruction for the evaluation of bone tissue.
- *Nanocomposite*: solid multiphasic material in which one of the phases has one, two or three dimensions of less than 100 nanometers.

Summary

- Subcutaneous adipose tissue is a valuable source of adult stem cells (ASCs).
- ASCs can be easily harvested from adipose tissue. Using this attractive cell population, several studies have explored the efficacy of implanted/administrated ASCs in various animal models.
- ASCs can undergo osteogenesis without *in vitro* pre-differentiation and/or any stimulation when loaded on an osteoconductive scaffold *in vivo*.
- ASCs provide opportunity of autologous transplant, making them an appealing cell source and promising candidate for skeletal tissue repair.
- A greater understanding of the mechanism(s) of interactions among ASCs, growth factors and biomaterials on tissue regeneration is needed to advance the clinical utility of this therapy.

Acknowledgment

The authors would like to thank Dr B. Levi for help in graphics.

List of Abbreviations

ASCs	:	adipose-derived stem cells
BCP/PCL -nHA	:	biphasic calcium phosphates/polycaprolactone- hydroxyapatite nanoparticles
BM	:	bone marrow
BMP-2, -6 -7-	:	bone morphogenetic protein-2, -6 -7
DAPI	:	4′,6-diamidino-2-phenylindole
ERK	:	extracellular signal-regulated kinase
FGF-2	:	fibroblast growth factor-2
HA-/β-TCP	:	hydroxyapatite/beta-tricalcium phosphate
HGF	:	hepatocyte growth factor
nHA	:	hydroxyapatite nanoparticles
MSCs	:	mesenchymal stem cells
μCT	:	micro-computed tomography
Oct-4	:	octamer-binding transcription factor-4
PCNA	:	proliferation cell nuclear antigen
PDGF	:	platelet derived growth factor
PLC	:	polycaprolactone
PLGA	:	polylactic-co-glycolic acid
Rex-1	:	reduced expression gene-1
RUNX2	:	Runt-related transcription factor 2
Sox-2	:	SRY (sex determining region Y)-box 2
SVF	:	stromal vascular fraction
TGF-β1	:	transforming growth factor β1
VCFs	:	vertebral compression fractures
VEGF	:	vascular endothelial growth factor

References

Aalami, O.O., R.P. Nacamuli, K.A. Lenton, C.M. Cowan, T.D. Fang, K.D. Fong, Y.Y. Shi, H.M. Song, D.E. Sahar and M.T. Longaker. 2004. Applications of a mouse model of calvarial healing: differences in regenerative abilities of juveniles and adults. Plast Reconstr Surg. 114: 713–720.

Bacigalupo, A., J. Tong, M. Podesta, G. Piaggio, O. Figari, P. Colombo, G. Sogno, E. Tedone, F. Moro, M.T. Van Lint et al. 1992. Bone marrow harvest for marrow transplantation: effect of multiple small (2 ml) or large (20 ml) aspirates. Bone Marrow Transplant. 9: 467–470.

Basmanav, F.B., G.T. Kose and V. Hasirci. 2008. Sequential growth factor delivery from complexed microspheres for bone tissue engineering. Biomaterials. 29: 4195–4204.

Batinic, D., M. Marusic, Z. Pavletic, V. Bogdanic, B. Uzarevic, D. Nemet and B. Labar. 1990. Relationship between differing volumes of bone marrow aspirates and their cellular composition. Bone Marrow Transplant. 6: 103–107.

Behr, B., C. Tang, G. Germann, M.T. Longaker and N. Quarto. 2011. Locally applied vascular endothelial growth factor A increases the osteogenic healing capacity of human adipose-derived stem cells by promoting osteogenic and endothelial differentiation. Stem Cells. 29: 286–296.

Bohnenblust, M.E., M.B. Steigelman, Q. Wang, J.A. Walker and H.T. Wang. 2009. An experimental design to study adipocyte stem cells for reconstruction of calvarial defects. J Craniofac Surg. 20: 340–346.

Cai, X., Y. Lin, G. Ou, E. Luo, Y. Man, Q. Yuan and P. Gong. 2007. Ectopic osteogenesis and chondrogenesis of bone marrow stromal stem cells in alginate system. Cell Biol Int. 31: 776–783.

Caplan, A.I. 1991. Mesenchymal stem cells. J Orthop Res. 9: 641–650.

Cowan, C.M., Y.Y. Shi, O.O. Aalami, Y.F. Chou, C. Mari, R. Thomas, N. Quarto, C.H. Contag, B. Wu and M.T. Longaker. 2004. Adipose-derived adult stromal cells heal critical-size mouse calvarial defects. Nat Biotechnol. 22: 560–567.

Cui, L., B. Liu, G. Liu, W. Zhang, L. Cen, J. Sun, S. Yin, W. Liu and Y. Cao. 2007. Repair of cranial bone defects with adipose derived stem cells and coral scaffold in a canine model. Biomaterials. 28: 5477–5486.

de Girolamo, L., E. Arrigoni, D. Stanco, S. Lopa, A. Di Giancamillo, A. Addis, S. Borgonovo, C. Dellavia, C. Domeneghini and A.T. Brini. 2011. Role of autologous rabbit adipose-derived stem cells in the early phases of the repairing process of critical bone defects. J Orthop Res. 29: 100–108.

De Ugarte, D.A., K. Morizono, A. Elbarbary, Z. Alfonso, P.A. Zuk, M. Zhu, J.L. Dragoo, P. Ashjian, B. Thomas, P. Benhaim, I. Chen, J. Fraser and M.H. Hedrick. 2003. Comparison of multi-lineage cells from human adipose tissue and bone marrow. Cells Tissues Organs. 174: 101–109.

D'Ippolito, G., P.C. Schiller, C. Ricordi, B.A. Roos and G.A. Howard. 1999. Age-related osteogenic potential of mesenchymal stromal stem cells from human vertebral bone marrow. J Bone Miner Res. 14: 1115–1122.

Dragoo, J.L., J.Y. Choi, J.R. Lieberman, J. Huang, P.A. Zuk, J. Zhang, M.H. Hedrick and P. Benhaim. 2003. Bone induction by BMP-2 transduced stem cells derived from human fat. J Orthop Res. 21: 622–629.

Dragoo, J.L., J.R. Lieberman, R.S. Lee, D.A. Deugarte, Y. Lee, P.A. Zuk, M.H. Hedrick and P. Benhaim. 2005. Tissue-engineered bone from BMP-2-transduced stem cells derived from human fat. Plast Reconstr Surg. 115: 1665–1673.

Dudas, J.R., K.G. Marra, G.M. Cooper, V.M. Penascino, M.P. Mooney, S. Jiang, J.P. Rubin and J.E. Losee. 2006. The osteogenic potential of adipose-derived stem cells for the repair of rabbit calvarial defects. Ann Plast Surg. 56: 543–548.

Follmar, K.E., F.C. Decroos, H.L. Prichard, H.T. Wang, D. Erdmann and K.C. Olbrich. 2006. Effects of glutamine, glucose, and oxygen concentration on the metabolism and proliferation of rabbit adipose-derived stem cells. Tissue Eng. 12: 3525–3533.

Fraser, J.K., I. Wulur, Z. Alfonso and M.H. Hedrick. 2006. Fat tissue: an underappreciated source of stem cells for biotechnology. Trends Biotechnol. 24: 150–154.

Geiger, F., H. Bertram, I. Berger, H. Lorenz, O. Wall, C. Eckhardt, H.G. Simank and W. Richter. 2005. Vascular endothelial growth factor gene-activated matrix (VEGF165-GAM) enhances osteogenesis and angiogenesis in large segmental bone defects. J Bone Miner Res. 20: 2028–2035.

Gerber, H.P., T.H. Vu, A.M. Ryan, J. Kowalski, Z. Werb and N. Ferrara. 1999. VEGF couples hypertrophic cartilage remodeling, ossification and angiogenesis during endochondral bone formation. Nat Med. 5: 623–628.

Gimble, J. and F. Guilak. 2003. Adipose-derived adult stem cells: isolation, characterization, and differentiation potential. Cytotherapy. 5: 362–369.

Gimble, J.M., A.J. Katz and B.A. Bunnell. 2007. Adipose-derived stem cells for regenerative medicine. Circ Res. 100: 1249–1260.

Huibregtse, B.A., B. Johnstone, V.M. Goldberg and A.I. Caplan. 2000. Effect of age and sampling site on the chondro-osteogenic potential of rabbit marrow-derived mesenchymal progenitor cells. J Orthop Res. 18: 18–24.

Izadpanah, R., C. Trygg, B. Patel, C. Kriedt, J. Dufour, J.M. Gimble and B.A. Bunnell. 2006. Biologic properties of mesenchymal stem cells derived from bone marrow and adipose tissue. J Cell Biochem. 99: 1285–1297.

Jeon, O., S.J. Song, H.S. Yang, S.H. Bhang, S.W. Kang, M.A. Sung, J.H. Lee and B.S. Kim. 2008. Long-term delivery enhances *in vivo* osteogenic efficacy of bone morphogenetic protein-2 compared to short-term delivery. Biochem Biophys Res Commun. 369: 774–780.

Kanczler, J.M. and R.O. Oreffo. 2008. Osteogenesis and angiogenesis: the potential for engineering bone. Eur Cell Mater. 15: 100–114.

Kang, S.W., J.S. Kim, K.S. Park, B.H. Cha, J.H. Shim, J.Y. Kim, D.W. Cho, J.W. Rhie and S.H. Lee. 2011. Surface modification with fibrin/hyaluronic acid hydrogel on solid-free form-based scaffolds followed by BMP-2 loading to enhance bone regeneration. Bone. 48: 298–306.

Kempen, D.H., L. Lu, T.E. Hefferan, L.B. Creemers, A. Maran, K.L. Classic, W.J. Dhert and M.J. Yaszemski. 2008. Retention of *in vitro* and *in vivo* BMP-2 bioactivities in sustained delivery vehicles for bone tissue engineering. Biomaterials. 29: 3245–3252.

Kern, S., H. Eichler, J. Stoeve, H. Kluter and K. Bieback. 2006. Comparative analysis of mesenchymal stem cells from bone marrow, umbilical cord blood, or adipose tissue. Stem Cells. 24: 1294–1301.

Kilroy, G.E., S.J. Foster, X. Wu, J. Ruiz, S. Sherwood, A. Heifetz, J.W. Ludlow, D.M. Stricker, S. Potiny, P. Green, Y.D. Halvorsen, B. Cheatham, R.W. Storms and J.M. Gimble. 2007. Cytokine profile of human adipose-derived stem cells: expression of angiogenic, hematopoietic, and pro-inflammatory factors. J Cell Physiol. 212: 702–709.

Kwan, M.D., M.A. Sellmyer, N. Quarto, A.M. Ho, T.J. Wandless and M.T. Longaker. 2011. Chemical control of FGF-2 release for promoting calvarial healing with adipose stem cells. J Biol Chem. 286: 11307–11313.

Lan Levengood, S.K., S.J. Polak, M.J. Poellmann, D.J. Hoelzle, A.J. Maki, S.G. Clark, M.B. Wheeler and A.J. Wagoner Johnson. 2010. The effect of BMP-2 on micro- and macroscale osteointegration of biphasic calcium phosphate scaffolds with multiscale porosity. Acta Biomater. 6: 3283–3291.

Lee, R.J., Q. Fang, P.A. Davol, Y. Gu, R.E. Sievers, R.C. Grabert, J.M. Gall, E. Tsang, M.S. Yee, H. Fok, N.F. Huang, J.F. Padbury, J.W. Larrick and L.G. Lum. 2007. Antibody targeting of stem cells to infarcted myocardium. Stem Cells. 25: 712–717.

Lendeckel, S., A. Jodicke, P. Christophis, K. Heidinger, J. Wolff, J.K. Fraser, M.H. Hedrick, L. Berthold and H.P. Howaldt. 2004. Autologous stem cells (adipose) and fibrin glue used to treat widespread traumatic calvarial defects: case report. J Craniomaxillofac Surg. 32: 370–373.

Levi, B., A.W. James, E.R. Nelson, D. Vistnes, B. Wu, M. Lee, A. Gupta and M.T. Longaker. 2010. Human adipose derived stromal cells heal critical size mouse calvarial defects. PLoS One. 5: e11177.

Levi, B., J.S. Hyun, E.R. Nelson, S. Li, D.T. Montoro, D.C. Wan, F. Jia, J.C. Glotzbach, A.W. James, M. Lee, M. Huang, N. Quarto, G.C. Gurtner, J.C. Wu and M.T. Longaker. 2011. Non-Integrating Knockdown and Customized Scaffold Design Enhances Human Adipose Derived Stem Cells in Skeletal Repair. Stem Cells.

Lin, Y., T. Wang, L. Wu, W. Jing, X. Chen, Z. Li, L. Liu, W. Tang, X. Zheng and W. Tian. 2007. Ectopic and *in situ* bone formation of adipose tissue-derived stromal cells in biphasic calcium phosphate nanocomposite. J Biomed Mater Res A. 81: 900–910.

Lin, Y., W. Tang, L. Wu, W. Jing, X. Li, Y. Wu, L. Liu, J. Long and W. Tian. 2008. Bone regeneration by BMP-2 enhanced adipose stem cells loading on alginate gel. Histochem Cell Biol. 129: 203–210.

Lu, Z., S.I. Roohani-Esfahani, G. Wang and H. Zreiqat. 2011. Bone biomimetic microenvironment induces osteogenic differentiation of adipose tissue-derived mesenchymal stem cells. Nanomedicine.

Mesimaki, K., B. Lindroos, J. Tornwall, J. Mauno, C. Lindqvist, R. Kontio, S. Miettinen and R. Suuronen. 2009. Novel maxillary reconstruction with ectopic bone formation by GMP adipose stem cells. Int J Oral Maxillofac Surg. 38: 201–209.

Miranville, A., C. Heeschen, C. Sengenes, C.A. Curat, R. Busse and A. Bouloumie. 2004. Improvement of postnatal neovascularization by human adipose tissue-derived stem cells. Circulation. 110: 349–355.

Moon, M.H., S.Y. Kim, Y.J. Kim, S.J. Kim, J.B. Lee, Y.C. Bae, S.M. Sung and J.S. Jung. 2006. Human adipose tissue-derived mesenchymal stem cells improve postnatal neovascularization in a mouse model of hindlimb ischemia. Cell Physiol Biochem. 17: 279–290.

Muschler, G.F., C. Nakamoto and L.G. Griffith. 2004. Engineering principles of clinical cell-based tissue engineering. J Bone Joint Surg Am. 86-A: 1541–1558.

Peterson, B., J. Zhang, R. Iglesias, M. Kabo, M. Hedrick, P. Benhaim and J.R. Lieberman. 2005. Healing of critically sized femoral defects, using genetically modified mesenchymal stem cells from human adipose tissue. Tissue Eng. 11: 120–129.

Petite, H., V. Viateau, W. Bensaid, A. Meunier, C. de Pollak, M. Bourguignon, K. Oudina, L. Sedel and G. Guillemin. 2000. Tissue-engineered bone regeneration. Nat Biotechnol. 18: 959–963.

Pieri, F., E. Lucarelli, G. Corinaldesi, N.N. Aldini, M. Fini, A. Parrilli, B. Dozza, D. Donati and C. Marchetti. 2010. Dose-dependent effect of adipose-derived adult stem cells on vertical bone regeneration in rabbit calvarium. Biomaterials. 31: 3527–3535.

Pittenger, M.F., A.M. Mackay, S.C. Beck, R.K. Jaiswal, R. Douglas, J.D. Mosca, M.A. Moorman, D.W. Simonetti, S. Craig and D.R. Marshak. 1999. Multilineage potential of adult human mesenchymal stem cells. Science. 284: 143–147.

Planat-Benard, V., J.S. Silvestre, B. Cousin, M. Andre, M. Nibbelink, R. Tamarat, M. Clergue, C. Manneville, C. Saillan-Barreau, M. Duriez, A. Tedgui, B. Levy, L. Penicaud and L. Castellla. 2004. Plasticity of human adipose lineage cells toward endothelial cells: physiological and therapeutic perspectives. Circulation. 109: 656–663.

Prockop, D.J. 1997. Marrow stromal cells as stem cells for nonhematopoietic tissues. Science. 276: 71–74.

Quarto, N. and M.T. Longaker. 2006. FGF-2 inhibits osteogenesis in mouse adipose tissue-derived stromal cells and sustains their proliferative and osteogenic potential state. Tissue Eng. 12: 1405–1418.

Quarto, N., D.C. Wan and M.T. Longaker. 2008. Molecular mechanisms of FGF-2 inhibitory activity in the osteogenic context of mouse adipose-derived stem cells (mASCs). Bone. 42: 1040–1052.

Rehman, J., D. Traktuev, J. Li, S. Merfeld-Clauss, C.J. Temm-Grove, J.E. Bovenkerk, C.L. Pell, B.H. Johnstone, R.V. Considine and K.L. March. 2004. Secretion of angiogenic and antiapoptotic factors by human adipose stromal cells. Circulation. 109: 1292–1298.

Santos, M.I. and R.L. Reis. 2010. Vascularization in bone tissue engineering: physiology, current strategies, major hurdles and future challenges. Macromol Biosci. 10: 12–27.

Sawin, P.D., V.C. Traynelis and A.H. Menezes. 1998. A comparative analysis of fusion rates and donor-site morbidity for autogeneic rib and iliac crest bone grafts in posterior cervical fusions. J Neurosurg. 88: 255–265.

Scherberich, A., R. Galli, C. Jaquiery, J. Farhadi and I. Martin. 2007. Three-dimensional perfusion culture of human adipose tissue-derived endothelial and osteoblastic progenitors generates osteogenic constructs with intrinsic vascularization capacity. Stem Cells. 25: 1823–1829.

Sheyn, D., I. Kallai, W. Tawackoli, D. Cohn Yakubovich, A. Oh, S. Su, X. Da, A. Lavi, N. Kimelman-Bleich, Y. Zilberman, N. Li, H. Bae, Z. Gazit, G. Pelled and D. Gazit. 2011. Gene-modified adult stem cells regenerate vertebral bone defect in a rat model. Mol Pharm. 8: 1592–1601.

Street, J., M. Bao, L. deGuzman, S. Bunting, F.V. Peale, Jr., N. Ferrara, H. Steinmetz, J. Hoeffel, J.L. Cleland, A. Daugherty, N. van Bruggen, H.P. Redmond, R.A. Carano and E.H. Filvaroff. 2002. Vascular endothelial growth factor stimulates bone repair by promoting angiogenesis and bone turnover. Proc Natl Acad Sci USA. 99: 9656–9661.

Takahashi, K. and S. Yamanaka. 2006. Induction of pluripotent stem cells from mouse embryonic and adult fibroblast cultures by defined factors. Cell. 126: 663–676.

Thesleff, T., K. Lehtimaki, T. Niskakangas, B. Mannerstrom, S. Miettinen, R. Suuronen and J. Ohman. 2011. Cranioplasty with adipose-derived stem cells and biomaterial: a novel method for cranial reconstruction. Neurosurgery. 68: 1535–1540.

Tobita, M., A.C. Uysal, R. Ogawa, H. Hyakusoku and H. Mizuno. 2008. Periodontal tissue regeneration with adipose-derived stem cells. Tissue Eng Part A. 14: 945–953.

Wan, D.C., Y.Y. Shi, R.P. Nacamuli, N. Quarto, K.M. Lyons and M.T. Longaker. 2006. Osteogenic differentiation of mouse adipose-derived adult stromal cells requires retinoic acid and bone morphogenetic protein receptor type IB signaling. Proc Natl Acad Sci USA. 103: 12335–12340.

Yoon, E., S. Dhar, D.E. Chun, N.A. Gharibjanian and G.R. Evans. 2007. *In vivo* osteogenic potential of human adipose-derived stem cells/poly lactide-co-glycolic acid constructs for bone regeneration in a rat critical-sized calvarial defect model. Tissue Eng. 13: 619–627.

Younger, E.M. and M.W. Chapman. 1989. Morbidity at bone graft donor sites. J Orthop Trauma. 3: 192–195.

Zaragosi, L.E., G. Ailhaud and C. Dani. 2006. Autocrine fibroblast growth factor 2 signaling is critical for self-renewal of human multipotent adipose-derived stem cells. Stem Cells. 24: 2412–2419.

Zhang, S., M.R. Doschak and H. Uludag. 2009. Pharmacokinetics and bone formation by BMP-2 entrapped in polyethylenimine-coated albumin nanoparticles. Biomaterials. 30: 5143–5155.

Zuk, P.A., M. Zhu, H. Mizuno, J. Huang, J.W. Futrell, A.J. Katz, P. Benham, H.P. Lorenz and M.H. Hedrick. 2001. Multilineage cells from human adipose tissue: implications for cell-based therapies. Tissue Eng. 7: 211–228.

Zuk, P.A., M. Zhu, P. Ashjian, D.A. De Ugarte, J.I. Huang, H. Mizuno, Z.C. Alfonso, J.K. Fraser, P. Benhaim and M.H. Hedrick. 2002. Human adipose tissue is a source of multipotent stem cells. Mol Biol Cell. 13: 4279–4295.

Section 2
Cellular and Molecular Aspects

CHAPTER 7

The Modulation of Skeletal Stem Cell Function Through Nanoscale Topography

Manus J.P. Biggs[1],* and Laura E. McNamara[2]

ABSTRACT

Skeletal stem cells have the capacity to differentiate into various lineages, and the ability to reliably direct stem cell fate would have tremendous potential for basic research and clinical therapy. Recent advances in microelectronic engineering techniques have paved the way for the development of nanoscale cellular technologies in medicine and basic science. In particular nanotopographical modification of an orthopedic device may provide valuable stimulation for guiding differentiation and cellular function, presenting specific cues which are more durable than surface chemistry and can be modified in size, shape and density to suit the desired application. Furthermore, due to a continuously advancing state of the art, nanofabrication methods are quickly giving rise to an ability to faithfully produce feature of sub

[1]Network of Excellence for Functional Biomaterials (NFB), National University of Ireland, Galway, IDA Business Park, Dangan, Galway, Ireland.
E-mail: manus.biggs@nuigalway.ie
[2]Centre for Cell Engineering, Institute of Molecular, Cell & Systems Biology, College of Medical, Veterinary and Life Sciences, Joseph Black Building, University of Glasgow, Glasgow, G12 8QQ, UK.
E-mail: Laura.McNamara@glasgow.ac.uk
*Corresponding author

List of abbreviations given at the end of the text.

5 nm, the scale of a single molecule. In this chapter, nanotopography is examined as a means to guide stem cell differentiation, with a particular focus on skeletal (mesenchymal) stem cells. To address the mechanistic basis underlying the topographical effects on stem cells, the likely contributions of indirect (biochemical signal-mediated) and direct (force-mediated) mechanotransduction are discussed, with emphasis on mechanotransduction through focal adhesion complexes and the cytoskeletal network. Finally, the technological advances in nanofabrication of materials and devices with the potential for clinical translation will be addressed.

Introduction

It is becoming increasingly evident that stem cells are highly sensitive to their environment and will respond to cues provided by chemistry, stiffness in two- and three-dimensional (3D) culture, and topography. This chapter will focus on the responses of skeletal stem cells to nanotopography and its mechanistic basis and is adapted from McNamara et al. (2010).

The extracellular environment presents complex chemical and topographical cues, which differs significantly from the uncharacterized surfaces normally used for *in vitro* culture systems. Cells may encounter topographies, ranging from macro- (bone, ligaments or vessels), to micro- (cellular extensions, protein fibrils) and nanoscale features (such as collagen banding, protein conformation, and ligand presentation), each of which has the potential to influence cell behaviour and functionality. An early study showed that cells were responsive to structural cues (Carrel and Burrows 1911), and over the last decade, the effects of microtopography have been well documented. Microtopographies, which include pits, grooves, pillars or meshes, are observed to guide cellular orientation by physical confinement or alignment. These substrata can induce changes in cell attachment, spreading, contact guidance, cytoskeletal architecture, nuclear shape, nuclear orientation, programmed cell death, macrophage activation, transcript levels and protein abundance (Ross et al. 2011).

Critically, evidence is also gathering on the importance of nanotopographical features in the design of smart biomimetic materials, capable of modulating the cell response and facilitating functional tissue genesis. Interaction with nanotopographies has also been observed to modulate cell morphology, adhesion, motility, proliferation, endocytotic activity, protein abundance, and gene regulation, and nanotopographical responsiveness has been observed in numerous and diverse cell types. This is intriguing from a biomaterials perspective as it demonstrates that intrinsic surface features inherent to a material or derived from the methods of manufacton can modulate the cellular response. To date, the smallest feature size shown to affect cell behaviour is 10 nm (Dalby et al. 2004), which illustrates the importance of considering the topographical cues deliberately or inadvertently presented to cells during *in vitro* culture and implantation of devices. As an increasing range of precision nanofabrication techniques become available for biological assesment, including electron beam lithography (Fig. 7.1), colloidal/nanorod self-assembly and polymer phase separation (For a review of current methodology see Table 7.1), it becomes possible to dissect out the effects of nanoscale features and particles on stem cell function and implement these materials as noninvasive tools to investigate cellular functioning, with an aim to developing functional biomaterials.

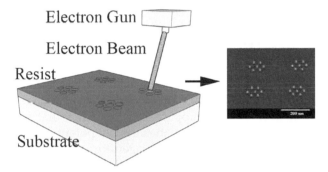

Figure 7.1 Schematic of electron-beam lithographic nanofabrication. Electron-beam lithography relies on the emission of a beam of electrons across a surface covered with a sensitive film termed a resist, which selectively removes either exposed (positive resist) or non-exposed (negative resist) regions of the resist to form a desired pattern. Currently the lowest x-y resolution obtainable with this technology lies in the 5–10 nm range. Here a negative resist has been exposed to form a hexagonal configuration of 10 nm wide pillars (Image unpublished data, courtesy of M. Schvartzman).

Table 7.1 Key Features of Nanofabrication. Nanotechnology is concerned with the fabrication, study or implementation of components and phenomena with dimensions ranging from 1–100 nm. This table lists some of the most prevalent fabrication methods for nanoscale structures and topographies (Unpublished data).

i) *Electron-beam, X-ray & ion-beam lithography* all utilize a directed beam of radiated energy to delineate patterns within an energy sensitive "resist".

ii) *Colloidal lithography* utilizes self-assembled monolayers of polymeric or metallic nanospheres. Spacing can be tightly regulated by the addition of spacing molecules.

iii) *Anodising* is an electrolytic process used to grow a natural oxide topography on the surface of metallic substrates.

iv) *Polymer demixing* refers to the process of phase separation in a polymer blend resulting in the formation of discrete polymer islands.

v) *Self-assembly* represents the processes by which molecules and particles (E.G nanorods) adopt a defined arrangement without guidance or management from an outside source.

vi) *Rapid prototyping* techniques generate nanostructures in a layer-by-layer approach, i.e., repeated deposition of a polymer or molecular ink.

The general protocols for nanomanufacturing require high resolution and throughput coupled with low cost. With respect to biological investigations, nanotopographies should occur across a large surface area (ensuring repeatability of experiments and patterning of implant surfaces), be reproducible (allowing for consistency in experiments), and preferably, be accessible (limiting the requirement for specialized equipment). The extent to which nanotopography influences cell behaviour within an *in vitro* environment remains unclear, and investigation into this phenomenon is still ongoing. A question being asked in the field of medical device manufacture however, is whether nanofeatures offer any relevant stimuli to the cellular component of the surrounding tissues *in vivo* and if so, whether implants could be fabricated to also include these topographical structures.

Skeletal Stem Cells and Topography

The adult stem cell, first described in the haematopoietic system following an investigation sparked by the detonation of atomic devices in Nagasaki and Hiroshima has been isolated from virtually every tissue of the body. Of greater interest however is the apparent phenomenon that these adult cells, arising from different sources have the inherent potential to spontaneously trans-differentiate into other tissue progenitor cells. Skeletal stem cells, found in the bone marrow, have been shown to differentiate into various cell types including osteoblasts, adipocytes, chondrocytes, smooth muscle cells, and, controversially, neurons. In most of these cases, however, differentiation of the stem cells has required the use of differentiation factors, such as dexamethasone for osteogenic differentiation, insulin for adipogenic differentiation, and hydrocortisone for smooth muscle cell differentiation. There is now compelling evidence that topography alone can produce the same effects as chemical agents. This is particularly attractive given that induction by some of these medium supplements, although successful, is not physiologically relevant and offers the possibility for development of improved clinical prostheses with topographies that can directly modulate stem cell fate.

Nanoscale pits

Nanopits and pores are identified as common constituents of tissues *in vivo*, notably basement membrane of the cornea, the aortic heart valve and the vascular system, and may be implicated in the regulation of cell behaviour and function. Pitted topographies have been shown to produce differing effects on cellular adhesion *in vitro*, depending on pit diameter and the spacing and symmetry of pit positioning. One important study examined STRO-1 enriched skeletal stem cells cultured on nanopits (120 nm diameter,

100 nm depth) with varying degrees of disorder and geometry, ranging from an absolute square and controlled disorder to a random arrangement embossed into the polymer poly(methylmethacrylate) (PMMA) (Dalby et al. 2007). Key observations of this study were that when cultured specifically on nanopits with a controlled disorder of ±50 nm, skeletal stem cells expressed particularly high levels of bone cell markers. This was comparable with the results from stem cells cultured on control planar substrates in medium supplemented with dexamethasone and ascorbic acid, a chemical enhancer of skeletal stem cell differentiation down the osteogenic lineage, but in contrast to that of the square or random pit topographies. Skeletal stem cells cultured on planar control substrates without osteogenic medium had negligible amounts of bone cell markers and appeared to have a fusiform fibroblastic-like appearance. Further evidence gathered from quantitative real-time PCR (polymerase chain reaction) and microarray data revealed an increase in expression of genes associated with bone cell development, comparable with those in cells on the flat surface in osteogenic medium, and considerably higher than expression levels from cells on the supplement-free planar controls.

In a follow-up study (McMurray, unpublished data), a temporal differentiation profile of skeletal stem cells cultured on the osteoinductive controlled disorder topography was carried out in reference to the classical osteogenic differentiation profile. This study examined the skeletal stem cell markers STRO-1 and ALCAM (activated leukocyte cell adhesion molecule), together with the bone cell markers osteocalcin and osteopontin, to examine the progression from an undifferentiated stem cell towards a committed bone cell. It was found that the skeletal stem cells cultured on the controlled disorder topography had a normal differentiation profile in line with the proposed osteogenic differentiation model. This provides further evidence that differentiation of skeletal stem cells cultured on nanotopography can produce an equally effective, if not superior, method for differentiation than chemical induction.

Nanoscale protrusions

Nanoprotrusions and raised topographical features have been reported within the extracellular matrix (ECM) in a large number of tissues. Carbon nanotubes have recently been fabricated into nanoprotrusion topographies and are attractive candidates for osteospecific studies due to their biomimetic morphology. A study of the effect of carbon nanotube dimensions on skeletal stem cell fate revealed that by increasing the diameter size of the nanotubes from a range of 30 nm up to 100 nm, it was possible to alter the focal adhesion size and differentiation of these stem cells (Oh et al. 2009). It was shown that the 30 nm nanotubes had a higher number of adherent cells with a more rounded morphology, in contrast to that of the stem cells cultured on the 100 nm nanotubes, which developed highly elongated morphologies with a low level of cell adhesion.

Osteogenic differentiation of the skeletal stem cells in this study was observed to occur on the carbon nanotubes with a 100 nm diameter, with negligible amounts of osteogenic markers observed on carbon nanotubes of 50 nm or less. Interestingly, the authors also investigated the relationship between pore size and initial cell density, which indicated that a lower seeding density led to skeletal stem cell differentiation into osteoblasts, with a higher seeding density predisposing differentiation into adipocytes. It was also discovered that there was an inverse correlation between pore size and cell density as found on the 100 nm nanotubes, which ultimately led to an increase in osteogenic gene expression. Further to this, studies have investigated the modulation of STRO+ skeletal stem cell adhesion and behaviour on surfaces with 45 nm-high "islands" manufactured by polymer phase separation. These STRO+ skeletal stem cells were shown to upregulate the synthesis of osteospecific proteins critical for bone formation (Biggs et al. 2009).

Nanoscale grooves/ridges

Nanogrooved topographies consisting of alternating grooves and ridge features differ from both nanopits and nanoprotrusions in that they produce very notable effects on cellular morphology—which, it can be argued, are directly related to cellular alignment through contact guidance.

A key fabrication tenet of nanogroove substrates for the study of cell-interface interactions is that of biomimetic ECM design, an attempt to mimic the topographical cues imparted by the fibrous nature of ECM. ECM components include both individual fibril elements, which have been reported to measure ~20–30 nm in diameter in vasculature basement membrane, and fibril bundles which range from 15–400 μm in diameter in tendon tissue. Key to this is the observation that nanogrooved surfaces may induce enhanced tissue organization and facilitate active self-assembly of ECM molecules to further mediate cell attachment and orientation. Indeed, the elongated morphology and alignment induced by grooved substrates may resemble the natural state of many cell populations *in vivo* and is observed to occur in a wide range of cell types, including osteoblasts, skeletal stem cells and nerve cells which respond profoundly to grooved substrates and have been shown to upregulate the expression of components of the ECM (Chou et al. 1995) as well as proteins key in cellular adhesion (Dalby et al. 2008).

Interestingly, skeletal stem cells cultured on nanoscale gratings have been shown to differentiate down a non-skeletal lineage, known as transdifferentiation—a somewhat controversial process. Studies have reported neurospecific differentiation of skeletal stem as indicated by the upregulation of mature neuronal cell markers, MAP2 (microtubule-associated protein 2) and β-tubulin III, when cultured on nanogratings of 350 nm depth and 700 nm pitch, in the absence of neuronal differentiation medium (Yim et al. 2007). Quantitative real-time PCR showed that expression of MAP2 was consistently higher in skeletal stem cells cultured over 14 d on the nanotopography without retinoic acid, a neuronal differentiation factor, than that of skeletal stem cells cultured on planar PDMS poly(dimethylsiloxane) supplemented with retinoic acid. Such studies indicate that nanotopography alone might have a stronger influence upon skeletal stem cell differentiation than chemical induction alone. Interestingly, a later study showed that embryonic stem cells could also be prompted to differentiate along a neuronal lineage on nanogrooved substrates of 500 nm depth and 700 nm pitch (Lee et al. 2010), confirming that nanogrooved cues may be useful for stimulating the development of neural tissue.

The Mechanistic Basis of Stem Cell Response to Topography

The ability to control stem cell differentiation using topography alone has focused attention on elucidating the mechanism by which a cell perceives these topographical cues and relays this information into the nucleus to initiate a cellular response. A key idea that appears to be corroborated in each of the chapters outlined in the previous section is the influence of cell adhesion to the topography. It has been proposed, for example, that the effects of nanoscale features on skeletal stem cell adhesion and function were due to alterations in the surface area available for protein adsorption, restricting ECM deposition and therefore the size of the cell-substrate adhesion sites that can form, specialized structures termed focal adhesions.

Focal adhesions

Focal adhesions, the sites of cell attachment to the underlying substrate, play a pivotal role in all subsequent cell actions in response to nanotopography. These dynamic adhesions are subject to complex regulation, involving integrin binding to ECM components, and the reinforcement of the adhesion plaque by recruitment of additional proteins. The process of integrin-mediated mechanotransduction relies on the ability of proteins of the focal adhesion to change chemical activity state when physically distorted, exploiting mechanical energy to drive biochemical processes by modulating the kinetics of intracellular protein-protein or protein-ligand interactions within the cell (Fig. 7.2). The ability of proteins to translate the mechanical forces observed at the site of focal adhesion to nuclear activity facilitates bidirectional signalling between the cell and the ECM, activating both direct mechanotransductive signalling and indirect molecular cascades that regulate transcription factor activity, gene and protein expression, and ultimately growth and differentiation, as will be discussed.

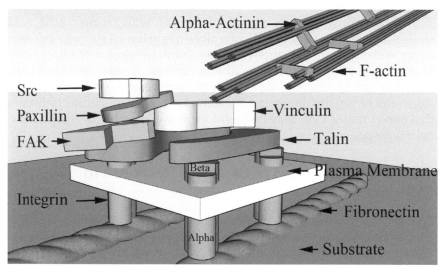

Figure 7.2 A simplified overview of the focal adhesion. Focal adhesions are macromolecular structures which serve as mechanical linkages of the cell cytoskeleton (F-actin) to the ECM, and as a biochemical signalling hubs involved with the transmission of external mechanical forces through numerous signalling proteins which interact at sites of integrin binding and clustering (Unpublished data).

The initiating event in tissue neogenesis is the transition of a pluri- or multipotent cell population into tissue-forming, differentiated cells—a process controlled by sequential activation of diverse signalling pathways and transcription factors that regulate the expression of specific genes. It is becoming increasingly clear that epigenetic modulation of cellular behaviour and subsequent cellular differentiation is highly regulated by mechanotransductive events, and that topographical modification may be a viable strategy to regulate this process.

Studies to date indicate that integrin clustering and the formation of focal adhesions are modulated by nanofeatures *in vitro* and that subsequent changes in both focal adhesion density and length are linked to changes in stem cell function and differentiation (Biggs et al. 2009). Studies also indicate that features such as pillars, grooves, and pits, with a critical feature size greater than ~73 nm modulate focal adhesion formation and influence the cell response (Figs. 7.3 and 7.4). Typically, the focal adhesion plaque undergoes anisotropic growth with increased intracellular tension leading to elongation and integrin clustering. Critically, this elongation is associated with both an increase in cytoskeletal strengthening and the recruitment of focal adhesion-associated signalling molecules.

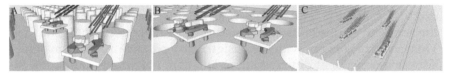

Figure 7.3 Critical nanofeature dimensions and spacing reduce focal adhesion formation. (A) Nanoprotrusions with an interfeature dimension > 73 nm and a z dimension >34 nm reduce focal adhesion reinforcement and cellular spreading through a reduction in integrin activation and clustering. Integrin clustering can still occur on the protrusion surface if the x-y dimension is > 73 nm. Conversely, nanopits (B) with an x-y dimension > 73 nm and a z dimension >34 nm reduce focal adhesion reinforcement and cellular spreading. Similarly, integrin clustering and focal adhesion reinforcement can still occur at the interpit regions, providing the pit edge-edge spacing is >73 nm. Nanoscale groove/ridges © may be considered as anisotropic dual pit/protrusion topographies, containing elements of both. Reducing the ridge width to <73 restricts focal adhesion formation to within the nanoscale ridges (Unpublished data).

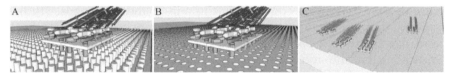

Figure 7.4 Critical nanofeature dimensions and spacing facilitate focal adhesion formation. (A) Nanoprotrusions with an interfeature dimension <73 nm facilitate integrin clustering and focal adhesion reinforcement irrespective of feature diameter. Similarly, nanoscale pits (B) with a diameter <73 do not disrupt integrin clustering and focal adhesion formation. Groove/ridge topographies © do not modulate either focal adhesion formation or cellular contact guidance when the groove/ridge z-dimension is < 34 nm (Unpublished data).

Mechanotransduction through focal adhesions

The integrin-dependent signalling pathway is mediated by non-receptor tyrosine kinases, most notably focal adhesion kinase (FAK), which is constitutively associated with the β-integrin subunit. FAK localizes at focal adhesions or focal complexes and can influence cellular transcriptional events through adhesion-dependent phosphorylation of downstream signalling molecules. In particular, the extracellular signal-regulated kinase (ERK) signalling cascade, a member of the MAPK family of pathways, is activated by signaling molecules downstream to FAK and acts as a mediator of cellular differentiation (Ge et al. 2007). Furthermore, it has been shown that ERK 1/2 is translocated to the nucleus in cells cultured on topographical features and that this affects the expression of cellular transcription factors and modulates differential function (Hamilton and Brunette 2007).

Primitive cells of the embryonic inner cell mass undergo integrin-dependent activation of ERK1/2 during early gastrulation, inducing cellular differentiation and the formation of the primitive endoderm (Liu et al. 2009). Yee et al. (2008) proposed an elegant model to explain how ERK signalling might become activated through increased binding of $a_5\beta_1$ integrins to extracellular matrix proteins (Yee et al. 2008), whereby FAK recruitment to the adhesion plaque induces downstream ERK-dependent differentiation. This model not only provides an insight into a possible mechanism of focal adhesion-dependent differentiation, but more importantly, also indicates that nanotopographical surface modification may directly regulate stem cell differentiation (Fig. 7.5 and Table 7.2).

Functional differentiation in skeletal stem cell populations is highly dependent on focal adhesion formation and cellular spreading, processes which are to a degree, dependent on nanotopographical cues. Indeed, it has been reported that adipogenic differentiation of skeletal stem cells versus osteospecific differentiation is directly related to cellular spreading (McBeath et al. 2004). This is possible because adipogenic and osteogenic cells share part of the early differentiation cascade followed by other, partly unknown, signals that determine commitment to one of these lineages. Studies suggest that a reduction in cellular adhesion, cytoskeletal development, and deactivation of FAK induce adipospecific differentiation, perhaps indicating that anti-adhesive nanotopographies may be employed to direct progenitor cell differentiation and reduce osteospecific differentiation. Conversely, nanotopographical modification that induces an increase in integrin-substratum interaction and cellular spreading may be employed to induce osteospecific differentiation. Again, mesenchymal stem cells have been shown to undergo osteospecific differentiation and functional tissue formation when cultured on nanoscale topographies, which increase focal adhesion frequency and reinforcement (Biggs et al. 2009).

Signalling cascades are initiated by exertion of mechanical forces at focal adhesions, leading to sequential post-translational modification (principally phosphorylation) events that transfer the message into the nucleus, and stimulate transcriptional effects. Physical pulling on focal adhesions can promote their anisotropic growth in response to tyrosine phosphorylation of the GTPase Rho (Riveline et al. 2001), which illustrates the mechanosensitivity of these sites. Tension can also induce conformational changes in focal adhesion proteins. It has been shown, for example, that tensile force modifies the physical conformation of the adaptor protein talin, uncovering cryptic binding sites and allowing binding to vinculin

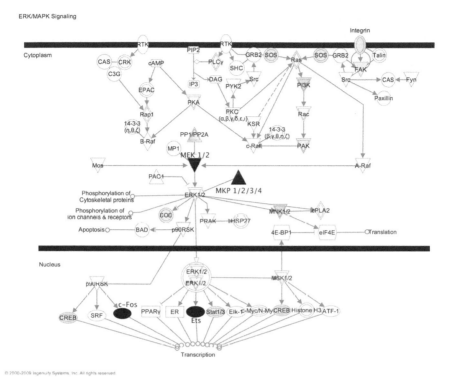

Figure 7.5 Nanoscale topography modulates integrin mediated signalling pathways in skeletal stem cells. Microarray data of transcriptional changes in a population of human skeletal stem cells cultured on a nanogrooved topography. Extracellular forces are translated through transmembrane integrin receptors into differential cellular function through FAK recruitment and subsequent activation of Grb2-Sos-Ras pathway. This in turn leads to activation of ERK 1/2 signalling which acts on the nuclear machinery of cellular transcription. Nanotopographies which modulate integrin clustering and focal adhesion reinforcement, induce downstream changes in ERK 1/2 mediated cellular transcription and subsequent cellular differentiation and function (grey = upregulated, Black = downregulated) (Unpublished data).

(del Rio et al. 2009), and the functionality of the adapter protein p130Cas in the promotion of integrin signalling (Sawada et al. 2006).

It can be inferred that nanoscale features influence differential pathways in adherent stem cell populations by modulating integrin clustering and adhesion formation, and that subsequent activation of FAK acts to regulate the ERK signalling pathway and influence stem cell differentiation and tissue neogenesis. Indeed, many recent studies suggest that focal adhesion formation and the phosphorylation of both FAK and ERK are influenced by nanoscale topography in skeletal stem cell populations (Biggs et al. 2009), and that FAK activity indirectly regulates stem cell differentiation.

Since such a wide variety of signals affect bone marrow stromal cell differentiation, it is very likely that no single signalling pathway is

Table 7.2 The influence of nanoscale features on skeletal stem cell function. Nanotopographical modification induces functional changes in skeletal stem cell populations. To date, nanoscale protrusions and pits have principally been shown to upregulate osteospecific markers. Nanoscale grooves / ridge substrates however have been shown to induce neuro and adipospecific differentiation in skeletal stem cells (Unpublished data).

Topography	Study	Cell type	Chemistry	Width	Edge-edge	Depth/ Height	Functional modification
Nanoscale protrusions	(Sjostrom et al. 2009)	Human skeletal stem cells	Ti	40 nm	28 nm	15 nm	Upregulation of osteospecific markers
	(McNamara et al. 2011)	Human skeletal stem cells	Ti	22 nm	41 nm	15 nm	Upregulation of osteospecific markers
Nanoscale pits	(Dalby et al. 2007)	Human skeletal stem cells	poly(methyl methacrylate)	120 nm	180 nm	100 nm	Upregulation of osteospecific markers
	(Lim et al. 2007)	Human fetal osteoblastic	poly(L-lactic acid)/ poly(styrene) blend	500 nm	variable	29 nm	Increase in FAK phosphorylation
Nanoscale Groove/ridges	(Biggs et al. 2009)	Human skeletal stem cells	poly(methyl methacrylate)	10 μm	10 μm	330 nm	Upregulation of adipospecific signalling pathways
	(Yim et al. 2007)	Human skeletal stem cells	Poly(dimethylsiloxane)	350 nm	700 nm	350 nm	Upregulation of neuronal markers- MAP2 and GFAP

responsible for the regulation of early osteoprogenitor differentiation. Rather, a network of signalling pathways is likely to be invloved, and FAK is highly suited to integrate these signalling activities. These pathways are intimately related and activation may be mediated by adhesive mechanisms or influenced further by soluble signalling factors. There is growing appreciation that simultaneous activation and repression of gene expression is a feature of multiple developmental signalling pathways, and importantly, it is this crosstalk between biochemical and other mechanotransductive pathways that regulates transcriptional events.

Direct mechanotransduction

Direct mechanotransduction requires the simple transmission of extracellular forces into the nucleus in order to exert a physical change in nuclear conformations or loads. In tensegrity theory, the cytoskeletal elements are described in terms of their likely contribution to force transmission. In this model, actin stress fibres generate cellular tension while microtubules resist compression. These opposing forces, together with a contribution from the intermediate filaments (such as vimentin and the nucleoskeletal lamins) provide "prestress" to the cellular tensegrity structure. The cytoskeleton circumscribes and is physically linked to the nucleoskeleton via bridging proteins including SUN (Sad1p-Unc-84) and the nesprins (Haque et al. 2006). On the nucleoplasmic side, DNA is physically associated with the lamina, at sites called matrix attachment regions (MARs). Kilian et al. (2010) used chemical surface patterning to confine cell morphology in an analogous manner to surface topography, with angular patterns inducing osteogenic differentiation, and rounded adhesive patterns increasing adipospecific differentiation (Kilian et al. 2010). Myosin II staining suggested that cytoskeletal tension had been decreased in cells on rounded patterns and enriched in cells on the more angular shapes. Such differentials of tensile force could impact upon the positioning of chromosomes, by unequal distribution of mechanical forces onto the nucleus via attachment to the nucleoskeleton. Similarly, the enclosure of multipotent stem cells within 3D gels of variable stiffness containing adhesive peptides modulated fate determination by affecting the extent to which the cells were able to cluster adhesive ligands, and exert a contractile force upon the substrata (Huebsch et al. 2010). Such tension would be expected to affect the nucleus, and ablation experiments have suggested that heterochromatin is important in the structural maintenance of nuclear architecture, also contributing in the transmission of forces from the cytoskeleton into the nucleus, via its association with the nuclear lamina (Fig. 7.6).

It is generally accepted that peripheral nuclear DNA is comprised of rarely transcribed—or transcriptionally silent—heterochromatin, while

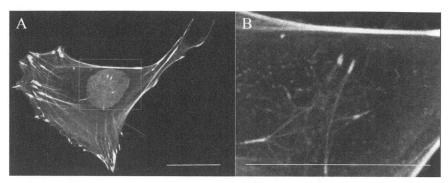

Figure 7.6 Direct mechanotransduction through the peri-nuclear cytoskeletal network.
Fluorescent image of a human skeletal stem cell. Tensile forces from the ECM are relayed through the focal adhesions to a specialized perinuclear actin cytoskeleton. Interactions between actin, the intermediate filaments and the lamins of the nucleoskeleton induce nuclear distortion, and DNA rearrangement, resulting in direct mechanotransductive effects. Together with signalling inputs, these changes may include redistribution of chromosomes, nucleoli, and other nuclear components. High mag. image of boxed area with nucleus removed for clarity (B) Bar, 20 µm (Unpublished data).

centrally the more frequently expressed sequences are typically found. Extraction experiments indicate a mechanical link between chromosomes and nucleoli, supported by studies of cellular micromanipulation, where pulling at the peripheral cytoplasm induced a relocation of the nucleus and concurrent redistribution of nucleoli (Maniotis et al. 1997). It was concluded that the force had been transmitted between the interconnected cytoskeleton and nucleoskeleton. The use of nanotopography as a mechanical modulator of cellular function has the additional advantage that it is a noninvasive stimulus, and thus cells are not damaged, unlike traditional interventional approaches. Experimentally, spatial chromosomal modulations have also been observed with FISH (Fluorescence *in situ* hybridisation) which has been employed to examine the centromeric positioning of chromosome 3 in the interphase nuclei of fibroblasts cultured on nanocolumns (Dalby et al. 2007). The authors proposed that observed alterations were consistent with topography-induced direct mechanotransduction, as decreased cell spreading, resulting in less taut cytoskeletons, lowered tensile forces and reduced pulling on the nuclear chromatin. Chromosomal redistribution also has the potential to affect the transcription of multiple genes if the associated genes undergo a spatial relocation from for example, hetero- to euchromatic regions of the nucleus.

Recently, a number of models have been proposed to explain how tension from focal adhesions could be converted into gene- or protein-level effects. These are based on the hypothesis that DNA might be subject to local dissociation at MARs, facilitating the action of transcription factors. Alternatively, further models indicate that tensile strain may directly modulate the assembly of transcription factor complexes, or that nuclear pores could be affected, leading to alteration of mRNA transport. The mechanical changes exerted by nanotopography will most likely have multiple, synergistic effects on the nucleus for the enforcement of gene-level changes. It is also probable that topography-invoked mechanotransduction will have various outputs, acting at both the mRNA and protein levels (Fig. 7.5): (1) transcriptional, with the physical accessibility of genes to transcription factors being altered, (2) post-transcriptional, with effects on mRNA splicing, editing or transport, (3) translational, by altering protein production, (4) post-translational, by rapidly altering the activity state of proteins, and (5) conformational, with changes in protein structure, composition of protein assemblies, and exposure of cryptic binding sites.

In this complex interplay, the complementary nature of direct and indirect (chemical signal-mediated) mechanotransduction can be seen. It can be inferred, although research is still ongoing, that direct mechanotransduction plays a significant role in the physical changes in the cell resulting from external mechanical stimuli. Such modulations include nuclear morphology, redefinition of chromosomal territories, force-induced changes in protein conformation, and nuclear membrane distortions (e.g., in nuclear pore diameter) that might manipulate mRNA or protein transport. Conversely, chemical signals (such as integrin-linked signal cascades) would be expected to predominate in inducing the rapid-response modifications in protein activity state observed in skeletal cells cultured on nanoscale structures. Although one type of mechanotransduction may dominate for a particular functional response, it seems highly likely that the two pathways ultimately act to reinforce each other to generate the complete and apposite cellular response to nanotopography.

Clinical Relevance: Challenges and Applications

Nanotopographies can be designed to promote or reduce cell adhesion and alter stem cell fate, all of which would be useful attributes for a range of applications in regenerative medicine, including, but not limited to orthopaedics and dentistry. Chemical surface patterning of implants could induce these effects, but topographies should persist more efficiently than chemical modification on the devices. Indented features such as nanopits would also be advantageous, as these would be less easily abraded during surgery or *in situ* by temporal wear and tear. A combination of different topographies could be used to differentially functionalize implants for distinct applications, or demarcate particular "zones" within a single device.

Cell-substrate interactions can be regarded as the defining factors for the long-term performance and biofunctionality of many devices *in situ*. It can be reasoned that the integration of exogenous materials can be regulated by controlling the associated interfacial reactions, in an attempt to minimize non-functional tissue generation or aseptic loosening. Materials that promote osteoblast specific adhesion may enhance functional differentiation resulting in the neogenesis of mineralized matrix, bony tissue formation and deposition.

Conversely, many functional biomaterials require minimal protein and/ or cellular interaction *in vivo* for optimal device function or to facilitate future device removal. For example a body of research suggests that permanent device retention following orthopaedic fixation is not ideal and may present future implant site morbidity. Increased osteoblast adhesion and bony tissue mineralization can, for example, complicate the removal of plating systems, increasing removal torque and predisposing screw damage and bone re-fracture during the removal procedure. It follows that selective adhesion of specific cellular phenotypes is crucial to regulate optimal tissue specific integration while preventing inflammatory cell adhesion and scar tissue formation.

For surgical applications, nanofeatures would need to be patterned into clinically relevant materials, which can be technically demanding. Scaling up production of ordered nanotopographies for the generation of whole devices presents technical challenges, particularly if the nanofeatures consist of complex or highly ordered features, requiring precision techniques such as electron beam and nanoimprint lithography. More readily applied surface modification technologies, such as implant roughening by sand-blasting, acid etching, or plasma immersion ion implantation, offer a means to generate less defined arrangements of features of mixed sizes. While this can be a useful means of modulating cell adhesion and osseointegration, the irreproducible nature of the features is likely to contribute to interimplant discrepancies and variable success rates. Use of complementary micro- and nanotopographies should be valuable in the development of more efficacious prostheses.

Techniques such as colloidal and nanoimprint lithography could be useful for nanostructuring of large areas of devices with more consistent topographies. In addition, chemical vapour deposition (CVD) can be utilized to deposit nanoscale features over large areas, in biocompatible materials such as silicon and titanium. With a growing number of studies indicating that topographical modification of the cell-substrate interface is a significant regulator of cellular adhesion and function, we may see modified biomaterials in clinical use in the near future. In particular, biodegradable devices may be functionally modified to control cellular interactions, with an aim to enhance tissue regeneration.

Conclusions

There is a wealth of accumulating evidence to demonstrate that maintenance of the undifferentiated state of stem cells and the direction of stem cell fate can be modified by the topographic substratum. Surface features are detected by a variety of mechanosensors, including integrin-linked focal adhesions, which respond to the mechanical constraints of the surface by inducing signalling cascades, such as the ERK-MAPK pathway. Physical pulling of the rearranging cytoskeleton on the nucleus complements the chemical signalling, and, together, such direct and indirect mechanotransduction can modulate nuclear components, altering gene expression to direct stem cell responses and modifying the cellular protein profile concomitantly with the state of differentiation.

The potential for the use of nanotopography as a means of influencing cell behaviour and lineage commitment for scientific studies has obvious advantages in reducing, abolishing, or even enhancing chemical inducers and feeder layers. As our understanding of the molecular and whole-cell responses of stem cells to topography increases, there will be enormous scope for the creation of next-generation materials possessing defined features to tailor stem cell fate and functioning to specific laboratory, industrial, and therapeutic applications.

Summary Points

- Nanotopography modulates focal adhesion formation by promoting or disrupting integrin clustering and focal adhesion associated protein reinforcement.
- Skeletal stem cells respond to critical dimensions of 73 nm in the x-y and 34 nm in the z.
- Nanotopographically medicated modulation of skeletal stem cell function may be as a result of direct or indirect mechanotransductive processes.
- Indirect mechanotransduction (biochemical signalling) in skeletal stem cells is mediated by the integrin and ERK 1/2 signalling pathways (although not exclusively).
- Direct mechanotransduction (nuclear rearrangement by external forces) in skeletal stem cells is mediated by crosstalk between focal adhesions and the cytoskeletal-nucleoskeletal networks.
- Nanotopographical modification may be exploited to regulate cellular function at the device/tissue interface.

Dictionary of Key Terms

- *Topography*: Used to indicate the geometric conformation or general organization of three dimensional features distributed on a solid surface.
- *Biomimetic*: A term applied to materials and devices which have been engineered to mimic the structure and function of biological systems.
- *Haematopoietic*: Pertaining to, haematopoiesis, which is the formation of both myeloid and lymphoid blood cells through multipotent progenitor cells.
- *Dexamethasone*: A synthetic glucocorticoid steroid which acts via RUNX2 to induce osteoblast differentiation in osteospecific progenitor cells.
- *Adipogenic*: A term used to describe the condition of being related to, or characteristic of fat production.
- *Inner cell mass*: The mass of cells inside the blastocyst that will eventually give rise to all cell types of the definitive structures of the fetus.
- *Focal adhesion kinase*: A protein tyrosine kinase involved in cellular adhesion which is phosphorylated in response to integrin engagement, eliciting intracellular signal transduction pathways.
- *ERK*: An intracellular protein cascade that communicates a signal from surface receptors to the machinery of nuclear transcription.
- *Nesprins*: A family of proteins that are found in the outer nuclear membrane and which primarily act to couple the nuclear membrane to the cellular cytoskeleton.
- *Nuclear lamina*: The fibrillar network of intermediate filaments found inside the nucleus which provides nuclear integrity and regulates nuclear processes such as transcription.

List of Abbreviations

ALCAM	:	activated leukocyte cell adhesion molecule
Cdks	:	cyclin-dependent kinases
CVD	:	chemical vapor deposition
DIGE	:	(2D)-fluorescence difference gel electrophoresis
ECM	:	extracellular matrix
ERK	:	extracellular signal-regulated kinase
FAK	:	focal adhesion kinase
MAP2	:	microtubule-associated protein 2
MAPK	:	mitogen-activated protein kinase
MARs	:	matrix attachment regions

PDMS : poly(dimethylsiloxane)
PMMA : poly(methylmethacrylate)
SUN : Sad1p-Unc-84

References

Biggs, M.J., R.G. Richards, N. Gadegaard, R.J. McMurray, S. Affrossman, C.D. Wilkinson, R.O. Oreffo and M.J. Dalby. 2009. Interactions with nanoscale topography: adhesion quantification and signal transduction in cells of osteogenic and multipotent lineage. J Biomed Mater Res A. 91: 195–208.

Biggs, M.J., R.G. Richards, N. Gadegaard, C.D. Wilkinson, R.O. Oreffo and M.J. Dalby. 2009. The use of nanoscale topography to modulate the dynamics of adhesion formation in primary osteoblasts and ERK/MAPK signalling in STRO-1+ enriched skeletal stem cells. Biomaterials. 30: 5094 5103.

Carrel, A. and M. Burrows. 1911. Culture *in vitro* of malignant tumors. The Journal of Experimental Medicine. 12: 571–575.

Chou, L., J.D. Firth, V.J. Uitto and D.M. Brunette. 1995. Substratum surface topography alters cell shape and regulates fibronectin mRNA level, mRNA stability, secretion and assembly in human fibroblasts. J Cell Sci. 108 (Pt 4): 1563–1573.

Dalby, M.J., M.O. Riehle, H. Johnstone, S. Affrossman and A.S. Curtis. 2004. Investigating the limits of filopodial sensing: a brief report using SEM to image the interaction between 10 nm high nano-topography and fibroblast filopodia. Cell Biol Int. 28: 229–236.

Dalby, M.J., M.J. Biggs, N. Gadegaard, G. Kalna, C.D. Wilkinson and A.S. Curtis. 2007. Nanotopographical stimulation of mechanotransduction and changes in interphase centromere positioning. J Cell Biochem. 100: 326–338.

Dalby, M.J., N. Gadegaard, R. Tare, A. Andar, M.O. Riehle, P. Herzyk, C.D. Wilkinson and R.O. Oreffo. 2007. The control of human mesenchymal cell differentiation using nanoscale symmetry and disorder. Nat Mater. 6: 997–1003.

Dalby, M.J., A. Hart and S.J. Yarwood. 2008. The effect of the RACK1 signalling protein on the regulation of cell adhesion and cell contact guidance on nanometric grooves. Biomaterials. 29: 282–289.

del Rio, A., R. Perez-Jimenez, R. Liu, P. Roca-Cusachs, J.M. Fernandez and M.P. Sheetz. 2009. Stretching single talin rod molecules activates vinculin binding. Science. 323: 638–641.

Ge, C., G. Xiao, D. Jiang and R.T. Franceschi. 2007. Critical role of the extracellular signal-regulated kinase-MAPK pathway in osteoblast differentiation and skeletal development. J Cell Biol. 176: 709–718.

Hamilton, D.W. and D.M. Brunette. 2007. The effect of substratum topography on osteoblast adhesion mediated signal transduction and phosphorylation. Biomaterials. 28: 1806–1819.

Haque, F., D.J. Lloyd, D.T. Smallwood, C.L. Dent, C.M. Shanahan, A.M. Fry, R.C. Trembath and S. Shackleton. 2006. SUN1 interacts with nuclear lamin A and cytoplasmic nesprins to provide a physical connection between the nuclear lamina and the cytoskeleton. Mol Cell Biol. 26: 3738–3751.

Huebsch, N., P.R. Arany, A.S. Mao, D. Shvartsman, O.A. Ali, S.A. Bencherif, J. Rivera-Feliciano and D.J. Mooney. 2010. Harnessing traction-mediated manipulation of the cell/matrix interface to control stem-cell fate. Nat Mater. 9: 518–526.

Kilian, K.A., B. Bugarija, B.T. Lahn and M. Mrksich. 2010. Geometric cues for directing the differentiation of mesenchymal stem cells. Proc Natl Acad Sci USA. 107: 4872-4877.

Lee, M.R., K.W. Kwon, H. Jung, H.N. Kim, K.Y. Suh, K. Kim and K.S. Kim. 2010. Direct differentiation of human embryonic stem cells into selective neurons on nanoscale ridge/groove pattern arrays. Biomaterials. 31: 4360–4366.

Lim, J.Y., A.D. Dreiss, Z. Zhou, J.C. Hansen, C.A. Siedlecki, R.W. Hengstebeck, J. Cheng, N. Winograd and H.J. Donahue. 2007. The regulation of integrin-mediated osteoblast focal adhesion and focal adhesion kinase expression by nanoscale topography. Biomaterials. 28: 1787–1797.

Liu, J., X. He, S.A. Corbett, S.F. Lowry, A.M. Graham, R. Fassler and S. Li. 2009. Integrins are required for the differentiation of visceral endoderm. J Cell Sci. 122: 233–242.

Maniotis, A.J., C.S. Chen and D.E. Ingber. 1997. Demonstration of mechanical connections between integrins, cytoskeletal filaments, and nucleoplasm that stabilize nuclear structure. Proc Natl Acad Sci USA. 94: 849–854.

McBeath, R., D.M. Pirone, C.M. Nelson, K. Bhadriraju and C.S. Chen. 2004. Cell shape, cytoskeletal tension, and RhoA regulate stem cell lineage commitment. Dev Cell. 6: 483–495.

McNamara, L.E., R.J. McMurray, M.J.P. Biggs, F. Kantawong, R.O.C. Oreffo, M.J. Dalby, Nanotopographical control of stem cell differentiation. JTE. Volume 2010, Article ID 120623. doi: 10.4061/2010/120623.

McNamara, L.E., T. Sjostrom, K.E. Burgess, J.J. Kim, E. Liu, S. Gordonov, P.V. Moghe, R.M. Meek, R.O. Oreffo, B. Su and M.J. Dalby. 2011. Skeletal stem cell physiology on functionally distinct titania nanotopographies. Biomaterials. 32: 7403–7410.

Oh, S., K.S. Brammer, Y.S. Li, D. Teng, A.J. Engler, S. Chien and S. Jin. 2009. Stem cell fate dictated solely by altered nanotube dimension. Proc Natl Acad Sci USA. 106: 2130–2135.

Riveline, D., E. Zamir, N.Q. Balaban, U. S. Schwarz, T. Ishizaki, S. Narumiya, Z. Kam, B. Geiger and A.D. Bershadsky. 2001. Focal contacts as mechanosensors: externally applied local mechanical force induces growth of focal contacts by an mDia1-dependent and ROCK-independent mechanism. J Cell Biol. 153: 1175–1186.

Ross, A.M., Z. Jiang, M. Bastmeyer and J. Lahann. 2011. Physical Aspects of Cell Culture Substrates: Topography, Roughness, and Elasticity. Small.

Sawada, Y., M. Tamada, B.J. Dubin-Thaler, O. Cherniavskaya, R. Sakai, S. Tanaka and M.P. Sheetz. 2006. Force sensing by mechanical extension of the Src family kinase substrate p130Cas. Cell. 127: 1015–1026.

Sjostrom, T., M.J. Dalby, A. Hart, R. Tare, R.O. Oreffo and B. Su. 2009. Fabrication of pillar-like titania nanostructures on titanium and their interactions with human skeletal stem cells. Acta Biomater. 5: 1433–1441.

Yee, K.L., V.M. Weaver and D.A. Hammer. 2008. Integrin-mediated signalling through the MAP-kinase pathway. IET Syst Biol. 2: 8–15.

Yim, E.K., S.W. Pang and K.W. Leong. 2007. Synthetic nanostructures inducing differentiation of human mesenchymal stem cells into neuronal lineage. Exp Cell Res. 313: 1820–1829.

CHAPTER 8

Myostatin (GDF-8) Signaling in Progenitor Cells and Applications to Bone Repair

Moataz N. Elkasrawy[1] and Mark W. Hamrick[2,*]

ABSTRACT

Myostatin is a member of the TGF-β superfamily of growth and differentiation factors and is widely recognized as a potent regulator of skeletal muscle mass and regeneration. It has, however, recently been shown that myostatin is expressed in tissues aside from muscle including fat and tendon. In addition, studies monitoring gene expression in the fracture callus following injury have shown that myostatin is highly expressed in the early phases of fracture healing. Furthermore, other research has demonstrated that mice lacking myostatin show increased bone density and a marked increase in fracture callus size following fibula osteotomy. Yet, the basic mechanisms by which myostatin may impact bone healing after injury has not been fully elucidated. Our *in vitro* studies using 3D aggregate culture to induce chondrogenic differentiation of bone marrow derived stromal cells (BMSCs) reveal that

[1]University of Colorado-Denver, School of Dental Medicine, Mail Stop F838, 13065 East 17th Avenue, Room 104, Aurora, CO 80045. USA.
E-mail: moataz.elkasrawy@ucdenver.edu
[2]Department of Cellular Biology & Anatomy, Georgia Health Sciences University, CB1116 Laney Walker Blvd, Augusta, GA 30912 USA.
E-mail: mhamrick@georgiahealth.edu
*Corresponding author

List of abbreviations given at the end of the text.

myostatin suppresses the proliferation and chondrogenic differentiation of BMSCs by altering the expression of Sox-9 and various Wnt-related factors. These *in vitro* studies suggest that myostatin may directly impair bone regeneration, a hypothesis that is also supported by our recent finding that exogenous myostatin reduces fracture callus bone volume. Together, the data suggest that myostatin may represent a novel therapeutic target for management of orthopedic trauma where both bone and muscle are damaged and, furthermore, that myostatin inhibitors may enhance fracture healing and improve recovery following musculoskeletal injury.

Introduction

Myostatin (GDF-8) was first identified as a new member of the transforming-beta growth factor (TGF-beta) superfamily of growth and differentiation factors that was highly expressed in skeletal muscle progenitors (McPherron et al. 1997). The functional characterization of myostatin using a myostatin-knockout mouse led to the surprising observation that loss of normal myostatin signaling produced a dramatic increase in muscle mass (McPherron et al. 1997). Since the original discovery of myostatin, its role in muscle development and muscle regeneration has been explored using a variety of *in vivo* and *in vitro* approaches (Zimmers et al. 2002; Reis-Porszaz et al. 2003; Li et al. 2008; McFarlane et al. 2011). These studies have generally shown that myostatin plays an important role in the proliferation and differentiation of myoblasts, and that inhibiting myostatin expression improves muscle regeneration and increases muscle fiber size and number (Lee 2004; Burks and Cohn 2011).

It is now clear, however, that myostatin is expressed in many tissues aside from muscle (Allen et al. 2008; Mendias et al. 2008), and that it has important functions in cells ranging from fibroblasts to adipocytes. This raises the possibility that myostatin is involved in the differentiation of mesenchymal stem cells (MSCs), since mice lacking myostatin show significant alterations in tissues of mesenchymal origin. Specifically, myostatin deficient mice show decreased body fat (McPherron and Lee 2002; Hamrick et al. 2006), increased bone mineral density (Elkasrawy and Hamrick 2010), and tendons and ligaments that are relatively weak and exhibit cellular hypoplasia (Mendias et al. 2008; Fulzele et al. 2010). In addition, other studies show that the myostatin receptor (Acvr2b, or ActRIIB) is expressed in bone marrow-derived stromal cells (Hamrick et al. 2007), and myostatin can regulate the differentiation of C3H 10T(1/2) mouse mesenchymal multi-potent cells (Rebbapragada et al. 2003; Artaza et al. 2005). Finally, GDF-11, a member of the TGF-β superfamily that is structurally and functionally very similar to myostatin (McPherron et al. 2009), has been shown to decrease chondrogenic and myogenic differentiation of MSCs in micromass cultures (Gamer et al. 2001).

Previous work therefore suggests that myostatin is an important factor in regulating not only the fate of mesenchymal progenitors but also the proliferation, differentiation, and function of mesenchymal derivatives such as adipocytes, chondrocytes, myoblasts, and fibroblasts. These observations have obvious implications for bone repair, since myostatin is also expressed in bone during the early phases of fracture healing (Cho et al. 2002). This chapter reviews the basic biology of myostatin signaling, its role in mediating the proliferation and differentiation of mesenchymal progenitors, and the effects of myostatin and myostatin inhibition on bone repair.

Myostatin Signaling Through the Type IIB Activin Receptor (Acvr2b, or ActRIIB)

Myostatin circulates in a latent form bound to a propeptide, which is cleaved by a BMP1/Tolloid matrix metalloproteinase, releasing the active form capable of binding the type IIB activin receptor. Follistatin and follistatin-related gene (FLRG) are other proteins that can bind and inhibit the activity of myostatin by maintaining its latency (Lee 2004; Tsuchida 2004). Myostatin has previously been shown to bind the type IIB activin receptor (Acvr2b, or ActRIIB) with high affinity (Lee 2004), with activin itself having a higher affinity for the type IIa activin receptor (Donaldson et al. 1992). Other ligands are, however, also known to bind Acvr2b along with myostatin. In mice these include BMP-9, -10, and -11 along with Inhibinβ and TGF-β1. In human serum the same ligands were found bound to Acvr2b, with the exception of TGF-β1 (Souza et al. 2008).

Acvr2b is a serine/threonine kinase receptor, which recruits and phosphorylates activin-like kinase receptor 5 (ALK5) or ALK4 (Rebbapragada et al. 2003). This consequently activates and phosphorylates Smad2/3, which dissociate from the ligand/receptor complex to bind with the co-Smad, Smad4, allowing translocation of the Smad complex to the nucleus where it targets several DNA-binding proteins to regulate transcriptional response (Lee 2004) (Fig. 8.1). Smad4 also activates Smurf-mediated ubiquitination of the ActRIIB/ALK4-5 receptor complex, and promotes its proteosomal degradation (Bradley et al. 2008). Although myostatin, like other TGF-β family members, is most well known to activate Smad signaling, it can also activate the p38 MAPK and ERK1/2 pathways (Allendorph et al. 2006; Steelman et al. 2006; Ekaza and Cabello 2007). The p53 MAPK and ERK1/2 pathways work on promoting the survival of mature muscle synthetia by the action of p53, which blocks apoptosis.

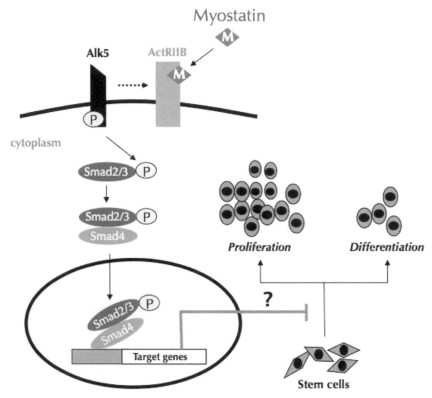

Figure 8.1 The myostatin signaling pathway. Myostatin (M) binds the type IIB activin receptor (ActRIIB) which recruits the type I co-receptor ALK5, phosphorylating Smad2/3 which translocate to the nucleus where they regulate target gene expression. A number of factors thought to be involved in the proliferation and differentiation of mesenchymal stem cells are regulated through this TGF-beta signaling pathway.

ERK1/2 also activates p21, which arrests myoblast proliferation (Bradley et al. 2008). Autocrine myostatin signaling is regulated in part through a negative feedback loop, where myostatin activation of smad2/3 stimulates expression of the inhibitory smad Smad7, which in turn inhibits myostatin signal transduction (Zhu et al. 2004).

A number of studies now provide evidence that myostatin may also signal through the Wnt pathway. For example, Wnt4 expression is significantly elevated in skeletal muscle from myostatin-deficient mice (Steelman et al. 2006), and myostatin enhances nuclear translocation of beta-catenin and formation of the Smad3-beta-catenin-TCF4 complex in human mesenchymal stem cells (Guo et al. 2008). These findings are significant from the perspective of bone healing and regeneration, since components of the Wnt pathway such as Wnt4, Wnt5a, and Wnt5b, are known to be upregulated during the process of fracture repair (Hadjiagyrou et al. 2002;

Zhong et al. 2006). In addition, β-catenin-mediated Wnt pathway signaling is also involved in endochondral bone formation (Zhong et al. 2006; Chen et al. 2007a, 2007b). The Wnt pathway ultimately regulates the levels of β-catenin in the cytoplasm and promotes its nuclear translocation, where it acts in conjunction with other transcription factors to promote different gene transcription (Akiyama 2000).

A Role for Myostatin in Regulating Stem Cell Differentiation

A growing body of evidence reveals that myostatin plays a role in regulating myogenesis, adipogenesis and tissue fibrosis by altering the differentiation of progenitor cells of mesenchymal origin. Given the double-muscled phenotype of mice lacking myostatin, it is not surprising that myostatin suppresses cell proliferation during myogenic differentiation of C2C12 progenitors (Taylor et al. 2001; Rios et al. 2001). Myostatin also induces the differentiation of myofibroblasts in C3H 10T1/2 mesenchymal stem cells (Artaza et al. 2008). *In vivo*, inhibition of myostatin using an antibody decreases apoptosis and caspase-3 expression in skeletal muscle (Murphy et al. 2010). Myostatin likewise increases proliferation and extracellular matrix synthesis of primary muscle fibroblasts (Li et al. 2008), as well as fibroblasts from human knee ligaments (Fulzele et al. 2010). These data indicate that myostatin is a pro-fibrotic factor that contributes to the accumulation of collagen-rich fibrous tissue after muscle injury (Zhu et al. 2007).

While myostatin appears to inhibit myogenesis and stimulate fibrosis, it also inhibits adipogenic differentiation of both human bone marrow-derived stromal cells (Guo et al. 2008) and adipogenic differentiation of mouse pre-adipocytes (Kim et al. 2001) and mesenchymal stem cells (Rebbapragada et al. 2003). In human stromal cells, myostatin induced nuclear translocation of beta-catenin, altered the expression of several Wnt/beta-catenin pathway genes, and suppressed expression of the adipogenic transcription factors PPAR gamma and C/EBP alpha (Guo et al. 2008). In mouse MSCs and pre-adipocytes, adipogenesis was induced with BMP-7, and myostatin treatment suppressed this BMP7 induced effect (Rebbapragada et al. 2003). The inhibitory effect of myostatin on adipogenesis in these experiments could be reversed by blocking ALK5. In contrast, other studies have shown that myostatin can directly stimulate adipogenesis in C3H 10T1/2 cells (Artaza et al. 2005; Feldman et al. 2006). These apparently contradictory findings are difficult to reconcile, particularly when congenital absence of myostatin is associated with a dramatic decrease in body fat (Lin et al. 2002; McPherron and Lee 2002). Further investigation into myostatin's role in adipocyte differentiation is clearly needed.

We have previously shown that absence of myostatin increases the osteogenic differentiation of primary bone marrow stromal cells from mice

(Hamrick et al. 2007). This finding is perhaps consistent with some of the *in vitro* experiments above demonstrating that myostatin may favor the adipogenic differentiation of mesenchymal progenitors; hence, absence of myostatin or myostatin inhibition would be expected to direct MSCs toward the osteogenic or myogenic lineages. This hypothesis is supported by the finding that myostatin inhibitors increase bone formation and bone mass in mice (Bialek et al. 2008; Ferguson et al. 2009), and also improve bone repair following osteotomy (Hamrick et al. 2010). We have also found that myostatin treatment inhibits the chondrogenic differentiation of bone marrow-derived stromal cells by suppressing the expression of Sox-9, and by altering the expression of a number of Wnt-related genes. Specifically, exogenous myostatin reduced the expression of type II collagen and the proliferation of growth plate chondrocytes as well as bone marrow stromal cells *in vitro* (Elkasrawy et al. 2011a). The Wnt signaling factor Dkk1 is significantly upregulated with myostatin treatment during TGF-β1-induced chondrogenesis, and GSK3A (responsible for beta-catenin instability) is also upregulated, whereas the Wnt ligand binding protein Fzd3 is significantly downregulated with myostatin treatment (Fig. 8.2). Dkk1 is a molecule that inhibits the Wnt signaling pathway by binding to and antagonizing LRP5/6, forming a ternary complex with LRP6 that induces its rapid endocytosis and removal from the plasma membrane (Mao et al. 2002). GSK3A is another factor that is involved in Wnt signaling, which is a component of the GSK3/Axin/APC proteosomal complex responsible for ubiquitination of β-catenin to prevent its accumulation in the cytoplasm and translocation into the nucleus. Fzd3, a transmembrane protein that can bind Wnt ligands including Wnt5a, and Wnt5a is known to be a molecule that plays a key role in chondrogenesis (Church et al. 2002; Yang et al. 2003), is downregulated in the presence of myostatin during TGF-β1-induced chondrogenesis. In addition, as noted above, GDF-11, a member of the TGF-β superfamily that is structurally very similar to myostatin (McPherron et al. 2009), has been shown to decrease chondrogenic and myogenic differentiation of MSCs in micromass cultures (Gamer et al. 2001).

TGF-β1 is a growth factor that is well known to enhance several key processes in the chondrogenic pathway, including cellular condensation, adhesion, and extracellular matrix production during chondrogenesis (James et al. 2009). TGF-β1 and myostatin can both bind the same co-receptor, ALK5 (Rebbapragada et al. 2003; Derynck and Feng 1997). We have shown that Sox-9 was downregulated with myostatin treatment, a treatment effect that was also seen with ALK5 inhibition during TGF-β1-induced chondrogenesis (Elkasrawy et al. 2011a). We also showed that GSK3a, responsible for beta-catenin instability, is upregulated with myostatin treatment during TGF-β1-induced chondrogenesis (Fig. 8.2). These data suggest that myostatin could be competitively inhibiting ALK5-mediated TGF-β1 signaling

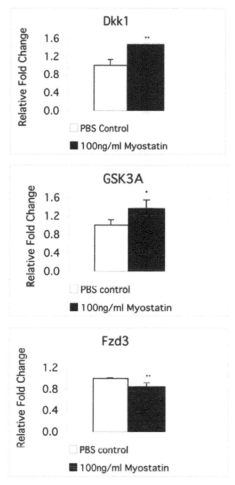

Figure 8.2 Effects of myostatin treatment on gene expression. Exposure of bone marrow-derived mesenchymal stem cells to exogenous myostatin alters the expression of several factors involved in the Wnt signaling pathway.

(Fig. 8.3). TGF-β signaling upregulates Sox-9 and Wnt5a, and increases nuclear accumulation and stability of beta-catenin during chondrogenesis (Zhou et al. 2004; Kawakami et al. 2006; Lorda-Diez 2009). Fzd3 is a transmembrane proteins that bind Wnt ligands (Chen and Struhl 1999). These experimental findings indicate that myostatin might be capable of competitively inhibiting ALK5-mediated TGF-β1 signaling, or alternatively that myostatin may stimulate the expression of certain factors that could actively inhibit TGF-β1, such as Smad7 (e.g., Zhu et al. 2004; Iwai et al. 2008).

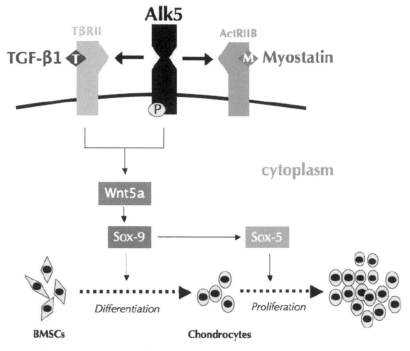

Figure 8.3 Hypothetical model showing mechanisms by which myostatin may mediate TGF-beta-induced chondrogenesis. TGF-beta 1 and myostatin compete for recruitment of ALK5. TGF-beta stimulates Wnt5a activation of the chondrogenic cascade, whereas myostatin may attenuate this effect.

Implications for Bone Repair

The first indication that myostatin might play a key role in bone repair and regeneration came from our work in myostatin-knockout mice, which revealed that fracture callus bone volume was significantly increased in mice lacking myostatin following fibular osteotomy (Fig. 8.4; Kellum et al. 2009). The idea that myostatin might in some way suppress or inhibit bone healing was further supported by another study in which we showed that a myostatin inhibitor (propeptide) increased fracture callus bone volume in mice two weeks following osteotomy (Hamrick et al. 2010a). Fracture healing in rodents involves three key phases that have been well-described by numerous authors: an initial inflammatory phase, a chondrogenic phase, and an osteogenic phase. The *in vivo* findings in mice lacking myostatin and in mice treated with a myostatin inhibitor are generally supportive of the *in vitro* data referenced above, suggesting that myostatin acts to suppress chondrogenesis directly (Table 8.1). Our immunolocalization studies showed that myostatin is highly expressed in the region of the wound blastema 12

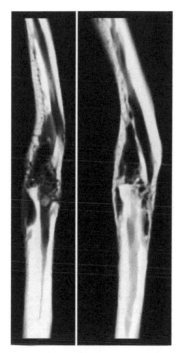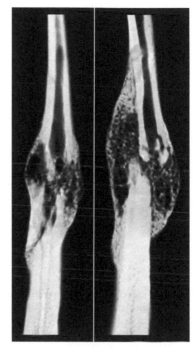

Figure 8.4 Effects of myostatin deficiency on fracture healing. MicroCT images of the fibula fracture callus in normal mice (left two panels) and mice lacking myostatin (right two panels).

Table 8.1 Key Facts Related to Myostatin and Bone Healing. This table lists key facts that are related to the role of myostatin in bone repair, including direct roles for myostatin on osteo- and chondroprogenitors, and indirect roles mediated by muscle-derived factors.

1. The receptor for myostatin is expressed in bone marrow stromal cells, and chondrocytes as well as muscle cells express myostatin.
2. Following traumatic musculoskeletal injury, myostatin is most highly expressed by injured muscle cells in the first 24–48 hr following the injury.
3. Congenital absence of myostatin increases the volume of cartilage and bone in the fracture callus, whereas delivery of exogenous myostatin impairs muscle and bone healing.
4. *In vitro* experiments show that myostatin can inhibit the chondrogenic differentiation of bone marrow stromal cells.
5. Blocking myostatin signaling immediately following musculoskeletal injury is therefore a potential therapeutic strategy for improving the healing of both muscle and bone.

and 24 hr following osteotomy, and that exogenous myostatin reduces fracture callus bone volume (Elkasrawy et al. 2011b). This is one mechanism by which myostatin may influence bone repair (Fig. 8.5; Table 8.1), but we believe there are also several others which we describe below.

In addition to suppressing chondrogenesis during fracture healing, myostatin may also mediate the differentiation of muscle-derived progenitors that contribute to callus formation. It has been recognized for

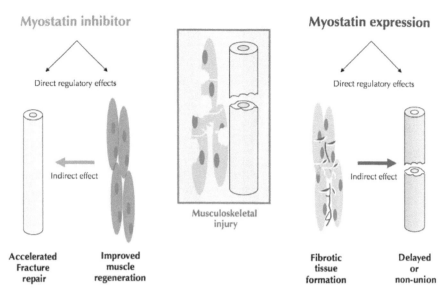

Myostatin inhibitor   **Myostatin expression**

Direct regulatory effects   Direct regulatory effects

Indirect effect   Indirect effect

Musculoskeletal
injury

Accelerated	Improved	Fibrotic	Delayed
Fracture	muscle	tissue	or
repair	regeneration	formation	non-union

Figure 8.5 General model summarizing effects of myostatin inhibition (left) and myostatin expression (right) on musculoskeletal repair and regeneration. Blocking myostatin signaling following injury of muscle and bone is likely to enhance bone healing in two ways (left side of figure): directly, because suppression of myostatin activity will enhance chondrogenesis and indirectly, because enhancing muscle regeneration is likely to restore the normal secretion of paracrine, osteogenic trophic factors (e.g., IGF-1) from muscle. Myostatin expression following musculoskeletal injury inhibits bone healing in two ways (right side of figure), first by suppressing chondrogenesis directly and second by increasing fibrosis in injured muscle thereby reducing secretion of osteogenic paracrine factors from muscle tissue.

Color image of this figure appears in the color plate section at the end of the book.

decades that the muscle bed itself plays a crucial role in fracture healing (Pritchard and Ruzicka 1950), and that cells derived from muscle may in fact migrate into the site of injury and promote bone healing (Schindeler et al. 2009; Liu et al. 2010). Cells harvested from muscle after exposure to an adjacent fracture are highly osteogenic and form bone nodules *in vitro* (Glass et al. 2011). Thus, muscle tissue not only facilitates bone repair by providing a vascular bed and perhaps trophic factors (see below), but it is a possible source of progenitor cells that can differentiate into osteoblasts, further improving bone healing (Glass et al. 2011). Given that myostatin is a potent factor that can mediate the differentiation of muscle-derived cells, it is likely that elevated levels of myostatin following injury may drive muscle-derived stromal cells toward a myofibroblast lineage rather than toward an osteogenic fate. In this way local or systemic myostatin could inhibit bone healing by limiting the supply of osteoprogenitors available from the pool of muscle-derived stromal cells (Fig. 8.5).

A third manner in which myostatin may alter bone healing is by regulating muscle regeneration and hence the secretion of muscle-derived paracrine factors ("myokines"; Hamrick 2011). Studies in which skeletal muscle is transplanted into cardiac muscle support the notion that myofibers secrete a number of paracrine factors that are not only osteogenic (Hamrick 2011) but also promote the survival of neighboring muscle fibers ("paracrine theory of transplantation"; Perez-Ilzarbe et al. 2008). Factors that are actively secreted by myotubes *in vitro* include insulin-like growth factor 1 (IGF-1), osteonectin (SPARC), and basic fibroblast growth factor (FGF-2) (Hamrick, 2011). All of these factors have been observed to increase bone formation when injected peripherally, and deficiency of these factors is associated with low bone mass (Hamrick 2011). Elevated levels of myostatin impair muscle regeneration and induce muscle atrophy, which would in turn be expected to reduce the secretion of factors normally produced by healthy myofibers such as IGF-1 and FGF-2 (Hamrick et al. 2010b). Recently it was also shown that mice lacking myostatin have elevated serum levels of IGF-1, and that this was associated with elevated expression of IGF-1 in the liver (Williams et al. 2011). Thus, elevated secretion of myostatin, which is known to occur with disuse, infection, burns, or AIDS- and cancer-related cachexia, may indirectly impair bone healing and bone formation by decreasing circulating levels of liver-derived IGF-1 (Williams et al. 2010).

Conclusion

Since the discovery of myostatin in 1997 our understanding of its biology and function has grown considerably so that we now recognize a broad range of potential roles for myostatin in musculoskeletal growth, development, and regeneration. These include effects of myostatin on tendon and ligament development and healing (Mendias et al. 2008; Eliasson et al. 2009; Fulzele et al. 2010), adipogenesis and fat deposition (McPherron and Lee 2002; Hamrick et al. 2006), and bone density and bone healing (Morissette et al. 2009; Kellum et al. 2009), in addition to the well-documented effects of myostatin on muscle mass and regeneration (Lee 2004). These and other *in vivo* and *in vitro* studies now point to a role for myostatin in the proliferation and differentiation of mesenchymal progenitors. These effects are mediated by TGF-beta signaling through the type IIB activin receptor, involving phosphorylation of Smad2/3, but our own work in chondrocytes and the work of others in MSCs and myoblasts suggest that myostatin also signals through the Wnt pathway. Various myostatin inhibitors have, not surprisingly, been observed to increase muscle mass, decrease fat mass, and increase bone formation. These findings suggest that myostatin represents a potential therapeutic target for improving tissue repair and regeneration in musculoskeletal diseases and following musculoskeletal injury.

Summary Points

- Myostatin, also known as growth and differentiation factor-8 (GDF-8), was discovered in 1997.
- Myostatin plays a number of roles in musculoskeletal growth, development, and regeneration.
- These include effects of myostatin on tendon and ligament development and healing, adipogenesis and fat deposition, and bone density and bone repair.
- Blocking myostatin signaling is well-known to increase muscle mass and improve muscle regeneration.

Dictionary

- *Wnt pathway*: The wnt signaling pathway is most well known for its role in regulating cell fate and cell-cell interactions during embryonic development in both vertebrates and invertebrates. The term "wnt" refers to the name of the gene implicated in the wingless mutation in fruit flies (Drosophila).
- *Endochondral ossification*: Long bones such as the humerus and femur of the mammalian skeleton first form as condensations of cartilage, which then undergo a process of ossification to form a bony skeleton. This process is referred to as endochondral ossification, and a similar process occurs during bone healing, where a "soft" callus of cartilage is replaced by a "hard" bony callus.
- *Bone marrow stromal cells*: Bone marrow includes a variety of cell types, including cells that can give rise to tissue types of mesodermal origin (e.g., muscle, fat, bone). These cells in bone marrow are referred to as bone marrow stromal cells.
- *Myokines*: It is now known that muscle can secrete a number of different growth factors and cytokines that can affect other organs, and these muscle-derived factors are referred to as myokines.

Acknowledgments

Many of the findings discussed in this chapter are derived from our research on myostatin which has been supported by the National Institutes of Health (AR049717), the Office of Naval Research (N000140810197), and the Department of the Army (USAMRMC PR093619). We are grateful to Drs. Paul Yaworsky, Alexandra McPherron, and Li-Fang Liang for many helpful discussions regarding the biology of myostatin. Ethan Kellum, Phonepasong Arounleut, Matthew Bowser, David Immel, Craig Byron, Penny Roon, Donna Kumiski, and Cathy Pennington provided valuable technical assistance and support for many of the studies reviewed in this chapter.

List of Abbreviations

ActRIIB	:	type IIB activin receptor
Acvr2b	:	type IIB activin receptor
Alk4/5	:	activin-like kinase receptor 4 and 5
BMP-9, -10, 11	:	bone morphogenetic protein-9, -10, -11
BMSC	:	bone marrow stromal cell
Dkk1	:	gene encoding Dickkopf-related protein 1
ERK	:	extracellular signal-regulated kinase
FLRG	:	follistatin-related gene
Fzd3	:	gene encoding the frizzled-3 protein
GDF-8	:	growth and differentiation factor-8, or myostatin
GDF-11	:	growth and differentiation factor-11
Gsk3a	:	gene encoding glycogen synthase kinase 3 alpha
MAPK	:	mitogen-activated protein kinase
microCT	:	micro-computed tomography
TGF-β	:	transforming growth factor-beta
Wnt4	:	Wingless-type MMTV integration site family, member 4
Wnt5a	:	Wingless-type MMTV integration site family, member 5a

References

Akiyama, T. 2000. Wnt/beta-catenin signaling. Cytokine Growth Factor Rev. 11: 273–82.

Allen, D.L., A. Cleary, K. Speaker, S.F. Lindsay, J. Uyenishi, J.M. Reed, M.C. Madden and R.S. Mehan. 2008. Myostatin, activin receptor IIB, and follistatin-like-3 gene expression are altered in adipose tissue and skeletal muscle of obese mice. Am J Physiol Endocrinol Metab. 294: E918–27.

Allendorph, G., W. Vale and S. Choe. 2006. Structure of the ternary signaling complex of a TGF-beta superfamily member. Proc Natl Acad Sci USA. 103: 7643–7648.

Artaza, J., S. Bhasin, T. Magee, S. Reisz-Porszasz, R. Shen, N. Groome, M. Fareez and N. Gonzalez-Cadavid. 2005. Myostatin inhibits myogenesis and promotes adipogenesis in C3H 10T(1/2) mesenchymal multi-potent cells. Endocrinology. 146: 3547–57.

Bialek, P., J. Parkington, L. Warner, M. St. Andre, L. Jian, D. Gavin, C. Wallace, J. Zhang, G. Yan, A. Root, H. Seeherman and P. Yaworsky. 2008. Mice treated with a myostatin/GDF-8 decoy receptor, ActRIIB-Fc, exhibit a tremendous increase in bone mass. Bone. 42: S46.

Bradley, L., P. Yaworsky and F. Walsh. 2008. Myostatin as a therapeutic target for musculoskeletal disease. Cell Mol Life Sci. 65: 2119–24.

Burks, T. and R. Cohn. 2011. Role of TGF-beta signaling in inherited and acquired myopathies. Skelet. Muscle 1: 19.

Chen, C. and G. Struhl. 1999. Wingless transduction by the Frizzled and Frizzled2 proteins of Drosophila. Development. 126: 5441–42.

Chen, Y. and B.A. Alman. 2009. Wnt pathway, an essential role in bone regeneration. J Cell Biochem. 106: 353–62.

Chen, Y., H.C. Whetstone, A. Youn, P. Nadesan, E.C. Chow, A.C. Lin and B.A. Alman. 2007a. Beta-catenin signaling pathway is crucial for bone morphogenetic protein 2 to induce new bone formation. J Biol Chem. 282: 526–33.

Chen, Y., H.C. Whetstone, A.C. Lin, P. Nadesan, Q. Wei, R. Poon and B.A. Alman. 2007b. Beta-catenin signaling plays a disparate role in different phases of fracture repair: implications for therapy to improve bone healing. PLoS Med. 4(7): e249.

Cho, T., L. Gerstenfeld and T. Einhorn. 2002. Differential temporal expression of members of the transforming growth factor beta superfamily during murine fracture healing. J Bone Miner Res. 17: 513–520.

Church, V., T. Nohno, C. Linker, C. Marcelle and P. Francis-West. 2002. Wnt regulation of chondrocyte differentiation. J. Cell Sci. 115: 4809.

Derynck, R. and X. Feng. 1997. TGF-beta receptor signaling. Biochim Biophys Acta. 1333: F105–50.

Donaldson, C., L. Mathews and W.W. Vale. 1992. Molecular cloning and binding properties of the human type II activin receptor. Biochem Biophys Res Commun. 194: 310–6.

Ekaza, D. and G. Cabello. 2007. The myostatin gene: physiology and pharmacological relevance. Curr Opin Pharmacol. 7: 310–5.

Eliasson, P., T. Andersson, J. Kulas, P. Seemann and P. Aspenberg. 2009. Myostatin in tendon maintenance and repair. Growth Factors. 27: 247–54.

Elkasrawy, M.N. and M.W. Hamrick. 2010. Myostatin (GDF-8) as a key factor linking muscle mass and bone structure. J Musculoskelet Neuronal Interact. 10: 56–63.

Elkasrawy, M.N., S. Fulzele, M. Bowser, K. Wenger and M.W. Hamrick. 2011a. Myostatin (GDF-8) inhibits chondrogenesis and chondrocyte proliferation in vitro by suppressing Sox-9 expression. Growth Factors. July 15 (ePub ahead of print).

Elkasrawy M., D. Immel, L.-F. Liang, X. Wen and M.W. Hamrick. 2011b. Immunolocalization of myostatin (GDF-8) following musculoskeletal injury and the effects of exogenous myostatin on muscle and bone healing. J Histochem. Cytochem (in press).

Feldman, B.J., R. Streeper, R. Farese and K. Yamamoto. 2006. Myostatin modulates adipogenesis to generate adipocytes with favorable metabolic effects. Proc Natl Acad Sci USA. 103: 15675–80.

Ferguson, V., R. Paietta, L. Stodieck, A. Hanson, M. Young, T. Bateman et al. 2009. Inhibiting myostatin prevents microgravity-associated bone loss in mice. J Bone Miner Res. 24 (supp 1): 1288.

Fulzele, S., P. Arounleut, P. Cain, S. Herberg, M. Hunter, K. Wenger and M.W. Hamrick. 2010. Role of myostatin (GDF-8) signaling in the human anterior cruciate ligament. J Orthop Res. 28: 1113–1118.

Gamer, L.W., K.A. Cox, C. Small and V. Rosen. 2001. Gdf11 is a negative regulator of chondrogenesis and myogenesis in the developing chick limb. Dev Biol. 229: 407–20.

Glass, G., J. Chan, A. Freidin, M. Feldmann, N. Horwood and J. Nanchacal. 2011. TNF-alpha promotes fracture repair by augmenting the recruitment and differentiation of muscle-derived stromal cells. Proc Natl Acad Sci USA. 108: 1585–90.

Guo, W. et al. 2008. The effects of myostatin on adipogenic differentiation of human bone marrow-derived mesenchymal stem cells are mediated through cross-communication between Smad3 and Wnt/beta-catenin signaling pathways. J Biol Chem. 283: 9 136–9145.

Hadjiargyrou, M., F. Lombardo, S. Zhao, W. Ahrens, J. Joo, H. Ahn, M. Jurman, D.W. White and C.T. Rubin. 2002 Transcriptional profiling of bone regeneration. Insight into the molecular complexity of wound repair. J Biol Chem. 277: 30177–82.

Hamrick, M.W. 2011. A role for myokines in muscle-bone interactions. Exercise & Sports Science Reviews. 39: 43–47.

Hamrick, M.W., C. Pennington, C. Webb and C.M. Isales. 2006. Resistance to body fat gain in 'double-muscled' mice fed a high-fat diet. Int J Obes. 30: 868–870.

Hamrick, M.W., X. Shi, W. Zhang, C. Pennington, B. Kang, H. Thakore, M. Haque, C.M. Isales, S. Fulzele and K. Wenger. 2007. Loss of myostatin function increases osteogenic differentiation of bone marrow-derived mesenchymal stem cells but the osteogenic effect is ablated with unloading. Bone. 40: 1544–1553.

Hamrick, M.W., P. Arounleut, E. Kellum, M. Cain, D. Immel and L. Liang. 2010a. Recombinant myostatin (GDF-8) propeptide enhances the repair and regeneration of both muscle and bone in a model of deep penetrant musculoskeletal injury. Journal of Trauma. 69: 579–83.

Hamrick, M.W., P.L. McNeil and S.L. Patterson. 2010b. Role of muscle-derived growth factors in bone formation. J Musculoskel. Neuronal Interact. 10: 64–70.

Iwai, T., J. Murai, H. Yoshikawa and N. Tsumaki. 2008. Smad7 inhibits chondrocyte differentiation at multiple steps during endochondral bone formation and down-regulates p38 MAPK pathways. J Biol Chem. 283: 27154–64.

James, A.W., Y. Xu, J.K. Lee, R. Wang and M.T. Longaker. 2009. Differential effects of TGF-beta1 and TGF-beta3 on chondrogenesis in posterofrontal cranial suture-derived mesenchymal cells *in vitro*. Plast Reconstr Surg. 123: 31–43.

Kawakami, Y., J. Rodriguez-Leon and J.C. Izpisua Belmonte. 2006. The role of TGFbetas and Sox9 during limb chondrogenesis. Curr Opin Cell Biol. 18: 723–9.

Kellum, E., H. Starr, D. Immel, P. Arounleut, S. Fulzele, K. Wenger and M.W. Hamrick. 2009. Myostatin (GDF-8) deficiency increases fracture callus size, Sox-5 expression, and callus bone volume. Bone. 44: 17–23.

Kim, H., L. Liang, R. Dean, D. Hausman, D. Hartzell and C. Baile. 2001. Inhibition of preadipocyte differentiation by myostatin treatment in 3T3-L1 cultures. Biochem Biophys Res Commun. 281: 902–906.

Lee, S.J. 2004. Regulation of muscle mass by myostatin. Annu Rev Cell Dev Biol. 20: 61–86.

Li Z., H. Kollias and K.R. Wagner. 2008. Myostatin directly regulates skeletal muscle fibrosis. J Biol Chem. 283: 19371–1938.

Lin, J., H. Arnold, M. Della-Fera, M. Azain, D. Hartzell and C.A. Baile. 2002. Myostatin knockout in mice increases myogenesis and decreases adipogenesis. Biochem Biophys Res Commun. 291: 701–6.

Liu, R., A. Schindeler and D. Little. 2010. The potential role of muscle in bone repair. J. Musculoskel. Neuronal Interact. 10: 71–6.

Lorda-Diez, C.I., J.A. Montero, C. Martinez-Cue, J.A. Garcia-Porrero and J.M. Hurle. 2009. Transforming growth factors beta coordinate cartilage and tendon differentiation in the developing limb mesenchyme. J Biol Chem. 284(43): 29988–96.

Mao, B., W. Wu, G. Davidson, J. Marhold, M. Li, B.M. Mechler, H. Delius, D. Hoppe, P. Stannek, C. Walter, A. Glinka and C. Niehrs. 2002. Kremen proteins are Dickkopf receptors that regulate Wnt/beta-catenin signalling. Nature. 417: 664.

McPherron, A.C., A.M. Lawler and S.-J. Lee. 1997. Regulation of skeletal muscle mass in mice by a new TGF-ß superfamily member. Nature. 387: 83–90.

McPherron, A.C. and S.-J. Lee. 2002. Suppression of body fat accumulation in myostatin-deficient mice. J Clin Invest. 109: 595–601.

McPherron, A.C., T.V. Huynh and S.J. Lee. 2009. Redundancy of myostatin and growth/differentiation factor 11 function. BMC Dev Biol. 9: 24.

Mendias, C., K. Bakhurin and J. Faulkner. 2008. Tendons of myostatin-deficient mice are small, brittle, and hypocellular. Proc Natl Acad Sci USA. 105: 388–393.

McFarlane, C., G. Hui, W. Amanda, H. Lau, S. Lokireddy, G. Xiaojia, V. Mouly, G. Butler-Browne, P. Gluckman, M. Sharma and R. Kambadur. 2011. Human myostatin negatively regulates human myoblast growth and differentiation. Am J Physiol Cell Physiol. 301: C195–203.

Morissette, M., J. Stricker, M. Rosenberg, C. Buranasombati, E. Levitan, M. Mittleman and A. Rosenzweig. 2009. Effects of myostatin deletion in aging mice. Aging Cell. 8: 573–83.

Perez-Ilzarbe, M., B. Agbulut, C. Pelacho, E. Ciorba, S. Jose-Eneriz, M. Desnos, A. Hagege, P. Aranda, E. Andreu, P. Menasche and F. Prosper. 2008. Characterization of the paracrine effects of human skeletal myoblasts transplanted in infracted myocardium. Eur J Heart Fail. 10: 1065–72.

Phiel, C.J. and P.S. Klein. 2001 Molecular targets of lithium action. Annu Rev Pharmacol Toxicol. 41: 789–813.

Pritchard, J. and A. Ruzicka. 1950. Comparison of fracture repair in the frog, lizard, and rat. J Anat. 84: 236–261.

Rebbapragada, A., H. Benchabane, J. Wrana, A. Celeste and L. Attisano. 2003. Myostatin signals through a transforming growth factor beta-like signaling pathway to block adipogenesis. Mol Cell Biol. 23: 7230–7242.

Reisz-Porszasz, S., S. Bhasin, J. Artaza, R. Shen et al. 2003. Lower skeletal muscle mass in male transgenic mice with muscle-specific overexpression of myostatin. Am J Physiol Endocrinol Metab. 285: E876–88.

Rios, R., I. Carniero, V. Arce and J. Devesa. 2001. Myostatin regulates cell survival during C2C12 myogenesis. Biochem Biophys Res Commun. 280: 561–66.

Schindeler, A., R. Liu and D. Little. 2009. The contribution of different cell lineages to bone repair: exploring a role for muscle stem cells. Differentiation. 77: 12–18.

Souza, T. et al. 2008. Proteomic identification and functional validation of activins and bone morphogenetic protein 11 as candidate novel muscle mass regulators. Mol Endocrinol. 22: 2689–702.

Steelman, C.A., J.C. Recknor, D. Nettleton and J.M. Reecy. 2006. Transcriptional profiling of myostatin-knockout mice implicates Wnt signaling in postnatal skeletal muscle growth and hypertrophy. Faseb J. 20(3): 580–2.

Taylor, W.E., S. Bhasin, J. Artaza, F. Byhower, M. Azam, D. Willard, F. Kull and N. Gonzalez-Cadavid. 2001. Am. J. Physiol. Endocrinol. Metab. E221–8.

Tsuchida, K. 2004. Activins, myostatin and related TGF-beta family members as novel therapeutic targets for endocrine, metabolic and immune disorders. Curr Drug Targets Immune Endocr Metabol Disord. 4: 157–66.

Williams, N., J. Interlichia, M. Jackson, D. Hwang, P. Cohen and B. Rodgers. 2011. Endocrine actions of myostatin: systemic regulation of the IGF and IGF binding protein axis. Endocrinology. 152: 172–80.

Yang, Y., L. Topol, H. Lee and J. Wu. 2003. Wnt5a and Wnt5b exhibit distinct activities in coordinating chondrocyte proliferation and differentiation. Development. 130: 1003.

Zhang, F., C.J. Phiel, L. Spece, N. Gurvic and P.S. Klein. 2003. Inhibitory phosphorylation of glycogen synthase kinase-3 (GSK-3) in response to lithium. Evidence for autoregulation of GSK-3. J Biol Chem. 278: 33067–77.

Zhong, N., R.P. Gersch and M. Hadjiargyrou. 2006. Wnt signaling activation during bone regeneration and the role of Dishevelled in chondrocyte proliferation and differentiation. Bone. 39: 5–16.

Zhou, S., K. Eid and J. Glowacki. 2004. Cooperation between TGF-beta and Wnt pathways during chondrocyte and adipocyte differentiation of human marrow stromal cells. J Bone Miner Res. 19: 463–70.

Zhu. J., Y. Li, W. Shen, C. Qiao, F. Ambrosio, M. Lavasani, M. Nozak, M. Branca and J. Huard. 2007. Relationships between transforming growth factor-beta1, myostatin, and decorin: implications for skeletal muscle fibrosis. J Biol Chem. 282: 25852–63.

Zhu, X., S. Topouzi, L.F. Liang and R.L. Stotish. 2004. Myostatin signaling through Smad2, Smad3 and Smad4 is regulated by the inhibitory Smad7 by a negative feedback mechanism. Cytokine. 26: 262–72.

Zimmers, T.A., M.V. Davie, L.G. Koniaris, P. Haynes, A.F. Esquela, K.N. Tomkinson, A.C. McPherron, N.M. Wolfman and S.J. Lee. 2002. Induction of cachexia in mice by systemically administered myostatin. Science. 296: 1486–1488.

c-Jun-centred Regulatory Network of Signalling Pathways in Bone Stem Cells and Applications to Bone Repair

Jon Fernández-Rueda,[#] Emma Jakobsson,[#] Akaitz Dorronsoro and César Trigueros[a,*]

ABSTRACT

Activator protein-1 (AP-1) is a transcription factor composed of a core dimer of Jun and Fos protein families and several coactivators, and is essential for the regulation of a variety of fundamental cellular processes such as proliferation, differentiation, apoptosis and cell transformation in almost all cell types. In bone tissue, AP-1 is involved in the activation of the expression of a number of genes that modulate bone construction and remodelling, mainly by driving the differentiation process of the precursors of bone modelling cells, osteoblasts and osteoclasts. Several signalling pathways that lead to bone cell differentiation regulate the expression, phosphorylation and binding capacity of the constituents

Fundación Inbiomed, Hematopoietic and Mesenchymal Stem Cell Department, Paseo Mikeletegi, 61 20009 San Sebastián, Spain.
[a]E-mail: ctrigueros@inbiomed.org
*Corresponding author
[#]These authors contributed equally to this work.

List of abbreviations given at the end of the text.

of the AP-1 dimer and thereby their activity. These pathways include paracrine signalling pathways that arise from stromal cells of the bone tissue and hormone signalling that often modulates these pathways. Regarding bone development, specific functions have been linked to different components of the AP-1 core dimer, and among them, c-Jun has emerged as an important regulator of differentiation signalling networks in both osteoblasts and osteoclasts. Dysfunction of these networks results in the development of a number of bone diseases, such as osteoporosis, osteopetrosis or periodontal diseases, and thus, they constitute promising targets for novel therapies.

Introduction

The development and maintenance of the skeleton depends on the equilibrated action of bone constructing cells and bone resorbing cells. The balance between bone formation by osteoblasts and bone resorption by osteoclasts is essential for bone homeostasis. Impairment of this balance results in bone pathologies such as osteoporosis, Paget's disease and osteopetrosis, as well as diseases of immune nature, such as rheumatoid arthritis and periodontal disease (Rodan and Martin 2000). Much of this balance lies in the scheduled expression of specific gene sets that control proliferation and differentiation of osteoblast and osteoclast precursor cells. c-Jun protein, a member of the AP-1 transcription factor, is one of the key regulators of the expression of these genes.

The AP-1 Transcription Factor

The transcription factor AP-1 consists of a variety of core dimers composed by members of the Fos and Jun families and some members of the activating transcription factor (ATF) and cAMP response element binding (CREB) families. Fos family, namely c-Fos, FosB, Fra-1 and Fra-2, can only heterodimerize with Jun proteins, while Jun proteins (c-Jun, JunB and JunD) can also dimerize with themselves. These dimers bind primarily to promoters containing AP-1 consensus sites; nevertheless, ATF can target dimers to CRE-like sites (Hai and Curran 1991). AP-1 is activated by signalling pathways such as WNT, transforming growth factor beta (TGF-β) or mitogen-activated protein kinases (MAPK), and regulates essential cell processes such as proliferation, differentiation, apoptosis and cell transformation by regulating the expression of genes that contain AP-1 binding site. The activity of AP-1 is regulated by controlling the expression and phosphorylation of dimerization partners, and by interactions of AP-1 with other transcription factors (Jochum et al. 2001).

Specific functions of AP-1 proteins have been dissected mainly from loss-of-function and gain-of-function mouse models. These analyses have shown that AP-1 proteins are essential in embryonic development and play important roles as regulators of bone development (Table 9.1).

Table 9.1 AP-1 components from the Fos and Jun family. The Fos family members form dimers together with a member of the Jun family, whereas Jun proteins can form both homo- and heterodimers. Knock-out models has been done in mouse for all the Fos and Jun family members.

Family	Protein	Dimerization	KO result in mouse
FOS	c-Fos	Heterodimer with Jun protein	Lack osteoclasts, osteopetrotic phenotype
	FosB	Heterodimer with Jun protein	Nurturing defect
	Fra-1	Heterodimer with Jun protein	Aberrant placenta and yolk sac, embryonic death
	Fra-2	Heterodimer with Jun protein	Growth retardation, die within a week after birth
JUN	c-Jun	Homodimer or heterodimer	Defects in liver, heart and artery development leads to death of the embryo
	JunB	Homodimer or heterodimer	Die early due to defective vascularization of the placenta
	JunD	Homodimer or heterodimer	Impaired growth and spermatogenesis

Fos family

Members of the Fos family are essential transcription factors for both osteoblast and osteoclast development. Most Fos proteins play unique roles in this process, and their loss is associated with bone developmental diseases.

c-Fos is considered an osteoclast differentiation-specific transcription factor. Mice deficient of c-Fos lack osteoclasts and develop an osteopetrotic phenotype. Transgenic mice overexpressing c-Fos undergo normal cell differentiation, but suffer transformation of osteoblasts leading to osteosarcomas (Wang et al. 1995).

Mice lacking FosB display a severe nurturing defect, while FosB overexpression in transgenic mice causes no evident phenotype. Overexpression of ΔFosB, an alternatively spliced form of FosB that occurs in osteoblasts, leads to increased bone formation resulting in osteosclerosis, while inhibiting adipogenesis. Therefore, ΔFosB is likely to have a positive role in osteoblastic differentiation (Jochum et al. 2001).

Loss of Fra-1 and Fra-2 is lethal in mice. Fra-1 overexpressing transgenic mice exhibit osteosclerosis of entire skeleton due to increased osteoblast number, as well as enhanced osteoblast differentiation and osteoclastogenesis *in vitro* (Jochum et al. 2000). Mice lacking Fra-2 show increased size and number of osteoclasts, probably due to impaired LIF signalling (Bozec et al.

2008). Fra-2-deficient osteoblasts fail to differentiate *in vitro* and acquire an adipogenic phenotype. Fra-2 transgenic mice exhibit defective osteoblast differentiation and develop an osteosclerotic phenotype. These findings have revealed a role for Fra-2 as a positive regulator of bone and matrix formation (Bozec et al. 2010), while Fra-1 appears as an essential promoter of bone remodelling.

Jun family

The Jun family members, dimerization partners of Fos proteins, have not been so extensively linked to bone development, probably due to eventual functional compensation between family members in loss-of-function experimental approaches. However, careful analyses of osteogenic pathways show that this family plays indeed an essential role in osteogenesis both in early progenitor (mesenchymal stem cells, MSCs) fate and in late osteoblast differentiation, as will be described later.

Among Jun proteins, c-Jun is often considered an activator of the transcription of AP-1 target genes, although it can act as a repressor in certain contexts (Yang-Yen et al. 1990). Loss of c-Jun causes death of mouse embryos between day 12.5 and 13.5 of development, due to defects in liver, heart and artery development. These foetuses show impaired proliferation and increased apoptosis of hepatocytes (Jochum et al. 2001). Besides development, c-Jun also stimulates keratinocyte proliferation and differentiation (Zenz et al. 2003). On the other hand, JunD and JunB are generally considered transcriptional repressors. The inactivation of JunD causes only mild phenotypes, namely impaired growth and spermatogenesis, while JunB defective mice die between day 8.5 and 10 due to defective vascularization of the placenta (Jochum et al. 2001).

Transactivation activity of Jun proteins is enhanced by phosphorylation by c-Jun N-terminal kinases (JNK) (Whitmarsh and Davis 1996). Mice defective for JNK activity die during development (Kuan et al. 1999), but c-Jun phosphorylation site-defective mice show an apparently normal development (Behrens et al. 1999). While this suggests that functional compensation between Jun proteins may render specific modulation of c-Jun non-essential during embryogenesis, an active role in osteogenesis has been found for both c-Jun and JNK.

Osteogenic potential of human bone MSCs, precursors of both osteoblasts and adipocytes, is connected to cell cycle progression. It is believed that osteoblasts suffer concomitant proliferation and differentiation *in vivo*, while adipogenic fate requires absence of proliferation. c-Jun regulates *in vitro* proliferation and differentiation of bone MSCs. Overexpression of c-Jun in MSC cultures leads to an increase in cell proliferation and osteogenic potential at the expense of adipogenic differentiation, while depletion of c-Jun reduces proliferation and completely abolishes the osteogenic potential of these cells (Cárcamo Orive et al. 2010). Thus, c-Jun plays a decisive role in the balance of bone MSC differentiation pathways.

JNK/c-Jun Axis in Bone Paracrine Pathways

Bone morphogenetic protein pathway

Bone morphogenetic protein (BMP) pathway is one of the best-dissected pathways that lead to differentiation of precursor cells towards osteogenic cells. BMPs are members of the transforming growth factor β (TGF-β) superfamily and induce the differentiation of mesenchymal cells into osteoblasts by activating the SMAD pathway (Fujii et al. 1999). SMADs are serine/threonine kinases that phosphorylate specific substrates after being activated by phosphorylation. Among BMPs, BMP-2 is the most studied inducer of osteoblast differentiation. BMPs bind to specific cell surface type I and type II receptors; then, type II receptors phosphorylate type I receptors, which in turn phosphorylate SMAD1, SMAD5 and SMAD8 proteins. Each of these phosphorylated SMADs is able to form an active dimer with SMAD4, then translocate to the nucleus and modulate transcription either by direct binding to DNA or by associating with other transcription factors. SMAD1-5-8/SMAD4 dimers are able to bind AP-1, preferentially when it includes a phosphorylated form of c-Jun (Liberati et al. 1999; Guicheux et al. 2003) (Fig. 9.1).

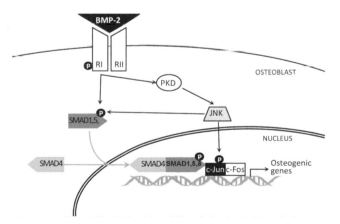

Figure 9.1 The bone morphogenetic protein pathway. When the secreted BMP binds to type I and type II receptors in the cell surface of osteoblast precursors, they activate the SMAD pathway, inducing the translocation of active SMAD dimers to the nucleus. Moreover, they activate JNKs via protein kinase D (PKD) activation, inducing phosphorylation of c-Jun. SMAD dimers bind AP-1 dimers containing phosphorylated c-Jun; then, this complex activates the expression of osteoblast-specific genes such as RUNX2.

In addition to the widely described activation of the SMAD pathway, BMP-2 is able to activate JNK and p38 kinases via protein kinase D during osteoblast differentiation. JNKs and p38 are also able to induce the expression of certain osteoblast-specific genes, e.g., osterix, alkaline phosphatase, osteocalcin (Guicheux et al. 2003) and RUNX2. This enhanced JNK activity contributes to the formation of SMAD-activable c-Jun /c-Fos dimers; moreover, direct activation of SMADs by JNK has also been described in osteoblasts (Fig. 9.1).

Either by association of SMADs with phosphorylated c-Jun or by direct JNK activation, JNKs and c-Jun have emerged as key factors in bone morphogenetic protein (BMP) family-induced osteoblastic differentiation of mesenchymal cells.

The WNT pathway

The WNT signalling pathway is an extensive network that plays a significant role in a broad variety of biological mechanisms, from embryonic development to adult physiological processes. The WNT pathway plays a wide role in bone development and homeostasis, and specifically, is an important regulator of osteogenic differentiation of MSCs.

In osteoblasts, the canonical WNT/β-catenin pathway acts synergistically with RUNX2, an osteoblast differentiation-specific gene, and promotes differentiation of osteoblast precursors. Non-canonical WNT pathway can induce RUNX2 transcription. Additionally, the cell polarity non-canonical WNT pathway, in which WNT proteins bind to Frizzled receptor activating rho and Rac GTPases, can activate JNK, which is able to induce c-Jun dependent expression of pro-osteogenic genes (Fig. 9.2). Some crosstalk has been found between canonical and non-canonical pathways in MSCs, as certain non-canonical WNT ligands are able to translocate β-catenin to the nucleus and to induce JNK-dependent differentiation of MSC towards osteoblastic lineage (Wagner et al. 2011).

In osteoclasts, non-canonical WNT pathway acts in synergy with RANK pathway (see Section below) to activate JNK/c-Jun activity, enhancing the expression of osteoclastogenic genes and ultimately favouring osteoclast differentiation.

The RANK pathway

Although osteoblast and osteoclast differentiation lineages are unrelated, JNK/c-Jun axis also plays an essential role in osteoclast differentiation, as targets of receptor activator of NF-κB (RANK) pathway signalling. The RANK signalling provides a pathway by which osteoblasts activate the differentiation of osteoclast precursors and promote bone remodelling.

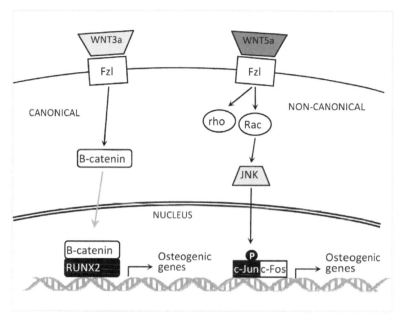

Figure 9.2 The WNT pathway. In osteoblasts, canonical WNT ligands, such as WNT3a, induce stabilization and translocation of β-catenin to the nucleus, where, in collaboration with RUNX2, it induces the expression of osteogenic genes. Non-canonical WNT ligands, such as WNT5a, induce JNK/c-Jun axis by activating rho and Rac GTPases.

Osteoblasts and stromal cells express the RANK ligand (RANKL) a member of TNF family. RANKL binds to its receptor, RANK, expressed in osteoclast precursors, which induces c-Fos expression and interacts with the intracellular TRAF6 protein. This, in turn, binds to SMAD3 and activates NF-κB, p38 and JNK pathways, all involved in osteoclast differentiation. JNK activation induces c-Jun/c-Fos dimer activity, and, together with NF-κB, upmodulates NFAT transcription factors. RANK-induced calcineurin pathway also upregulates NFAT activity. NFAT proteins form complexes with Fos/Jun dimers and are involved in regulating osteoclastogenesis by inducing the expression of osteoclast-specific genes such as TRAP, calcitonin receptor, cathepsin K, β3 integrin and OSCAR (Fig. 9.3) (Takanayagi 2007).

RANKL-deficient mice lack osteoclasts and suffer severe osteopetrosis. Similarly, c-Jun appears to be essential for their activation, and specific suppression of c-Jun in osteoclast precursors in mice impairs differentiation, leading to osteopetrosis (Teitelbaum 2004; Ikeda et al. 2004). Estrogen inhibits osteoclast differentiation by reducing c-Jun activity. Specifically, it inhibits RANKL-induced c-Jun expression and RANKL-induced JNK activation. The critical role of c-Jun in osteoclast differentiation is underscored by the

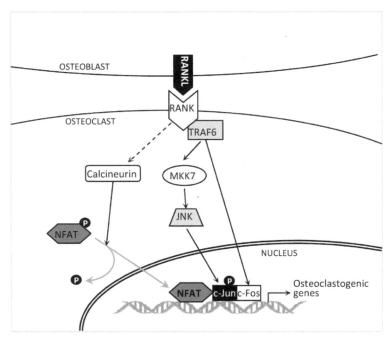

Figure 9.3 The RANK pathway. RANKL, expressed in osteoblasts and stromal cells, binds and activates RANK receptors. This induces enhancement of c-Fos expression and the induction of c-Jun dependent transcription via JNK activation. It is believed that induction of c-Fos expression and JNK activation are mediated by TRAF6, an adaptor protein. RANK also induces calcineurin activation, leading to NFAT dephosphorylation. Once activated, NFAT transcription factors are able to bind Fos/Jun dimers, inducing the expression of osteoclastogenic genes. Amongst them are NFAT transcription factors themselves.

fact that inactivation of c-Jun phosphorylation in macrophages prevents their differentiation into osteoclasts.

The PDGF pathway

The platelet-derived growth factor (PDGF) is an angiogenic factor and a potent inducer of cell division in mesenchymal cells, including bone MSCs. *In vitro*, PDGF treatment restores to a certain extent the reduction of proliferation and c-Jun expression that is seen under both *in vitro* adipocyte and osteoblast differentiation conditions. C-Jun NH$_2$-terminal kinase (JNK) and c-Jun are known targets of PDGF signalling and both are phosphorylated in MSCs upon treatment with the growth factor. *In vitro*, PDGF promotes both c-Jun phosphorylation and proliferation. Since the proliferative state of MSCs has been linked to differentiation towards osteoblasts at the expense of adipocytes, it is likely that PDGF induced c-Jun proliferative and differentiative networks coexist in this context (Cárcamo Orive et al. 2010).

JNK/c-Jun Axis in Hormone Signalling in Osteogenesis

Glucocorticoid signalling

Glucocorticoid hormones are widely used as anti-inflammatory therapy for chronic diseases and as prevention of transplant rejection. Among secondary effects that limit their use is the development of osteoporosis due to lengthened osteoclast lifespan, decreased osteoblast precursor proliferation and differentiation and increased osteoblast apoptosis. At high doses, dexamethasone, a synthetic glucocorticoid widely used in clinic, inhibits osteogenesis in human MSC cultures via glucocorticoid receptor (GR) pathway, which, in turn, reduces c-Jun levels in these cells. The way by which GR regulates c-Jun activity is not fully elucidated. Two mechanisms have been proposed. First, GR could inhibit AP-1 DNA binding by direct protein-protein interaction. Second, GR could exert the transcriptional repression of c-Jun by tethering to the nuclear isoform of the focal adhesion LIM domain protein Trip6, a coactivator of AP-1 (Kassel et al. 2004). While the second possibility remains as a hypothesis, direct GR/AP-1 binding has been found to be widely responsible for the glucocorticoid-mediated inhibition of AP-1 targets, including pro-osteogenic genes such as interleukin-11 or collagenase-1 (Fig. 9.4).

On the other hand, in physiological conditions, GR appears to be a positive regulator of osteogenesis by genotropic mechanisms in osteoblasts (Cárcamo Orive et al. 2010). In addition, binding of AP-1 to the chromatin has been linked to a greater accessibility for GR, promoting a wide co-occupancy of target genes (John et al. 2011).

Parathyroid hormone signalling

Parathyroid hormone (PTH) is an anabolic hormone that has shown to reverse bone loss in humans (Dempster et al. 1993) and to reduce the fracture risk of postmenopausal women (Neer et al. 2001). It stimulates the expression of certain osteogenic genes in osteoblast precursors. Although beneficial effects of PTH treatment are largely known, their molecular basis is complex and is not yet fully understood.

The intermittent administration of PTH elicits an anabolic response in bone due to antiapoptotic and prodifferentiative effects on osteoblasts (Jilka 2007). As PTH upregulates RUNX2 activity, antiapoptotic effects of PTH may be partially mediated by RUNX2-dependent expression of antiapoptotic genes. On the other hand, prodifferentiative effects rely on

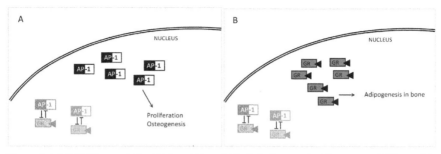

Figure 9.4 AP-1/GR crosstalk. AP-1 and ligand binded glucocorticoid receptor (GR) are able to inhibit reciprocally. Balance between these activities can determine MSC differentiation fate. (A) High AP-1 activity can inactivate GR activity and tilt the balance towards cell proliferation and osteoblastic differentiation. (B) An excess of activated GR can abrogate AP-1 activity and lead MSC to adipogenesis.

enhancement of RUNX2, c-Fos and c-Jun expression. PTH promotes c-Fos and c-Jun expression through protein kinase A and CREB pathway. Upon PTH treatment, c-Jun/c-Fos interact with Cbfa/runt and RUNX2 proteins to drive expression of osteogenic genes (Selvamurugan et al. 1998), and it has been shown that increase in bone mass after PTH treatment is dependent of c-Fos (Tanaka et al. 2004).

PTH also induces the expression of SMAD3 and SMAD3-related factor Tmem119 in osteoblasts. Induction of SMAD3, a TGF-β signalling molecule, by PTH stimulates βcatenin production, protects osteoblasts against apoptosis, and potentiates the anabolic effects of TGF-β, by modulating expression of certain genes (Tobimatsu et al. 2006). Amongst them, frizzled-related protein modulates WNT/β-catenin signalling, and Dlx3 upregulates RUNX2.

Sex steroid signalling

Postmenopausal estrogen deficiency has been linked to bone loss. Several studies have demonstrated that this effect is tied to the onset of cortical bone loss, but not to the trabecular bone loss. It is considered that estrogen inhibits the bone remodelling by reducing the initiation of new basic multicellular units, inhibits differentiation and promotes apoptosis of osteoclasts, and promotes differentiation of osteoblastic cells.

Estrogens and androgens exert their regulatory function by binding to specialized nuclear receptors: the estrogen receptors and the androgen receptor. These receptor/coactivator complexes attach to specific DNA response elements (estrogen response elements and androgen response elements), causing initiation of transcription in *cis*. The complexes can also bind and inactivate other transcription factors, preventing transcription in *trans*. Alternatively, they can bind and activate serine/threonine kinases, including ERK, JNK and p38 subfamilies in a so-called nongenotropic manner. Recent insights suggest that it is by this pathway that sex steroids exert some of their protective effect in bone. For instance, sex steroids attenuate the apoptosis of osteoblasts and osteocytes and by directly activating the Src/Shc/ERK pathway.

Clinical Applications

Bone remodelling is a dynamic process in which the skeleton is continuously regenerated. The same regenerative machinery is applied in the healing process at the site of the bone fracture upon injury and when bone grafts are incorporated after a surgical intervention. In this process a continuous supply of osteoblasts and osteoclasts, an appropriate blood flow to the site of injury and the presence of signalling molecules are crucial. When the body's own repair system fails these components or a substitution have to be added to aid the bone regeneration. Autogenous or allograft bone, calcium phosphate and collagen are materials that are used as a scaffold to provide a support where the bone-forming cells can attach. Growth factors are used to attract bone-precursor cells and stimulate their proliferation. Today there are a number of different treatments to aid bone healing. The applications in this chapter (Table 9.2), the growth factor rhPDGF-BB in combination with beta-tricalcium phosphate, rhBMP2, rhBMP7 and teriparatide, a 34 amino acid polypeptide from the parathyroid hormone, are all included in pathways that involve c-Jun.

Table 9.2 Clinical applications. Approved treatments based on molecules that participate in pathways including c-Jun.

Condition	Treatment	Status
Periodontal bone defects	rhPDGF-BB in combination with beta-tricalcium	FDA approved
Hindfoot and ankle fractures	rhPDGF-BB in combination with beta-tricalcium	Approved in Canada
Osteoporosis	Teriparatide	FDA approved
Osteoporosis	rhPTH	Approved in Europe
Fusion of the lumbar spine in degenerative disc disease	rhBMP-2	FDA approved
Certain oral maxillofacial and dental regenerative uses	rhBMP-2	FDA approved
Recalcitrant long bone nonunions where alternative treatments have failed	rhBMP-7	FDA approved
Lumbar spinal fusion where alternative treatments have failed	rhBMP-7	FDA approved

PDGF

An orchestra of growth factors and cytokines directs the bone remodelling by recruiting, initiating the differentiation and stimulating the proliferation of cell populations that are essential for the process. The growth factor PDGF is important for bone remodelling working as a chemotactic and mitogenic agent for a variety of mesenchymal-derived cells including the osteogenic progenitor cells (Hollinger et al. 2008). PDGF also acts indirectly by inducing the expression of proteins important for the healing process such as vascular endothelial growth factor (VEGF) (Bouletreau et al. 2002) involved in angiogenesis or c-Jun, which stimulates MSC expansion and promotes the mineralization of extracellular matrix during osteogenesis (Cárcamo Orive et al. 2010). The implications of PDGF in bone remodelling give this growth factor potential as a drug candidate in bone healing applications. rhPDGF-BB has been evaluated in clinical trials and has been approved by the Food and Drug Administration (FDA, USA) for use in repair of periodontal defects and is under evaluation for use in a bone graft product used in ankle fusion treatment.

rhPDGF-BB treatment in animal models

rhPDGF-BB treatment of bone fractures and periodontal defects has given improved regeneration in various models using several different delivery systems (Hollinger et al. 2008). Local application of the growth factor has shown to stimulate proliferation of periodontal cells, and rhPDGF-BB in combination with insulin-like growth factor 1 (IGF-1) resulted in an increase in bone and cementum, and in an improved osseointegration of dental implants in dog models (Lynch et al. 1991). The treatment has also been tried in nonhuman primates with promising results (Giannobile et al. 1996).

Collagen gels are used in the delivery of drugs to achieve a sustained release of therapeutic molecules. Rabbits with tibial osteotomies were treated with a collagen gel containing rhPDGF-BB and showed increased osteogenesis (Nash et al. 1994). In fact, the strength of the tibiae of rabbits treated with rhPDGF-BB was similar to untreated, non-fractured tibiae. In a rat model of osteoporosis a combination of rhPDGF-BB with a beta-tricalcium phosphate-collagen matrix was applied to fractures in tibiae. The torsional strength of the fractured bone was clearly weaker in control animals compared to animals treated with the combination of rhPDGF-BB with a beta-tricalcium phosphate-collagen matrix.

The half time of PDGF is relatively short, approximately 30 min. In an attempt to overcome the problem of the degradation of PDGF, gene delivery using adenoviral vectors was tried in rat models (Chang et al. 2009). This study showed that Ad-PDGF-B delivery is safe and improves new bone formation in peri-implant defects.

Clinical trials of rhPDGF-BB treatment

Eighty-seven percent of adults over 70 years of age suffer from loss of periodontal structures caused by chronic inflammation (Albandar and Kingman 1999). The periodontal tissues are the supporting structures of the teeth, a combination of soft and hard tissues that keep the tooth in place, feed it and protect it. The root of the tooth is covered by a layer of hard tissue, the cementum, and attached to the alveolar bone, the bone of the jaw, by the periodontal ligament. Gingiva is the soft tissue lining of the teeth in the mouth. rhPDGF-BB has been used in several clinical trials with the objective to restore periodontal tissue (Hollinger et al. 2008). The growth factor has been applied together with IGF-1 or in combination with a scaffolding material (bone allograft) and in the latter case they were able to regenerate complete attachment apparatus including bone, periodontal ligament and cementum (Howell et al. 1997).

Autogenous bone graft is regarded as the optimal solution when a graft material is needed to support bone healing in the treatment of fractures and non-unions (De Long Jr et al. 2007), but is often not readily available due to time, cost and increased risks associated with an additional surgical site (St John et al. 2003). A combination of rhPDGF-BB and beta-tricalcium phosphate is under evaluation for use in foot and ankle fusions. Clinical studies indicate that this combination gives results comparable to those achieved with autogenous bone graft (Hollinger et al. 2008).

PTH

PTH treatment has been used as an anabolic therapy: In contrast to anti remodelling agents, which primarily inhibit bone resorption, but subsequently also new bone formation, anabolic therapy increases bone remodelling rate with a positive net growth. As chronically elevated PTH level stimulates the expression of pro-osteoclastogenic RANKL in the surface of osteoblasts, administration must be intermittent to avoid the enhancement of bone resorption.

Teriparatide, a 1–34 amino acid fragment of the PTH, is the only anabolic agent approved for therapy by the Food and Drug Administration, while treatment with the full length PTH also is available in Europe (Canalis 2010). Teriparatide is under an ongoing clinical trial to evaluate its potential effect in treating symptoms of osteogenesis imperfecta.

BMP

BMPs play an important role in the differentiation of human bone MSCs to osteoblasts and are strong bone inducers. Two rhBMPs, rhBMP-2 and rhBMP-7, have been approved by FDA for use in the treatment of a number of well-defined conditions (Bishop and Einhorn 2007). The applications are effective but the current treatment setup with recombinant protein is accompanied by elevated costs due to high doses and problems with protein degradation. Moreover, the connection between treatments with high doses of BMP and cancer is under investigation. Therefore, attempts are being made to optimize the dose and delivery system to reduce the risks and costs of the treatment (Boerckel et al. 2011).

rhBMP-2 is currently approved for use in complicated cases of spinal fusion and tibial fractures and certain oral, maxillofacial and dental regenerative uses, whereas rhBMP-7 has received approval for limited use as an alternative to autograft in recalcitrant long bone nonunions and spinal fusions. rhBMP-7 is also currently under evaluation for treatment of osteoarthritis in the knee.

Key Facts of Osteoblast Differentiation

- Mesenchymal stem cells (MSC) are the precursors of osteoblasts.
- MSCs have the potential to differentiate into osteoblasts, adipocytes and chondrocytes *in vitro.*
- Differentiation towards osteoblastic cells requires paracrine pro-osteogenic signalling. The switch between osteogenesis and chondrogenesis is controlled by the WNT/β-catenin canonical pathway. WNT as well as BMP-2 pathways are important for the progression of osteoblastic differentiation.
- Osteoblastic differentiation can be assessed *in vitro* by analyzing the expression of osteoblastic markers, such as alkaline phosphatase, and the secretion of calcified matrix.
- Pro-osteogenic signalling leads to the activation of specific transcription factors, such as AP-1 or RUNX2, and thus to the expression of certain osteogenic genes, such as osterix, alkaline phosphatase, osteocalcin or the transcription factors themselves.
- Expression of these genes ultimately promotes the proliferation and differentiation of MSCs towards osteoblasts.
- As osteoblasts differentiate, they secrete bone matrix and are gradually buried in it.

Key Facts of c-Jun

- c-Jun is one of the components of the transcription factor AP-1.
- The protein can function as a homodimer or heterodimer with Fos or ATF proteins.
- c-Jun is mainly working as an activator of the AP-1 target genes, but can function as a repressor under certain conditions.
- The transactivation activity of c-Jun is modulated by phosphorylation by JNK.
- Lack of c-Jun is lethal in mouse.
- c-Jun is important for many crucial processes in the cell such as proliferation, differentiation and apoptosis.
- c-Jun is involved in the regulation of the switch between adipogenesis and osteogenesis in MSCs.

Key Facts of Platelet-derived Growth Factor (PDGF)

- PDGF is an important mitogen in hBMSCs and is implied in a number of signalling mechanisms in the cell.
- The protein functions as a dimer built up by a family of isoforms of the polypeptide chain including PDGF-A, -B, -C, and -D.
- The A and B polypeptide can form homodimers or heterodimers, while the C and D polypeptides can only form functional homodimers.
- The PDGF dimer interacts with different affinity to one or several of the dimeric tyrosine kinase receptors PDGFRα, PDGFRβ and PDGRFαβ.
- The homodimer PDGF-BB can bind all known PDGF receptors and is considered to be the universal PDGF.

Dictionary

- *Adipocyte*: Cells specialized in storage of fat.
- *Autogenous*: Generated by the organism itself, self generated.
- *Allograft*: A transplant of a tissue between different individuals of the same species.
- *Basic multicellular unit*: Temporary anatomical structure responsible of bone remodelling. It is composed of a leading front of osteoclasts, which advance through bone matrix by dissolving it, followed by a cluster of proliferating osteoblasts, which are gradually buried in the new matrix that they secrete. It permits temporal and spatial coupling of bone resorption and formation.
- *Chemotaxis*: The movement of a cell in a certain direction according to the concentration gradient of a molecule, called chemotactic factor.

- *Consensus site*: A DNA sequence containing a minimal nucleic acid pattern that can be recognized by a certain transcription factor. The representation of the consensus sequence shows which positions are conserved and which ones are changeable.
- *Genotropic*: Effect of a signalling molecule that is mediated by direct binding to gene promoters, modulating their transcription. On the other hand, non-genotropic effects do not involve direct DNA binding.
- *Mitogenic*: Any agent that stimulates the cells to enter into cell division or mitosis.
- *Osteopetrosis*: Literally "stone bone", is a hardening of the bones due to defects in the bone remodelling. It can cause osteosclerosis.
- *Osteosclerosis*: Meaning "bone hardening", is a pathogenic increase of bone density.
- *Mesenchymal stem cells (MSCs)*: Multipotent cells that can differentiate into, at least, osteoblasts, chondrocytes and adipocytes.
- *Paracrine*: Hormone signalling that affects cells in the immediate surroundings in contrast to endocrine signalling that reaches distant cells. An agent can act as a paracrine factor for a certain cell type and as an endocrine factor for another, more distant cell.
- *Transcription factor*: A protein that regulates the transcription of one or several genes by binding to specific DNA sequences and thereby promoting or blocking the binding of RNA polymerase.
- *Transgenic*: An organism in which a modified gene or a gene from another species has been artificially added.
- *Knockout*: Inactivate a gene in an organism by genetic methods.

Summary Points

- AP-1 is a transcription factor involved in the regulation of the proliferation and differentiation of osteoblast and osteoclast precursor cells, and has an essential role in the development of the bone tissue.
- c-Jun, a component of AP-1, is implicated in several pathways involved in the osteogenesis, such as BMP, WNT, RANK and PDGF.
- Glucocorticoid, sex steroid and parathyroid hormone signalling regulate bone development partially by modulating cJun activity.
- Most prevalent bone diseases involve dysregulation of osteoblast and/or osteoclast differentiation pathways.
- A variety of novel treatments of bone diseases involving stimulation of c-Jun-centred networks have been approved for clinical use.

List of Abbreviations

AP-1	:	Activator protein -1
ATF	:	Activating transcription factor
BMP	:	Bone morphogenetic protein
cAMP	:	Cyclic adenosine monophosphate
CREB	:	cAMP response element binding
ERK	:	Extracellular signal-regulated kinase
FDA	:	Food and Drug Administration
GR	:	Glucocorticoid receptor
GTP	:	Guanosine triphosphate
IGF	:	Insulin-like growth factor
JNK	:	c-Jun N-terminal kinase
LIF	:	Leukemia inhibitory factor
MAPK	:	Mitogen-activated protein kinase
MSC	:	Mesenchymal stem cell
NF-κB	:	Nuclear factor κ-light-chain-enhancer of activated B cells
NFAT	:	Nuclear factor of activated T-cells
OSCAR	:	Osteoclast-associated immunoglobulin-like receptor
PDGF	:	Platelet-derived growth factor
PTH	:	Parathyroid hormone
RANK	:	Receptor activator of NF-κB
RANKL	:	Receptor activator of NF-κB ligand
rh	:	Recombinant human
TGF-β	:	Transforming growth factor-β
VEGF	:	Vascular endothelial growth factor

References

Albandar, J. and A. Kingman. 1999. Gingival recession, gingival bleeding, and dental calculus in adults 30 years of age and older in the United States, 1988–1994. Journal of Periodontology. 70: 30–43.

Behrens, A., M. Sibilia and E.F. Wagner. 1999. Amino-terminal phosphorylation of c-Jun regulates stress-induced apoptosis and cellular proliferation. Nature Genetics. 21: 326–329.

Bishop, G.B. and T.A. Einhorn. 2007. Current and future clinical applications of bone morphogenetic proteins in orthopaedic trauma surgery. International Orthopaedics. 31: 721–727.

Boerckel, J.D., Y.M. Kolambkar, K.M. Dupont, B.A. Uhrig, E.A. Phelps, H.Y. Stevens, A.J. García and R.E. Guldberg. 2011. Effects of protein dose and delivery system on BMP-mediated bone regeneration. Biomaterials. 32(22): 5241–5251.

Bouletreau, P. J., S.M. Warren, J.A. Spector, D.S. Steinbrech, B.J. Mehrara and M.T. Longaker. 2002. Factors in the fracture microenvironment induce primary osteoblast angiogenic cytokine production. Plastic and Reconstructive Surgery. 110: 139.

Bozec, A., L. Bakiri, A. Hoebertz, R. Eferl, A.F. Schilling, V. Komnenovic, H. Scheuch, M. Priemel, C.L. Stewart and M. Amling. 2008. Osteoclast size is controlled by Fra-2 through LIF/ LIF-receptor signalling and hypoxia. Nature. 454: 221–225.

Bozec, A., L. Bakiri, M. Jimenez, T. Schinke, M. Amling and E.F. Wagner. 2010. Fra-2/AP-1 controls bone formation by regulating osteoblast differentiation and collagen production. The Journal of Cell Biology. 190: 1093.

Canalis, E. 2010. Update in new anabolic therapies for osteoporosis. Journal of Clinical Endocrinology & Metabolism. 95: 1496.

Cárcamo Orive, I., A. Gaztelumendi, J. Delgado, N. Tejados, A. Dorronsoro, J. Fernández-Rueda D.J. Pennington and C. Trigueros. 2010. Regulation of human bone marrow stromal cell proliferation and differentiation capacity by glucocorticoid receptor and AP-1 crosstalk. Journal of Bone and Mineral Research. 25: 2115–2125.

Chang, P.C., Y.J. Seol, J.A. Cirelli, G. Pellegrini, Q. Jin, L.M. Franco, S.A. Goldstein, L.A. Chandler, B. Sosnowski and W.V. Giannobile. 2009. PDGF-B gene therapy accelerates bone engineering and oral implant osseointegration. Gene therapy. 17: 95–104.

De Long Jr, W.G., T.A. Einhorn, K. Koval, M. Mckee, W. Smith, R. Sanders and T. Watson. 2007. Bone grafts and bone graft substitutes in orthopaedic trauma surgery. J Bone Joint Surg Am. 89: 649–658.

Dempster, D.W., F. Cosman, M. Parisien, V. Shen and R. Lindsay. 1993. Anabolic actions of parathyroid hormone on bone. Endocrine Reviews. 14: 690.

Fujii, M., K. Takeda, T. Imamura, H. Aoki, T.K. Sampath, S. Enomoto, M. Kawabata, M. Kato, H. Ichijo and K. Miyazono. 1999. Roles of bone morphogenetic protein type I receptors and Smad proteins in osteoblast and chondroblast differentiation. Molecular Biology of the Cell. 10: 3801.

Giannobile, W.V., R.A. Hernandez, R.D. Finkelman, S. Ryarr, C.P. Kiritsy, M.D'andrea and S.E. Lynch. 1996. Comparative effects of platelet derived growth factor BB and insulin like growth factor I, individually and in combination, on periodontal regeneration in Macaca fascicularis. Journal of Periodontal Research. 31: 301–312.

Guicheux, J., J. Lemonnier, C. Ghayor, A. Suzuki, G. Palmer and J. Caverzasio. 2003. Activation of p38 mitogen-activated protein kinase and c-Jun-NH2-terminal kinase by BMP-2 and their implication in the stimulation of osteoblastic cell differentiation. J Bone Miner Res. 18: 2060–8.

Hai, T. and T. Curran. 1991. Cross-family dimerization of transcription factors Fos/Jun and ATF/CREB alters DNA binding specificity. Proceedings of the National Academy of Sciences. 88: 3720.

Hollinger, J.O., C.E. Hart, S.N. Hirsch, S. Lynch and G.E. Friedlaender. 2008. Recombinant human platelet-derived growth factor: biology and clinical applications. Journal of Bone and Joint Surgery. American volume. 90: 48–54.

Howell, T.H., J.P. Fiorellini, D.W. Paquette, S. Offenbacher, W.V. Giannobile and S.E. Lynch. 1997. A phase I/II clinical trial to evaluate a combination of recombinant human platelet-derived growth factor-BB and recombinant human insulin-like growth factor-I in patients with periodontal disease. Journal of Periodontology. 68: 1186.

Ikeda, F., R. Nishimura, T. Matsubara, S. Tanaka, J. Inoue, S.V. Reddy, K. Hata, K. Yamashita, T. Hiraga and T. Watanabe. 2004. Critical roles of c-Jun signaling in regulation of NFAT family and RANKL-regulated osteoclast differentiation. Journal of Clinical Investigation. 114: 475–484.

Jilka, R.L. 2007. Molecular and cellular mechanisms of the anabolic effect of intermittent PTH. Bone. 40: 1434–1446.

Jochum, W., J.P. David, C. Elliott, A. Wutz, H. Plenk, K. Matsuo and E.F. Wagner. 2000. Increased bone formation and osteosclerosis in mice overexpressing the transcription factor Fra-1. Nature Medicine. 6: 980–984.

Jochum, W., E. Passegue and E.F. Wagner. 2001. AP-1 in mouse development and tumorigenesis. Oncogene. 20: 2401–12.

John, S., P.J. Sabo, R.E. Thurman, M.H. Sung, S.C. Biddie, T.A. Johnson, G.L. Hager and J.A. Stamatoyannopoulos. 2011. Chromatin accessibility pre-determines glucocorticoid receptor binding patterns. Nature Genetics. 43: 264–268.

Kassel, O., S. Schneider, C. Heilbock, M. Litfin, M. Göttlicher and P. Herrlich. 2004. A nuclear isoform of the focal adhesion LIM-domain protein Trip6 integrates activating and repressing signals at AP-1-and NF- B-regulated promoters. Genes & Development. 18: 2518.

Kuan, C.Y., D.D. Yang, D.R.S. Roy, R.J. Davis, P. Rakic and R.A. Flavell. 1999. The Jnk1 and Jnk2 protein kinases are required for regional specific apoptosis during early brain development. Neuron. 22: 667–676.

Liberati, N.T., M.B. Datto, J.P. Frederick, X. Shen, C. Wong, E.M. Rougier-Chapman and X.F. Wang. 1999. Smads bind directly to the Jun family of AP-1 transcription factors. Proceedings of the National Academy of Sciences. 96: 4844.

Lynch, S.E., G.R. De Castilla, R.C. Williams, C.P. Kiritsy, T.H. Howell, M.S. Reddy and H.N.Antoniades. 1991. The effects of short-term application of a combination of platelet-derived and insulin-like growth factors on periodontal wound healing. Journal of Periodontology. 62: 458–467.

Nash, T., C. Howlett, C. Martin, J. Steele, K. Johnson and D. Hicklin. 1994. Effect of platelet-derived growth factor on tibial osteotomies in rabbits. Bone. 15: 203–208.

Neer, R.M., C.D. Arnaud, J.R. Zanchetta, R. Prince, G.A. Gaich, J.Y. Reginster, A.B. Hodsman, E.F. Eriksen, S. Ish-Shalom and H.K. Genant. 2001. Effect of parathyroid hormone (1–34) on fractures and bone mineral density in postmenopausal women with osteoporosis. New England Journal of Medicine. 344: 1434–1441.

Rodan, G.A. and T.J. Martin. 2000. Therapeutic approaches to bone diseases. Science. 289: 1508.

Selvamurugan, N., W.Y. Chou, A.T. Pearman, M.R. Pulumati and N.C. Partridge. 1998. Parathyroid hormone regulates the rat collagenase-3 promoter in osteoblastic cells through the cooperative interaction of the activator protein-1 site and the runt domain binding sequence. Journal of Biological Chemistry. 273: 10647.

St John, T., A. Vaccaro, A. Sah, M. Sschaefer, S. Berta, T. Albert and A. Hilibrand. 2003. Physical and monetary costs associated with autogenous bone graft harvesting. American Journal of Orthopedics (Belle Mead, NJ). 32: 18.

Takanayagi, H. 2007. The role of NFAT in osteoclast formation. Annals of the New York Academy of Sciences. 1116: 227–237.

Tanaka, S., A. Sakai, M. Tanaka, H. Otomo, N. Okimoto, T. Sakata and T. Nakamura. 2004. Skeletal Unloading Alleviates the Anabolic Action of Intermittent PTH (1–34) in Mouse Tibia in Association With Inhibition of PTH Induced Increase in c fos mRNA in Bone Marrow Cells. Journal of Bone and Mineral Research. 19: 1813–1820.

Teitelbaum, S.L. 2004. RANKing c-Jun in osteoclast development. Journal of Clinical Investigation. 114: 463–464.

Tobimatsu, T., H. Kaji, H. Sowa, J. Naito, L. Canaff, G.N. Hendy, T. Sugimoto and K. Chihara. 2006. Parathyroid hormone increases -catenin levels through Smad3 in mouse osteoblastic cells. Endocrinology. 147: 2583.

Wagner, E.R., G. Zhu, B.Q. Zhang, Q. Luo, Q. Shi, E. Huang, Y. Gao, J.L. Gao, S.H. Kim and F. Rastegar. 2011. The therapeutic potential of the Wnt signaling pathway in bone disorders. Curr Mol Pharmacol. 4: 14–25.

Wang, Z.Q., J. Liang, K. Schellander, E.F. Wagner and A.E. Grigoriadis. 1995. c-fos-induced osteosarcoma formation in transgenic mice: cooperativity with c-jun and the role of endogenous c-fos. Cancer Res. 55: 6244–51.

Whimarsh, A. and R. Davis. 1996. Transcription factor AP-1 regulation by mitogen-activated protein kinase signal transduction pathways. Journal of Molecular Medicine. 74: 589–607.

Yang-Yen, H.F., J.C. Chambard, Y.L. Sun, T. Smeal, T.J. Schmidt, J. Drouin and M. Karin. 1990. Transcriptional interference between c-Jun and the glucocorticoid receptor: mutual inhibition of DNA binding due to direct protein-protein interaction. Cell. 62: 1205–1215.

Zenz, R., H. Scheuch, P. Martin, C. Frank, R. Eferl, L. Kenner, M. Sibilia and E.F. Wagner. 2003. c-Jun regulates eyelid closure and skin tumor development through EGFR signaling. Developmental Cell. 4: 879–889.

CHAPTER 10

Endothelial Progenitor Cells, Lnk and Bone Fracture Healing

Tomoyuki Matsumoto[1,2,a,*] and Takayuki Asahara[2,b,3]

ABSTRACT

Failures in fracture healing after conservative treatment or conventional open reduction and internal fixation combined with autologous/ allogenic bone grafting including vascularized bone grafting are recognized to be mainly due to poor tissue regeneration and vascularization. One emerging strategy in the regeneration and repair of bone and surrounding tissue is the use of stem cells, including bone marrow mesenchymal stem cells, which are the most investigated and reliable source for tissue engineering, as well as circulating skeletal stem/progenitor cells, which are recently receiving a lot of attention due to their ease of isolation and high osteogenic potential. In this chapter, we highlight the first proof-of-principle experiments that elucidate the collaborative multi-lineage differentiation of circulating CD34 positive cells—a cell-enriched population of endothelial/ hematopoietic progenitor cells—into not only endothelial cells but also osteoblasts. These cells include enriched population of endothelial

[1] Department of Orthopedic Surgery, Kobe University Graduate School of Medicine.
[a] E-mail: matsun@m4.dion.ne.jp
[2] Group of Vascular Regeneration Research, Kobe Institute of Biomedical Research and Innovation.
[b] E-mail: asa777@is.icc.u-tokai.ac.jp
[3] Department of Regenerative Medicine Science, Tokai University School of Medicine.
*Corresponding author

List of abbreviations given at the end of the text.

cells and develop a favorable environment for fracture healing via vasculogenesis/angiogenesis and osteogenesis, ultimately leading to functional recovery from fracture. Based on the *in vitro* experiments and pre-clinical studies using *in vivo* animal experiments, clinical trial of autologous transplantation of granulocyte colony-stimulating factor mobilized peripheral blood CD34+ cells for patients of tibial or femoral nonunion has been started with promising results. The use of cytokines such as granulocyte colony-stimulating factor, stromal cell-derived factor, and vascular endothelial growth factor has been reported to up-regulate or mobilize endothelial progenitor cells. Furthermore, to modify the current strategy and establish more efficient therapy for ischemic diseases including bone fracture, a new target such as the adaptor protein Lnk has received much attention. We proved Lnk inhibition up-regulated mobilization and incorporation of endothelial cells and provided a favorable environment for bone fracture healing. This chapter will highlight current concepts of circulating CD34 positive cell-based therapy, their potential application for bone repair, and novel molecule up-regulating endothelial progenitor cells, adopter protein Lnk.

Introduction

Embryonic stem cells in the blastocyst stage can generate into any differentiated cell type, while most adult stem cells have a limited potential for postnatal tissue/organ regeneration. Among the phenotypically characterized adult stem/progenitor cells, the hematopoietic system has traditionally been considered as an organized, hierarchical system that is spearheaded by multipotent, self-renewing stem cells at the top, followed by lineage-committed progenitor cells in the middle, and ends with lineage-restricted precursor cells—which give rise to terminally differentiated cells—at the bottom. Recently, however, a new population of stem cells has been added to this category, notably, the adult human peripheral blood (PB) CD34+ cells. These cells reportedly contain intensive endothelial progenitor cells (EPCs) as well as hematopoietic stem cells (HSCs) (Asahara et al. 1997), and were only recently discovered following the discovery of bone marrow (BM)-derived and circulating EPCs in adults—promotes embryonic vasculogenesis (Asahara et al. 1999; Asahara et al. 1997).

A fundamental aspect of tissue engineering research has been to identify various stimuli through which to direct stem cell activity towards tissue regeneration. To this end, recent research has demonstrated that EPCs respond to tissue ischemia as well as cytokines by mobilizing from BM into PB, ultimately migrating to regions of neovascularization to differentiate into mature endothelial cells and further promote vasculogenesis (Takahashi et al. 1999). As a result of this finding, many researchers have applied EPCs to promote therapeutic neovascularization in animal models of limb ischemia, myocardial infarction, and liver disorders. Promising results have been noticed particularly in the immunodeficient rat model of acute myocardial infarction following the transplantation of either human CD34+ cells or EPCs expanded *ex vivo* into the site of myocardial neovascularization. Following this implantation, these cells differentiate into mature endothelial cells, augmenting capillary density, inhibiting myocardial fibrosis and apoptosis, and ultimately preserving the left ventricular function (Iwasaki et al. 2006; Kawamoto et al. 2001; Kawamoto et al. 2003; Kocher et al. 2001). Based on these findings, clinical trials using PB CD34+ cells have been initiated and show promising results (Kawamoto et al. 2009; Losordo et al. 2007).

Circulating Endothelial/Osteo-progenitor Cells

Human PB CD34+ cells/EPCs are being investigated to specifically address the problem of delayed and atrophic non-unions in fracture healing, which has a significantly high (5–10 percent) annual incidence amongst all long bone fractures and results from an inadequate local blood supply around the injury zone. Because securing an adequate blood supply to this area is crucial for bone healing to occur, as would be evidenced radiographically by the formation of a bridging callus along a former fracture gap, an emerging focus in regenerative medicine is to develop EPCs to promote neo-angiogenesis. EPCs are appealing for this task in large part because the link between angiogenesis and the development of native bone on a larger scale has led to the discovery on a cellular level that there exists a developmental reciprocity between endothelial cells and osteoblasts. EPCs are also promising targets because the more traditional approach for enhancing local vascularity along a non-union or delayed union has been to perform vascular bone grafting, which requires painstaking microvascular surgical skills. Therefore, reduced invasion and more effective treatments for bone repair are still required. Our group has reported several successful outcomes when utilizing PB CD34+ cells/EPCs for fracture healing (Fukui et al. 2011; Matsumoto et al. 2006; Matsumoto et al. 2008; Mifune et al. 2008).

Systemic transplantation of PB CD34+ cells for bone healing

We first reported human PB CD34+ cells recruited to the fracture site following systemic delivery (Fig. 10.1A), developed a favorable environment for fracture healing by enhancing vasculogenesis/angiogenesis and osteogenesis (Fig. 10.1B), and finally led to functional recovery from the fracture (Fig. 10.1C) (Matsumoto et al. 2006). Briefly, we systemically

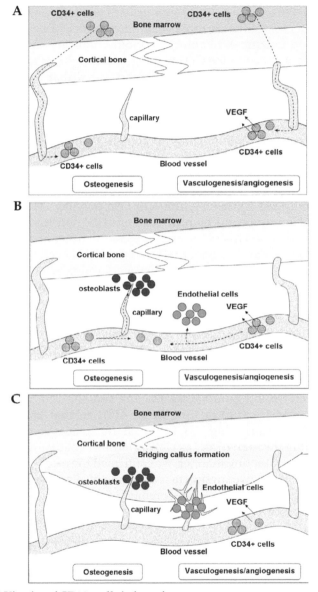

Figure 10.1 Kinetics of CD34+ cells in bone fracture.

A. Human peripheral blood CD34+ cells (endothelial progenitor cell-rich population) are mobilized from bone marrow as a result of fracture onset.

B. Mobilized peripheral blood CD34+ cells are recruited to the fracture site, developing a favorable environment for fracture healing by releasing vascular endothelial growth factor (VEGF) and differentiating osteoblasts and endothelial cells.

C. Enhanced vasculogenesis/angiogenesis and osteogenesis leads to functional recovery from fracture.

transplanted PB CD34+ cells, mononuclear cells (MNCs) or PBS into a non-healing femoral fracture model of immunodeficient rats. Bone healing assessed by radiological, histological, biomechanical examinations was significantly enhanced in the CD34+ cell group compared to the MNC and PBS group. Laser Doppler imaging demonstrated that fracture-induced ischemia was significantly recovered by CD34+ cell transplantation compared to the other groups. This healing potential by CD34+ cell transplantation was confirmed as mainly the result of two mechanisms. One is the osteogenic and endothelial differentiation potential of CD34+ cells, and the other is the paracrine effect of CD34+ cells by secreting vascular endothelial growth factor (VEGF). The former was first confirmed by the recruitment of human cells at the fracture site by labeling transplanted cells, and then confirmed by reverse-transcriptase polymerase chain reaction (RT-PCR) and immunohistochemical staining at the peri-fracture site, which showed molecular and histological expressions of human-specific markers for endothelial cells and osteoblasts 2 wk after CD34+ cell transplantation. Of note, our group sought to confirm the overlapping origin of endothelial and osteogenic markers by running single cell RT-PCR, and showed that ~20 percent of human peripheral blood CD34+ cells expressed the mRNA for osteocalcin. Previous reports showed that CD34+ and CD133+ cells were capable of differentiating into osteoblasts as well as hematopoietic and endothelial cells *in vitro* (Chen et al. 1997; Tondreau et al. 2005), supporting our findings. Interestingly, the loss of function test using sFlt1 (VEGF antagonist), following detection of human-specific VEGF expression in the CD34+ cell treated group at the fracture site, showed reduced angio/ osteogenesis and fracture healing in the sFlt1 treated group. This first series of study indicated that PB CD34+ cell transplantation contributed to fracture healing through direct as well as indirect cellular communication in an autocrine/paracrine manner; we confirmed the osteogenic properties of these cells via *in vitro* experiments consisting of alizarin red staining and RT-PCR (Mifune et al. 2008). These combined findings clearly highlight the therapeutic potential of PB CD34+ cells for functional bone healing.

Local transplantation of G-CSF mobilized CD34+ cells for bone healing

The therapeutic application of circulating CD34+ cells depends in large part on the availability of these cells in sufficient quantities. Our group has shown that while these cells can efficaciously heal skeletal defects after systemic transplantation, they also migrate to other tissues, including the lung, liver, thymus and brain, raising concerns of unforeseen side effects along these organs; this is particularly true given the large systemic doses that may be required for a clinical effect. In seeking alternative approaches to systemically infusing PB CD34+ cells for promoting skeletal healing, we seeded the femoral non-union site of immunodeficient rats with local CD34+ cell administration that were mobilized with granulocyte colony stimulating factor (G-CSF), and successfully promoted fracture union per radiographic and histologic assessment (Mifune et al. 2008). As expected, bone healing was significantly enhanced by high (10^5 cells) and middle (10^4 cells) dose CD34+ cell transplantation compared to animals receiving low (10^3 cells) dose CD34+ cells and PBS. In addition, we confirmed a dose-dependent response on molecular and histological levels via RT-PCR and immunohistochemical staining from tissue around the fracture site, which was most notable via detection of markers for endothelial cells and osteoblasts by 2-wk post-transplantation of CD34+ cells. The mechanism by which skeletal healing occurred in these experiments is likely related to dose-dependent vasculo/angiogenic effects of G-CSF-mobilized PB CD34+ cells. In one report, fracture-induced hypoxia and VEGF both upregulate the expression of bone morphogenetic protein (BMP)-2 in microvascular endothelial cells (Bouletreau et al. 2002). This is consistent with experimental results from our group in which we show an upregulation of rat VEGF and BMP-2 expression in the microvascular endothelial cells along a rat fracture site. We further showed that PB CD34+ cells have a paracrine effect through which the capillary and osteoblast density increases in a dose-dependent fashion. In this series of study, we confirmed that effective fracture healing occurred as long as there were more than 1×10^4 CD34+ cells per a rat implanted along the fracture site. We feel these findings provide feasible alternatives to current clinical strategies for addressing delayed unions and established non-unions.

Comparison between MNC and CD34+ cell transplantation for bone healing

In the field of revascularization, not only PB CD34+ cells but also PB total MNC transplantation for hind-limb ischemia or myocardial ischemia have shown their therapeutic efficiency in enhancing neovascularization at ischemic sites. In the clinical setting, the use of PB MNCs, which can be easily collected in a short time and at a low cost, is more attractive than PB CD34+ cells. However, some reports have shown a higher therapeutic potency for ischemic neovascularization in CD34+ cells than total MNC. Based on this controversy, we performed experiments to prove the hypothesis that PB G-CSF mobilized MNC transplantation may also contribute to fracture healing via vasculogenesis/angiogenesis and osteogenesis. In the series of study, FACS analysis showing about 1 percent ratio of CD34+ cells in PB MNCs led us to compare potential fracture healing with 1×10^7 MNCs and 1×10^5 CD34+ cells. Local transplantation of PB MNCs also contributes to fracture healing via vasculogenesis/angiogenesis and osteogenesis, confirmed by immunohistochemistry and RT-PCR analysis using human-specific marker. In addition, using rat specific marker, paracrine effect was also confirmed. This enhancement of angiogenesis/vasculogenesis and osteogenesis provided much more radiological and histological fracture healing compared to the PBS group. However, when compared with purified CD34+ cells, the therapeutic potential of PB MNCs for fracture healing could be inferior to that of purified CD34+ cells even if PB MNCs contained the same number of CD34+ cells (Fukui et al. 2011). The mechanism that causes MNC transplantation to be inferior to CD34+ cell in therapeutic effect for fracture healing is still being investigated. Total MNCs are mixed cell populations, and lymphocytes and monocytes/macrophages are the most prevalent mononuclear cells. These inflammatory cells could have a negative influence on osteogenesis. The recent report of our group about intramyocardial transplantation of G-CSF mobilized PB total MNCs showed the possible risk of severe hemorrhagic myocardial infarction in nude rats through the excessive inflammation induced by transplanted cells. We speculate that inflammation after transplantation of total MNC could interfere with the long term survival and differentiation of transplanted cells.

Physiological role of EPCs in fracture healing

While the physiological role of EPCs in fracture healing remains to be clarified, we have begun to characterize the role of mobilized BM-derived EPCs on fracture healing (Matsumoto et al. 2008). We have found that during the early phase of endochondral ossification, neovascularization peaks by 7 d after a fracture; this was noted via serial Laser Doppler perfusion imaging and histologic quantification of capillary density. Along this site, there is a significant increase in cell populations derived from the infiltrating EPC cells, most notably of BM cKit+Sca1+Lineage- (Lin-) and PB Sca1+ Lin-cells. The Sca1+ EPC cells contribute to vasculogenesis, as confirmed by double immunohistochemistry for CD31 and Sca1. We further showed that EPCs enhance neovascularization by transplanting BM that transcriptionally express Laz Z from transgenic donors into a fracture into wild type modes; these cells were regulated by the endothelial cell-specific Tie-2 promoter. We then confirmed that EPCs mobilize into a fracture site prior to healing by mapping this migration following the systemic administration of PB Sca+Lin-Green Fluorescent Protein (GFP)+ cells into an animal. These findings indicate that fracture may induce the mobilization of EPCs from the BM to the fracture sites, by way of transport through the PB as a way to augment neovascularization and ultimately bone healing.

Following our series of studies, some researchers have recently focused on the relationship between EPCs and bone fracture healing (Table 10.1). In mouse or rat fracture or destruction osteogenesis model, other groups reported the mobilization of EPCs for bone healing (Laing et al. 2007a; Lee et al. 2008). Moreover, recent reports have noted that there is a larger quantity of CD34+/AC133+ cells in the PB of patients with fractures (Laing et al. 2007b). Prior to our published reports, CD34+ and CD133+ cells were reported to be capable of differentiating into osteoblasts (Chen et al. 1997; Tondreau et al. 2005) and CD34+ osteoblastic cells were noted by Ford et al. to line cartilage cavities around the site of a tibial osteotomy in a rabbit model (Ford et al 2004), while intra-articularly injected human PB EPCs are reported to migrate to ischemic zones of nude rats undergoing distraction osteogenesis and participate in angiogenesis (Cetrulo et al. 2005). The latter phenomenon has been confirmed via Laser Doppler imaging to assess blood flow along the distraction zone as well as by labeling EPCs with a fluorescent dye, DiI, to track cell migration. In addition, other researchers confirmed that PB EPCs contribute to bone healing in rat segmental bone defect model (Atesok et al. 2010; Li et al. 2011). Further, Rozen et al reported that the *in vitro* expansion of autologous EPCs enhanced the healing of critical-sized bone defects in sheep (Rozen et al. 2009). These findings all indicate that EPCs heal large bone defects, and can potentially heal nonunions and delayed unions in both large and small animals, and thereby pave the way for the clinical use of these cells to enhance fracture repairs.

Clinical Application of Circulating CD34 Positive Cells for Bone Healing

From a clinical standpoint, PB cells receive interest because they can be isolated in a relatively minimally invasive, safe, and efficacious fashion. This gives PB cells an advantage over BM mesenchymal stem cells (MSCs), which have been used for bone healing with promising clinical results (Horwitz et al. 1999; Petite et al. 2000; Quarto et al. 2001), yet can only be isolated via BM aspiration under anesthesia and thus is considered a form of surgical intervention. Recently, autologous injection of cultured osteoblasts, which were isolated from BM and cultured for 4 wk in osteogenic conditioned medium, for long-bone fracture was reported by Kim et al. that the therapy showed fracture healing acceleration of statistical significance compared with the non-treated group (Kim et al. 2009). In contrast, PB cell aspiration does not require anesthesia. While we recognize that PB cells have advantages in that harvesting cells is minimally invasive, safe, and an easy procedure, we need further investigation to compare their efficacy in bone healing.

Table 10.1 Biological research of EPCs for bone healing. This table lists the biological research showing mobilization of EPCs in bone fracture healing or osteogenesis.

Author	Animal	model	Target materials	Biological Research Purpose
Cetrulo et al. (2005)	nude rat	**Systemic Transplantation** to tibial distraction osteogenesis model	Human cultured EPCs	Mobilization and incorporation of EPCs
Laing et al. (2007a)	mouse	Femoral fracture model	EPCs (Sca1+c-Kit+MNCs)	Mobilization of EPCs
Laing et al. (2007b)	human	Tibial fracture	CD34+/CD133+ cells	Mobilization of CD34+/CD133+ cells
Matsumoto et al. (2008)	mouse	Femoral fracture model	EPCs (KSL cells, Sca1+Lin- MNCs)	Mobilization and incorporation of EPCs
Lee et al. (2008)	rat	Tibial distraction osteogenesis model	EPCs (Dil-Ac-LDL/FITC-lectin double-stained MNCs)	Mobilization of EPCs

Table 10.2 Therapeutical research of EPCs for bone healing. This table lists the therapeutic research showing transplantation of various types of EPCs to bone fracture or nonunion animal model.

Author	Animal	model	Target materials	Therapeutical Research Purpose
Matsumoto et al. (2006)	nude rat	**Systemic Transplantation** to non-healing femoral fracture model	Human CD34+ cells	Mobilization and incorporation of CD34+ cells Enhanced angiogenesis/vasculogenesis and osteogenesis Radiological, histological, and biomechanical bone healing
Mifune et al. (2008)	nude rat	**Local Transplantation** to non-healing femoral fracture model	Human CD34+ cells	Enhanced angiogenesis/vasculogenesis and osteogenesis Radiological, histological, and biomechanical bone healing
Rozen et al. (2009)	sheep	**Local Transplantation** to tibial critical size defect model	Cultured EPCs	Radiological and histological bone healing
Atesok et al. (2010)	rat	**Local Transplantation** to femoral segmental bone defect model	Cultured EPCs	Radiological and histological bone healing
Li et al. (2011)	rat	**Local Transplantation** to femoral segmental bone defect model	Cultured EPCs	Radiological and biomechanical bone healing
Fukui et al. (2011)	nude rat	**Local Transplantation** to non-healing femoral fracture model	Human CD34+ cells and MNCs	Enhanced angiogenesis/vasculogenesis and osteogenesis Radiological, histological, and biomechanical bone healing

Based on these scientific backgrounds, we started phase I/ IIa clinical trial: autologous local transplantation of G-CSF mobilized peripheral blood CD34-positive cell for patients with the tibial or femoral nonunion. Inclusion criteria are (1) patient with tibial or femoral fracture, (2) patient with non-reactive and non-infectious nonunion, (3) 20 to 70 yr-old patient, (4) patient with informed consent by writing. Patients are selected and registered after informed consent and pre-registration examinations to check eligibility. After leukoapheresis following 5 d G-CSF injection, patient receives magnet sorting of CD34+ cells. Treatment is conventional surgery for nonunion combined with autologous transplantation of G-CSF mobilized peripheral blood CD34+ cells (10^6 cells/kg) suspended with atherocollagen gel. The schema is as shown in Fig. 10.2. The first case was a 42-yr-old male. He presented himself at our hospital complaining of tibial delayed union with pain at the fracture site and disability of life. He had had a closed tibial fracture and been treated by open reduction and internal fixation with plate at another hospital 9 mon before the initial presentation at our hospital. Then, autologeous local transplantation of G-CSF mobilized peripheral blood CD34+ cells with bone grafting was performed. Twelve weeks after the treatment, clinical and radiological healing was obtained (Kuroda et al. 2010).

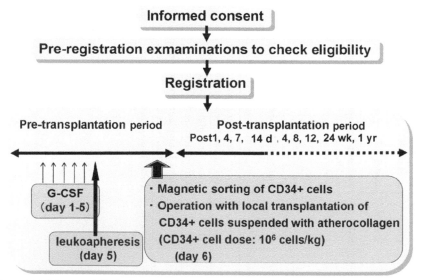

Figure 10.2 Schema of clinical trial. Patients are selected and registered after informed consent and pre-registration examinations to check eligibility. After leukoapheresis following 5 d G-CSF injection, patient receives magnet sorting of CD34+ cells. Conventional surgery for nonunion is performed at day 6, combined with autologous transplantation of G-CSF mobilized peripheral blood CD34+ cells (10^6 cells/kg) suspended with atherocollagen gel.

Lnk and Hematopoietic/Endothelial Progenitor Cells

G-CSF mobilized cell applications in the hematologic and cardiovascular fields have successfully involved the mobilization of EPCs to obtain sufficient therapeutic quantities of PB CD34+ cells (Losordo et al. 2007). Our group reported the effective application of G-CSF for bone healing in a canine bone defect model (Ishida et al. 2010), in which the local use of G-CSF combined with hydrogel exhibited enhancement of angiogenesis and osteogenesis at the fracture site compared with hydrogel only. G-CSF has been used in thousands of clinical cases, however severe complications such as spleen rupture, interstitial pneumonitis and acute coronary syndrome have been reported in a small number of cases. The rare but potential risks of using G-CSF as well as the high cost of CD34+ cell isolation need to be overcome before they can be considered a viable source. Many cytokines augment mobilization and/or recruitment of BM-derived EPCs, other than G-CSF, angiogenic growth factors such as VEGF and stromal cell–derived factor (SDF)-1; estrogen; and pharmaceutical drugs such as statins. However, these factors act not only on immature stem/progenitor cells but also on hematopoietic cells and mature endothelial cells. Thus, the identification of a novel molecule that specifically regulates immature populations involved in EPC kinetics in BM is warranted.

Lnk shares a pleckstin homology domain, a Src homology 2 domain, and potential tyrosine phosphorylation sites with APS and SH-2B. It belongs to a family of adaptor proteins implicated in integration and regulation of multiple signaling events (Ahmed and Pillay 2003; Takaki et al. 1997) and has also been suggested to act as a negative regulator in the stem cell factor (SCF)-c-Kit signaling pathway (Takaki et al. 2002; Takaki et al. 2000). Recently, Takaki et al. reported that Lnk is expressed in hematopoietic cell lineages, and BM cells of Lnk-deficient mice are competitively superior in hematopoietic population to those of wild-type mice (Takaki et al. 2002). They also clarified that not only HSC/hematopoietic progenitor cell (HPC) (c-Kit-positive, Sca-1-positive, lineage marker-negative (KSL) cell) numbers but also the self-renewal capacity of some HSCs/HPCs were markedly increased in Lnk-deficient mice (Ema et al. 2005). In addition, they identified the functional domains of Lnk and developed a dominant-negative Lnk mutant that inhibits the functions of Lnk endogenously expressed in the HSCs/HPCs and thereby potentiates the HPCs for engraftment (Takizawa et al. 2006).

The KSL cell population is also believed to be an attractive source for EPCs, with a portion of KSL cells able to differentiate into the endothelial lineage and to contribute to vasculogenesis. In our laboratory, the hypothesis was therefore first tested about a lack of Lnk signaling that may enhance postnatal neovascularization via specific control of the SCF–c-Kit–mediated

Table 10.3 Key Features of Lnk. This table lists the key facts of Lnk.

1. Lnk shares a pleckstin homology domain, a Src homology 2 domain, and potential tyrosine phosphorylation sites with APS and SH-2B.
2. Lnk belongs to a family of adaptor proteins implicated in integration and regulation of multiple signaling events.
3. Lnk act as a negative regulator in the stem cell factor-c-Kit signaling pathway
4. Lnk is expressed in hematopoietic cell lineages, and bone marrow cells of Lnk-deficient mice are competitively superior in hematopoietic population to those of wild-type mice.
5. Not only hematopoietic stem/progenitor cell (c-Kit-positive, Sca-1-positive, lineage marker-negative cell) numbers but also the self-renewal capacity of some these cells were markedly increased in Lnk-deficient mice.

regenerative potential of EPCs. We provide *in vitro* and *in vivo* evidence that Lnk plays a pivotal role in specific modulation of EPCs in terms of cell growth, commitment into endothelial lineage cell types, mobilization from BM into PB, and recruitment to ischemic sites for neovascularization (Kwon et al. 2009). In addition, in the field of neurovascular field, the role of Lnk was also investigated. We hypothesized that the transplantation of Lnk-deficient KSL cells might be more effective in promoting repair of injured spinal cord than wild type KSL cells. Our findings demonstrate that the transplantation of KSL cells promotes angiogenesis and astrogliogenesis in the acute phase of spinal cord injury and vascular stabilization, reduction of fibrous scar formation, axonal growth and functional recovery in the subacute phase and later. Lnk deletion enhances the effect of KSL cell transplantation even further, promoting regenerative changes in the injured spinal cord (Kamei et al. 2010).

Bone Fracture Healing in Lnk Deficient Mice

The concept of enhanced osteogenesis/angiogenesis by HSCs/EPCs, one of the novel factors responsible for stem/progenitor cell mobilization from BM, that is Lnk, attracted our research interests to develop a therapeutic strategy utilizing circulating EPCs for skeletal medicine. Thus, we have investigated the hypothesis that a lack of Lnk signaling, dependent on SCF-cKit signaling pathway, enhanced regenerative response via vasculogenesis and osteogenesis in fracture healing by HSC/EPC mobilization and recruitment to sites of fracture in Lnk-deficient mice. In our series of experiments, we clarified that negatively controlled Lnk system contributed via SCF-cKit signaling pathway to a favorable environment for fracture healing by enhancing vasculogenesis/angiogenesis and osteogenesis, which leads to prompt recovery from fracture (Matsumoto et al. 2010).

Takaki et al. reported that Lnk acts as a negative regulator in the SCF-cKit signaling pathway (Takaki et al. 2002; Takaki et al. 2000). Although the role of Lnk system has gradually been clarified in the field of hematology,

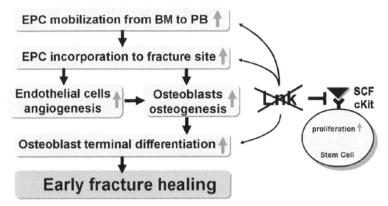

Figure 10.3 Effect of negatively regulated Lnk on fracture healing. Negatively controlled Lnk system contributed via SCF-cKit signaling pathway to a favorable environment for fracture healing by enhancing vasculogenesis/angiogenesis and osteogenesis, which leads to prompt recovery from fracture.

nothing is known on skeletal biology and bone regeneration and repair. Here, we have provided evidence that enhanced HSC/EPC mobilization into PB and its recruitment to the fracture site in Lnk-deficient mice are regulated, at least in part, by SCF-cKit signaling pathway, which was proved by gain and loss of function test using SCF and sKit. However, enhanced vasculogenesis and osteogenesis in Lnk-deficient mice was shown to be significantly superior to that of SCF-treated wild-type mice, suggesting that another mechanism other than SCF-cKit signaling pathway is involved in Lnk signal lacking accelerated bone healing. Indeed, molecular cDNA array analysis exhibited that BMPR2 mRNA expression was up-regulated in Lnk KO mice than in wild type mice. Moreover, lack of Lnk or SCF supplement in osteoblasts induced terminal differentiation and mineralized matrix formation with increased BMP2 gene expression, which was blocked by SCF antagonist. sKit. These findings are consistent with previous reports in which cKit expression is shown in osteoblasts and P-2 enhances osteoblast differentiation. (Bilbe et al. 1996; Hassel et al. 2006) Taken together, it is suggested that osteoblast differentiation and maturation is enhanced in Lnk-deficient mice via the mechanism of BMP2-involved SCF-cKit signaling pathway.

Conclusion and Perspective

Blood mesenchymal precursor cells (BMPCs) have been a central focus in regenerative medicine for bone regeneration ever since these cells were first discovered in the circulation of healthy patients. BMPCs were discovered by Zvaifler et al. who showed these cells adhere to plastic and glass and proliferate logarithmically in DMEM-20 percent fetal calf serum without growth factors, thereby suggesting that these cells relatively easily expand *in vitro* (Zvaifler et al. 2000). While the reproducibility of these experiments has been limited, there have been recent reports focusing on peripherally circulating skeletal-lineage cells from humans in which Eghbali-Fatourechi et al. demonstrate via flow cytometry that cells which positively stain with osteocalcin and alkaline phosphatase immunohistostaining are indeed present in this lineage of cells (Eghbali-Fatourechi et al. 2005; Khosla and Eghbali-Fatourechi 2006). Otsuru et al. have recently reported that osteoblast progenitor cells in circulation originated from BM and form ectopic bone after implanting these cells with BMP-2-containing collagen pellet into the skeletal muscle beds of mice (Otsuru et al. Kaneda 2007). In addition, they demonstrated that PB MNCs from the BMP-2-implanted mouse contained cells which could differentiate into osteoblasts *in vitro*. In a follow-up study, they proved that circulating BM-derived OB progenitor cells (MOPCs) were recruited to the bone-forming site by the CXCR4/SDF-1 pathway (Otsuru et al. Kaneda 2008).

Recently there has been increased interest in circulating stem/progenitor cells due to the ease of isolating them in a safe and efficacious manner, and their great potential for bone repair. Among them, we have investigated the potential of human PB CD34+ cells for bone healing and discussed their clinical feasibility. Our findings that PB CD34+ cells have a therapeutic potential via vasculogenesis/angiogenesis and osteogenesis for bone healing will provide an attractive strategy for cell-based therapy. Circulating EPCs, circulating osteoblast-lineage cells, and MOPCs are each isolated via different procedures, yet exhibit similar cell characteristics and behaviors for bone healing, indicating that there is some functional and perhaps phenotypic overlap in these populations. Taken together, recent advances in circulating stem/progenitor cell research provide new clinical applications for patients suffering from fracture non-unions and delayed unions. To further expand the clinical application of these cells and their therapeutic effects, we hope further in-depth investigations will focus on the inter-relationships of these three cell types and attempt to provide a greater understanding of their molecular mechanisms.

Harvest, isolation and transplantation of autologous, G-CSF mobilized CD34+ cells were first performed in a patient with nonunion/delayed union. Both clinical and radiological healing of the fracture was achieved 12 wk after the cell therapy and bone grafting, and no serious adverse events occurred during the 12 wk-follow up (Kuroda et al. 2010). These promising results of the first case in the phase I/ IIa clinical trial encourage us to use autologous circulating CD34+ cells for patients with nonunion of long bones. Efficacy of the cell therapy would be further elucidated in a randomized controlled clinical trial with an appropriate control group receiving G-CSF only or placebo in the future.

Summary Points

- Circulating CD34-positive cells—a cell-enriched population of endothelial/hematopoietic progenitor cells show the collaborative multi-lineage differentiation into not only endothelial cells but also osteoblasts.
- These cells include enriched population of endothelial cells and develop a favorable environment for fracture healing via vasculogenesis/ angiogenesis and osteogenesis, ultimately leading to functional recovery from fracture.
- Clinical trial of autologous transplantation of granulocyte colony-stimulating factor mobilized peripheral blood CD34+ cells for patients of tibial or femoral nonunion has been started with promising results.
- Inhibition of Lnk, a family of adaptor proteins and a negative regulator in the stem cell factor-c-Kit signaling pathway, exhibits enlarged number and high stem cell potential in hematopoietic cell lineages.
- Lnk inhibition up-regulated mobilization and incorporation of endothelial progenitor cells via stem cell factor-c-Kit signaling pathway and provided a favorable environment for bone fracture healing.

Dictionary

- *Endothelial progenitor cell (EPC)*: Bone marrow derived EPCs and circulating EPCs were discovered as a population of CD34-positive cells in adult human. These cells promote not only angiogenesis but also embryonic vasculogenesis.
- *Delayed union and non-union*: Five to ten percent of all long bone fractures result in delayed union and non-union mainly due to an inadequate local blood supply around the injury zone.
- *Vascular endothelial growth factor (VEGF)*: VEGF not only up-regulates angiogenesis in the various ischemic diseases but also mobilize hematopoietic stem/progenitor cells to peripheral blood.
- *Granulocyte colony stimulating factor (G-CSF)*: G-CSF mobilizing EPCs/ CD34+ cells to peripheral blood has often been applied for peripheral blood hematopoietic stem cell transplantation in the hematology fields.

- *Stem cell factor (SCF)-cKit pathway*: SCF stimulate proliferation and differentiation of hematopoietic stem cells and mobilize hematopoietic stem/progenitor cells into peripheral blood by binding with cKit.

Acknowledgements

The authors would like to thank Janina Tubby for her assistance with the preparation of this manuscript.

List of Abbreviations

BM	:	bone marrow
BMP	:	bone morphogenetic protein
BMPC	:	blood mesenchymal precursor cells
EPC	:	endothelial progenitor cells
G-CSF	:	granulocyte colony stimulating factor
GFP	:	green fluorescent protein
HPC	:	hematopoietic progenitor cell
HSC	:	hematopoietic stem cells
KSL	:	c-Kit-positive, Sca-1-positive, lineage marker-negative
MNC	:	mononuclear cell
MOPC	:	circulating bone marrow-derived osteoblast progenitor cell
MSC	:	mesenchymal stem cell
PB	:	peripheral blood
RT-PCR	:	reverse-transcriptase polymerase chain reaction
SCF	:	stem cell factor
SDF-1	:	stromal cell–derived factor -1
VEGF	:	vascular endothelial progenitor cell

References

Ahmed, Z. and T.S. Pillay. 2003. Adapter protein with a pleckstrin homology (PH) and an Src homology 2 (SH2) domain (APS) and SH2-B enhance insulin-receptor autophosphorylation, extracellular-signal-regulated kinase and phosphoinositide 3-kinase-dependent signalling. Biochem J. 371: 405–412.

Asahara, T., T. Murohara, A. Sullivan, M. Silver, R. van der Zee, T. Li, B. Witzenbichler, G. Schatteman and J.M. Isner. 1997. Isolation of putative progenitor endothelial cells for angiogenesis. Science. 275: 964–967.

Asahara, T., H. Masuda, T. Takahashi, C. Kalka, C. Pastore, M. Silver, M. Kearne, M. Magner and J.M. Isner. 1999. Bone marrow origin of endothelial progenitor cells responsible for postnatal vasculogenesis in physiological and pathological neovascularization. Circ Res. 85: 221–228.

Atesok, K., R. Li, D.J. Stewart and E.H. Schemitsch. 2010. Endothelial progenitor cells promote fracture healing in a segmental bone defect model. J Orthop Res. 28: 1007–1014.

Bilbe, G., E. Roberts, M. Birch and D.B. Evans. 1996. PCR phenotyping of cytokines, growth factors and their receptors and bone matrix proteins in human osteoblast-like cell lines. Bone. 19: 437–445.

Bouletreau, P.J., S.M. Warren, J.A. Spector, Z.M. Peled, R.P. Gerrets, J.A. Greenwald and M.T. Longaker. 2002. Hypoxia and VEGF up-regulate BMP-2 mRNA and protein expression in microvascular endothelial cells: implications for fracture healing. Plast Reconstr Surg. 109: 2384–2397.

Cetrulo, Jr., C.L., K.R. Knox, D.J. Brown, R.L. Ashinoff, M. Dobryansky, D.J. Ceradini, J.M. Capla, E.I. Chang, K.A. Bhatt, J.G. McCarthy and G.C. Gurtner. 2005. Stem cells and distraction osteogenesis: endothelial progenitor cells home to the ischemic generate in activation and consolidation. Plast Reconstr Surg. 116: 1053–1064; discussion 1065–1057.

Chen, J.L., P. Hunt, M. McElvain, T. Black, S. Kaufman and E.S. Choi. 1997. Osteoblast precursor cells are found in CD34+ cells from human bone marrow. Stem Cells. 15: 368–377.

Eghbali-Fatourechi, G.Z., J. Lamsam, D. Fraser, D. Nagel, B.L. Riggs and S. Khosla. 2005. Circulating osteoblast-lineage cells in humans. N Engl J Med. 352: 1959–1966.

Ema, H., K. Sudo, J. Seita, A. Matsubara, Y. Morita, M. Osawa, K. Takatsu, S. Takaki and H. Nakauchi. 2005. Quantification of self-renewal capacity in single hematopoietic stem cells from normal and Lnk-deficient mice. Dev Cell. 8: 907–914.

Ford, J.L., D.E. Robinson and B.E. Scammell. 2004. Endochondral ossification in fracture callus during long bone repair: the localisation of 'cavity-lining cells' within the cartilage. J Orthop Res. 22: 368–375.

Fukui, T., T. Matsumoto, Y. Mifune, T. Shoji, T. Kuroda, Y. Kawakami, A. Kawamoto, M. Ii, S. Kawamata, M. Kurosaka, T. Asahara and R. Kuroda. 2011. Local Transplantation of Granulocyte Colony Stimulating Factor-Mobilized Human Peripheral Blood Mononuclear Cells for Unhealing Bone Fractures. Cell Transplant.

Hassel, S., M. Yakymovych, U. Hellman, L. Ronnstrand, P. Knaus and S. Souchelnytskyi. 2006. Interaction and functional cooperation between the serine/threonine kinase bone morphogenetic protein type II receptor with the tyrosine kinase stem cell factor receptor. J Cell Physiol. 206: 457–467.

Horwitz, E.M., D.J. Prockop, L.A. Fitzpatrick, W.W. Koo, P.L. Gordon, M. Neel, M. Sussman, P. Orchard, J.C. Marx, R.E. Pyeritz and M.K. Brenner. 1999. Transplantability and therapeutic effects of bone marrow-derived mesenchymal cells in children with osteogenesis imperfecta. Nat Med. 5: 309–313.

Ishida, K., T. Matsumoto, K. Sasaki, Y. Mifune, K. Tei, S. Kubo, T. Matsushita, K. Takayama, T. Akisue, Y. Tabata, M. Kurosaka and R. Kuroda. 2010. Bone regeneration properties of granulocyte colony-stimulating factor via neovascularization and osteogenesis. Tissue Eng Part A. 16: 3271–3284.

Iwasaki, H., A. Kawamoto, M. Ishikawa, A. Oyamada, S. Nakamori, H. Nishimura, K. Sadamoto, M. Horii, T. Matsumoto, S. Murasawa, T. Shibata, S. Suehiro and T. Asahara. 2006. Dose-dependent contribution of CD34-positive cell transplantation to concurrent vasculogenesis and cardiomyogenesis for functional regenerative recovery after myocardial infarction. Circulation. 113: 1311–1325.

Kamei, N., S.M. Kwon, C. Alev, M. Ishikawa, A. Yokoyama, K. Nakanishi, K. Yamada, M. Horii, H. Nishimura, S. Takaki, A. Kawamoto, M. Ii, H. Akimaru, N. Tanaka, S. Nishikawa, M. Ochi and T. Asahara. 2010. Lnk deletion reinforces the function of bone marrow progenitors in promoting neovascularization and astrogliosis following spinal cord injury. Stem Cells. 28: 365–375.

Kawamoto, A., H.C. Gwon, H. Iwaguro, J.I. Yamaguchi, S. Uchida, H. Masuda, M. Silver, H. Ma, M. Kearney, J.M. Isner and T. Asahara. 2001. Therapeutic potential of ex vivo expanded endothelial progenitor cells for myocardial ischemia. Circulation. 103: 634–637.

Kawamoto, A., T. Tkebuchava, J. Yamaguchi, H. Nishimura, Y.S. Yoon, C. Milliken, S. Uchida, O. Masuo, H. Iwaguro, H. Ma, A. Hanley, M. Silver, M. Kearney, D.W. Losordo, J.M. Isner and T. Asahara. 2003. Intramyocardial transplantation of autologous endothelial progenitor cells for therapeutic neovascularization of myocardial ischemia. Circulation. 107: 461–468.

Kawamoto, A., M. Katayama, N. Handa, M. Kinoshita, H. Takano, M. Horii, K. Sadamoto, A. Yokoyama, T. Yamanaka, R. Onodera, A. Kuroda, R. Baba, Y. Kaneko, T. Tsukie, Y. Kurimoto, Y. Okada, Y. Kihara, S. Morioka, M. Fukushima and T. Asahara. 2009. Intramuscular transplantation of G-CSF-mobilized CD34(+) cells in patients with critical limb ischemia: a phase I/IIa, multicenter, single-blinded, dose-escalation clinical trial. Stem Cells. 27: 2857–2864.

Khosla, S. and G.Z. Eghbali-Fatourechi. 2006. Circulating cells with osteogenic potential. Ann N Y Acad Sci. 1068: 489–497.

Kim, S.J., Y.W. Shin, K.H. Yang, S.B. Kim, M.J. Yoo, S.K. Han, S.A. Im, Y.D. Won, Y.B. Sung, T.S. Jeon, C.H. Chang, J.D. Jang, S.B. Lee, H.C. Kim and S.Y. Lee. 2009. A multi-center, randomized, clinical study to compare the effect and safety of autologous cultured osteoblast(Ossron) injection to treat fractures. BMC Musculoskelet Disord. 10: 20.

Kocher, A.A., M.D. Schuster, M.J. Szabolcs, S. Takuma, D. Burkhoff, J. Wang, S. Homma, N.M. Edwards and S. Itescu. 2001. Neovascularization of ischemic myocardium by human bone-marrow-derived angioblasts prevents cardiomyocyte apoptosis, reduces remodeling and improves cardiac function. Nat Med. 7: 430–436.

Kuroda, R., T. Matsumoto, M. Miwa, A. Kawamoto, Y. Mifune, T. Fukui, Y. Kawakami, T. Niikura, S.Y. Lee, K. Oe, T. Shoji, T. Kuroda, M. Horii, A. Yokoyama, T. Ono, Y. Koibuchi, S. Kawamata, M. Fukushima, M. Kurosaka and T. Asahara. 2010. Local Transplantation of G-CSF-Mobilized CD34+ Cells in a Patient with Tibial Nonunion: A Case Report. Cell Transplant.

Kwon, S.M., T. Suzuki, A. Kawamoto, M. Ii, M. Eguchi, H. Akimaru, M. Wada, T. Matsumoto, H. Masuda, Y. Nakagawa, H. Nishimura, K. Kawai, S. Takaki and T. Asahara. 2009. Pivotal role of lnk adaptor protein in endothelial progenitor cell biology for vascular regeneration. Circ Res. 104: 969–977.

Laing, A.J., J.P. Dillon, E.T. Condon, J.C. Coffey, J.T. Street, J.H. Wang, A.J. McGuinness and H.P. Redmond. 2007a. A systemic provascular response in bone marrow to musculoskeletal trauma in mice. J Bone Joint Surg Br. 89: 116–120.

Laing, A.J., J.P. Dillon, E.T. Condon, J.T. Street, J.H. Wang, A.J. McGuinness and H.P. Redmond. 2007b. Mobilization of endothelial precursor cells: systemic vascular response to musculoskeletal trauma. J Orthop Res. 25: 44–50.

Lee, D.Y., T.J. Cho, J.A. Kim, H.R. Lee, W.J. Yoo, C.Y. Chung and I.H. Choi. 2008. Mobilization of endothelial progenitor cells in fracture healing and distraction osteogenesis. Bone.

Li, R., K. Atesok, A. Nauth, D. Wright, E. Qamirani, C.M. Whyne and E.H. Schemitsch. 2011. Endothelial progenitor cells for fracture healing: a microcomputed tomography and biomechanical analysis. J Orthop Trauma. 25: 467–471.

Losordo, D.W., R.A. Schatz, C.J. White, J.E. Udelson, V. Veereshwarayya, M. Durgin, K.K. Poh, R. Weinstein, M. Kearney, M. Chaudhry, A. Burg, L. Eaton, L. Heyd, T. Thorne, L. Shturman, P. Hoffmeister, K. Story, V. Zak, D. Dowling, J.H. Traverse, R.E. Olson, J. Flanagan, D. Sodano, T. Murayama, A. Kawamoto, K.F. Kusano, J. Wollins, F. Welt, P. Shah, P. Soukas, T. Asahara and T.D. Henry. 2007. Intramyocardial transplantation of autologous CD34+ stem cells for intractable angina: a phase I/IIa double-blind, randomized controlled trial. Circulation. 115: 3165–3172.

Matsumoto, T., A. Kawamoto, R. Kuroda, M. Ishikawa, Y. Mifune, H. Iwasaki, M. Miwa, M. Horii, S. Hayashi, A. Oyamada, H. Nishimura, S. Murasawa, M. Doita, M. Kurosaka and T. Asahara. 2006. Therapeutic potential of vasculogenesis and osteogenesis promoted by peripheral blood CD34-positive cells for functional bone healing. Am J Pathol. 169: 1440–1457.

Matsumoto, T., Y. Mifune, A. Kawamoto, R. Kuroda, T. Shoji, H. Iwasaki, T. Suzuki, A. Oyamada, M. Horii, A. Yokoyama, H. Nishimura, S.Y. Lee, M. Miwa, M. Doita,

M. Kurosaka and T. Asahara. 2008. Fracture induced mobilization and incorporation of bone marrow-derived endothelial progenitor cells for bone healing. J Cell Physiol. 215: 234–242.

Matsumoto, T., M. Ii, H. Nishimura, T. Shoji, Y. Mifune, A. Kawamoto, R. Kuroda, T. Fukui, Y. Kawakami, T. Kuroda, S.M. Kwon, H. Iwasaki, M. Horii, A. Yokoyama, A. Oyamada, S.Y. Lee, S. Hayashi, M. Kurosaka, S. Takaki and T. Asahara. 2010. Lnk-dependent axis of SCF-cKit signal for osteogenesis in bone fracture healing. J Exp Med. 207: 2207–2223.

Mifune, Y., T. Matsumoto, A. Kawamoto, R. Kuroda, T. Shoji, H. Iwasaki, S.M. Kwon, M. Miwa, M. Kurosaka and T. Asahara. 2008. Local delivery of granulocyte colony stimulating factor-mobilized CD34-positive progenitor cells using bioscaffold for modality of unhealing bone fracture. Stem Cells. 26: 1395–1405.

Otsuru, S., K. Tamai, T. Yamazaki, H. Yoshikawa and Y. Kaneda. 2007. Bone marrow-derived osteoblast progenitor cells in circulating blood contribute to ectopic bone formation in mice. Biochem Biophys Res Commun. 354: 453–458.

Otsuru, S., K. Tamai, T. Yamazaki, H. Yoshikawa and Y. Kaneda. 2008. Circulating bone marrow-derived osteoblast progenitor cells are recruited to the bone-forming site by the CXCR4/stromal cell-derived factor-1 pathway. Stem Cells. 26: 223–234.

Petite, H., V. Viateau, W. Bensaid, A. Meunier, C. de Pollak, M. Bourguignon, K. Oudina, L. Sedel and G. Guillemin. 2000. Tissue-engineered bone regeneration. Nat Biotechnol. 18: 959–963.

Quarto, R., M. Mastrogiacomo, R. Cancedda, S.M. Kutepov, V. Mukhachev, A. Lavroukov, E. Kon and M. Marcacci. 2001. Repair of large bone defects with the use of autologous bone marrow stromal cells. N Engl J Med. 344: 385–386.

Rozen, N., T. Bick, A. Bajayo, B. Shamian, M. Schrift-Tzadok, Y. Gabet, A. Yayon, I. Bab, M. Soudry and D. Lewinson. 2009. Transplanted blood-derived endothelial progenitor cells (EPC) enhance bridging of sheep tibia critical size defects. Bone. 45: 918–924.

Takahashi, T., C. Kalka, H. Masuda, D. Chen, M. Silver, M. Kearney, M. Magner, J.M. Isner and T. Asahara. 1999. Ischemia- and cytokine-induced mobilization of bone marrow-derived endothelial progenitor cells for neovascularization. Nature Medicine. 5: 434–438.

Takaki, S., J.D. Watts, K.A. Forbush, N.T. Nguyen, J. Hayashi, J. Alberola-Ila, R. Aebersold and R.M. Perlmutter. 1997. Characterization of Lnk. An adaptor protein expressed in lymphocytes. J Biol Chem. 272: 14562–14570.

Takaki, S., K. Sauer, B.M. Iritani, S. Chien, Y. Ebihara, K. Tsuji, K. Takatsu and R.M. Perlmutter. 2000. Control of B cell production by the adaptor protein lnk. Definition Of a conserved family of signal-modulating proteins. Immunity. 13: 599–609.

Takaki, S., H. Morita, Y. Tezuka and K. Takatsu. 2002. Enhanced hematopoiesis by hematopoietic progenitor cells lacking intracellular adaptor protein, Lnk. J Exp Med. 195: 151–160.

Takizawa, H., C. Kubo-Akashi, I. Nobuhisa, S.M. Kwon, M. Iseki, T. Taga, K. Takatsu and S. Takaki. 2006. Enhanced engraftment of hematopoietic stem/progenitor cells by the transient inhibition of an adaptor protein, Lnk. Blood. 107: 2968–2975.

Tondreau, T., N. Meuleman, A. Delforge, M. Dejeneffe, R. Leroy, M. Massy, C. Mortier, D. Bron and L. Lagneaux. 2005. Mesenchymal stem cells derived from CD133-positive cells in mobilized peripheral blood and cord blood: proliferation, Oct4 expression, and plasticity. Stem Cells. 23: 1105–1112.

Zvaifler, N.J., L. Marinova-Mutafchieva, G. Adams, C.J. Edwards, J. Moss, J.A. Burger and R.N. Maini. 2000. Mesenchymal precursor cells in the blood of normal individuals. Arthritis Res. 2: 477–488.

CHAPTER 11

Stromal Cell-derived Factor 1/CXCR4 Signaling, Stem Cells, and Fractures

Hiromu Ito

ABSTRACT

Stromal-derived factor-1 (SDF-1) was originally discovered as a unique growth factor and chemoattractant for lymphocytes expressed in bone marrow stromal cells. It acts through binding to its major receptor, CXC chemokine receptor-4 (CXCR4). Discovery of its crucial roles on hematopoietic, mesenchymal, and other types of cells has been broadened by vigorous investigation. Recent studies collectively indicate SDF-1 works as a crucial chemoattractant for mesenchymal stem cells (MSCs) migrating from surrounding tissues or circulation, or both, to repair tissue injury including fractures. While the roles of SDF-1 on tissue healing are elucidated, attempts to clinically apply SDF-1 to difficult tissue-healing problems become more plausible. The induction of MSC recruitment from surrounding tissues or from the circulation by SDF-1 may be a helpful modality for inducing or supporting cell-based therapy for tissue regeneration. Moreover, SDF-1 has been shown to play many other roles in MSC proliferation and differentiation, and in recruiting hematopoietic stem cells. These studies may expand the vista for elucidation of the tissue-healing mechanisms SDF-1 contributes to and potential clinical applications for SDF-1.

Department of the Control for Rheumatic Diseases, and Department of Orthopaedic Surgery, Kyoto University Graduate School of Medicine, Kyoto, Japan.
E-mail: hiromu@kuhp.kyoto-u.ac.jp

List of abbreviations given at the end of the text.

Keywords: Stromal cell-derived factor-1, CXCR4, mesenchymal stem cell, migration, bone repair, fracture, chemokine

Introduction

Stromal-derived factor-1 (SDF-1 or CXCL12) was originally discovered in a bone marrow-derived stromal cell line using a signal sequence trap, a unique cloning method (Tashiro et al. 1993). It was soon rediscovered as pre-B cell growth-stimulating factor (PBSF, Nagasawa et al. 1994). SDF-1 is broadly expressed in a variety of tissue types and acts as a potent chemoattractant for immature and mature hematopoietic cells. It has important roles in the homing of hematopoietic stem cells to the bone marrow and mediating the survival as well as the proliferation of progenitor/stem cells have been studied. SDF-1 has two major isoforms, α and β, but SDF-1α is the predominant isoform secreted by marrow stromal and endothelial cells and is identified in nearly all organs and tissues. SDF-1 plays many important roles through the activation of a G protein-coupled receptor CXCR4, and the interaction of SDF-1/CXCR4 and hematopoietic stem cells (HSCs) have been extensively reported. During the past decade, accumulating data have supported an emerging and fascinating hypothesis that SDF-1/CXCR4 also plays a pivotal role in the biologic and physiological functions of mesenchymal stem cells (MSCs) [Wynn et al. 2004; Dar et al. 2005]. SDF-1 is upregulated at injury sites and serves as a potent chemoattractant to recruit circulating or residing CXCR4-expressing MSCs, which are necessary for tissue-specific organ repair or the regeneration of many organs. As such, the involvement of the SDF-1/CXCR4 axis of MSCs in bone repair has been a focus in recent studies. Furthermore, there recently are many articles published on the crucial roles SDF-1 plays in cancer metastasis, leukemia, disease-related osteolysis, and other conditions. In this chapter the focus is on the unique and critical roles of SDF-1 on MSC migration, partially HSC migration, and possible roles on MSC proliferation and differentiation, especially for bone regeneration or fracture repair.

Bone-healing Mechanisms, Possible Sources of MSC, and its Migratory Factors

Bone-healing mechanisms and MSCs

Bone healing is thought to rely upon two types of healing mechanisms. Endochondral ossification is the major type of healing for most bone injuries. In this process, undifferentiated MSCs undergo proliferation, chondrogenic differentiation, hypertrophic change, and calcification, eventually being replaced by bone produced by osteoclasts and osteoblasts. This type of healing mechanism resembles that of development of the growth plate in long bones, and most researchers think this series of events "recapitulates" growth plate development, though some controversy certainly exists. The other type of healing is intramembranous ossification, in which MSCs or undifferentiated bone-forming cells directly differentiate into osteoblasts and efficiently form new bones. This process requires certain stable environmental and mechanical conditions. One of the most enigmatic yet fascinating questions is where the stem (or progenitor) cells come from and how the cells are recruited to the injured sites. Many researchers believe, and have proved in part, that residing stem or progenitor cells differentiate and form new bones. However, other progenitors provide another possible source of stem cells and they can come in from the circulation.

Possible sources of MSCs

The bone marrow is unanimously considered to be the most attractive source of stem cells. The existence of such cells in the bone marrow were reported as early as 1968. Moreover, clinical applications of bone marrow aspirates have been adopted for many years for the treatment of nonunion in orthopedic surgery. The favorable features of bone marrow for clinical application are its accessibility with little morbidity and the abundance of cells. However, many scientific and clinical questions remain to be investigated. Moreover, some strongly argue that human bone marrow MSCs represent a phenotypically homogeneous cell population that shares an identical phenotype with marrow adventitial reticular cells, stromal cells similar in nature to pericytes. This should be clarified.

From experience, the periosteum is unanimously considered by orthopedic surgeons to be one of the most crucial components for successful bone healing. If the periosteum is destroyed by the injury itself or surgery, the healing process will be much delayed, causing many problems. Furthermore, a number of studies have shown that when the periosteum is autografted at heterotopic sites, either as free grafts or in diffusion chambers, mineralized and cartilaginous tissue can be detected. However, the nature and kinetics of the cells involved have not been well studied. One of the major reasons for this is that the number of cells is limited and the extraction of the cells or of the periosteum itself is relatively difficult, especially in small animals such as rodents. By how much and by what way the periosteal cells contribute to actual bone healing in humans remains to be demonstrated, and the molecular and physiological mechanisms underlying their contributions are unclear.

Vessels are available virtually everywhere in the body except for certain "avascular" tissues such as cartilage, so they are easily envisaged as a possible source of stem cells. Some hypothesize that pericytes play a role in normal bone growth and development. Pericytes are an essential part of the angiogenic process, so these cells may directly contribute to skeletogenesis. Moreover, pericytes may serve as a reservoir of primitive precursor cells. Indeed, there are many phenotypic similarities between pericytes and stem cells isolated from adult tissues. However, isolation of distinct pericytes separate from other possible stem cells is required to prove this hypothesis *in vivo*, and it seems that some ambiguity and confusion exists in these experiments. It is also possible that pericytes and other types of stem cells such as "circulating" stem cells (described below) are unidentifiable in certain *in vivo* experiments. In addition, pericytes may not be good candidates for cell therapy because of their limited availability or difficulty in harvesting, with the exception of those in the bone marrow. Nevertheless, the potential of pericytes in bone-healing mechanisms is fascinating, and scientific investigations to elucidate their mechanism of action and possible therapeutic use are greatly sought.

Muscle has been one of the most plausible sources of cells for bone repair because of its proximity to bones and its ample blood supply. However, the lack of clear evidence of its contribution to bone repair has led to some doubt about its candidacy. There are some enthusiastic studies on muscle derived-stem cells on bone healing, but this work still seems to lack definite physiological evidence for the role of these cells in normal bone repair. With these limitations, the muscle is one of the most abundant tissues in our body, making it a good candidate for cell harvesting for cell-based therapy.

A breakthrough report was published that described the isolation of adherent and clonogenic cells from the whole blood of adult mammals of four different species, including humans. The study demonstrated that some polyclonal strains and several single colony-derived strains form bone upon *in vivo* transplantation (Kuznetsov et al. 2001). While the existence of circulating MSCs remains controversial, it is safe to say that recent investigations show their existence and that they have roles in bone healing, at least in rodents. Furthermore, the possibility has arisen that MSCs in circulation might only appear transiently after skeletal trauma (an unavoidable consequence of skeletal injuries), and one probable scenario is that the circulation of bone marrow MSCs helps the injured bone to heal.

Recent vigorous attempts to find any other sources of stem cells have led to the demonstration of several types of intriguing MSC-like cells, one of which is adipose tissue-derived stem cells. Another recent finding is that human and mouse tendons harbor a unique cell population that has universal stem cell characteristics such as clonogenicity, multipotency, and self-renewal capacity. Together with other types of stem cell-like cells such as synovium- and skin-derived stem cells, these have a certain potential for therapeutic applications, but it seems inconceivable that they would make a physiological contribution to bone healing. For clinical application of these cells in the treatment of bone diseases, their precise profiles and practical efficacy for bone regeneration should be demonstrated.

The possible sources of MSCs are summarized in another review (Ito 2011).

Migration of MSCs

During the course of organ regeneration, it has been demonstrated that both local MSCs derived from the injured tissue and circulating MSCs collaborate in the healing of damaged organs. A series of reports have supported the hypothesis that a particular set of molecules that are upregulated during bone injury are released around the injured site or into circulation, or both, stimulating MSCs to downregulate the adhesion molecules that hold them in their niche. Subsequently, resident or circulating MSCs "sense" a tissue injury, migrate to the sites of damage from surrounding tissues or from the circulation, and undergo tissue-specific differentiation. However, the mechanisms responsible for MSC migration to the site of bone injury have not been fully shown. Cytokines and chemokines probably play critical roles in these processes, and many of these factors are chemoattractants. Chemokines are potent chemoattractant cytokines that regulate trafficking of leukocytes and other types of cells in homeostatic and inflammatory processes. Depending on their topical concentration, chemokines directly recruit circulating leukocytes and other cells to the site of inflammation or injury. Many experimental and clinical studies have demonstrated that a substantial number of chemokines are involved in the pathophysiology of many diseases. In particular, chemokines have been shown to recruit MSC to inflammation or injury sites for tissue healing.

Possible factors inducing migration of MSCs

Chemokines (and their receptors) such as monocyte chemotactic protein-1 (MCP-1, CCL2), CCR2, macrophage inflammatory protein-1α (MIP-1α, CCL3), and MCP-2 (CCL8), and fractalkine (CX3CL1) are among the strong candidates for inducing migration of MSCs. MSCs have been shown to express a variety of chemokine receptors, and chemokine-mediated MSC migration has been demonstrated *in vitro* and *in vivo*. The importance of adhesion molecules such as very late antigen-1 (VLA-1), vascular cell adhesion molecule-1 (VCAM-1), intracellular adhesion molecule-1 (ICAM-1), VLA-4, β1-integrin, and P-selectin, and matrix-degrading enzymes such as matrix metalloproteinase-2 (MMP-2), membrane type-1 MMP (MT1-MMP), and tissue inhibitor of metalloproteinase-2 (TIMP-2) is undeniable. However, SDF-1 is one of the most investigated and potent factors for MSC migration. What is SDF-1? What can SDF-1 do for MSC migration? Are there any other possible roles that SDF-1 can play for bone healing? These issues will be discussed below. Other potent chemoattractants' roles in MSC migration processes are described in other chapters.

Stromal-derived Factor-1 (CXCL12) and its Receptors

What is SDF-1/CXCL12?

SDF-1 (also known as CXCL12 and PBSF) was the first soluble factor that was reported to be crucial for the earliest stage of B-cell lymphocyte development and one of the chemokines (Nagasawa 2006). Chemokines consist of a superfamily of chemoattracting, cytokine-like proteins that bind to and activate a family of heptahelical receptors coupled to heterotrimeric GTP-binding proteins (G-protein-coupled receptors). SDF-1 was isolated from stromal-cell lines used in *in vitro* B-cell development cultures and was first characterized as a growth-stimulating factor for a stromal-cell-dependent B-cell precursor clone. The main physiological receptor for SDF-1 is CXCR4, which also functions as an entry receptor for some strains of HIV-1. Studies using mutant mice with targeted gene disruption have shown that SDF-1 and CXCR4 are essential for various developmental processes including, not only B-cell development, but also angiogenesis, neurogenesis, and skeletogenesis. It has been shown that SDF-1 is involved in the colonization of the bone marrow by hematopoietic cells, including HSCs during ontogeny. The roles of SDF-1 on HSC migration for tissue repair will be discussed later. Recently numerous studies have demonstrated that SDF-1 acts not only as one of the most primary chemoattractants for migration of many types of cells, but also as a cell survival, proliferation, and even differentiation factor.

SDF-1 binds to CXCR4, a G-protein-coupled receptor with seven transmembrane domains. Once SDF-1 binds to CXCR4, the receptor forms a complex with the Gα1 subunit of G protein, resulting in inhibition of adenylyl cyclase-mediated cyclic adenosine monophosphate production and mobilization of intracellular calcium. Dissociation of the Gα1 subunit from G$\alpha\delta$ leads to activation of multiple downstream targets, including the focal adhesion kinase, extracellular signal-regulated kinases, protein kinase C, ERK1/2, MAPK, JNK, AKT, JAK/Stat, and NF-κB transcription pathways (Tavor and Petit 2010; Sun et al. 2010). Ligand-stimulated chemotaxis is accompanied by cytoskeletal rearrangements, actin polymerization, polarization, pseudopodia formation, and integrin-dependent adhesion to endothelial cells, and other biologic substrates. Those pathways are documented mostly in leukocytes, but MSCs and bone-forming cells are believed to have similar pathways. The interaction between SDF-1 and CXCR4 was previously exclusive , but it was recently identified that SDF-1 can also bind with CXCR7, which also binds with low affinity to CXCL11. CXCR7 expression has been found in T lymphocytes, B cell development, tumor cell lines, activated endothelium, fetal liver cells, and others. Several mechanisms underlying CXCR7 have been proposed as CXCR7 functions as a decoy receptor, scavenges or sequesters SDF-1 (generating gradients of SDF-1), and serves as a coreceptor for CXCR4 to enhance SDF-1-mediated G-protein signaling.

What can SDF-1 do for MSC migration?

During the past decade, accumulating data have supported an emerging hypothesis that SDF-1/CXCR4 plays a pivotal role in the biologic and physiological functions of MSCs (reviwed by Ito 2011). SDF-1 is upregulated at injury sites and serves as a potent chemoattractant to recruit circulating or residing CXCR4-expressing MSCs, which are necessary for tissue-specific organ repair or the regeneration of the many organs such as liver, heart, kidney and skin. Moreover, the local delivery of SDF-1 into injured tissue promotes the recruitment of circulating mesenchymal stromal and progenitor cells to lesions in the heart, brain and lung. Among these, myocardial infarction is one of the most plausible targets that SDF-1 can be used for in MSC treatment (Takahashi 2010). A few potential roles of the SDF-1/CXCR4 system in myocardial infarction have been suggested. At the site of tissue ischemia, HIF-1α is induced in response to reduced oxygen tension and then stimulates SDF-1 expression. Activation of the SDF-1/CXCR4 axis protects cardiomyocytes from apoptotic cell death. SDF-1 also recruits bone marrow-derived stem/progenitor cells and induces the production of angiogenic factors, thereby leading to angiogenesis. Moreover, SDF-1 stimulates migration of endothelial cells and enhances angiogenesis. As such, several therapeutic strategies using the SDF-1/CXCR4 axis have been proposed. First, SDF-1 can be locally delivered to induce cell migration or SDF-1 cleavage can be inhibited to sustain SDF-1 concentration, or both. Second, CXCR4 expression can be upregulated in cardiomyocytes and cardiac stem cells by inducing gene transfer or other methods. Third, CXCR4+ stem cells can be mobilized from the bone marrow into peripheral circulation. Lastly, CXCR4 expression can be upregulated in bone marrow-derived cells and MSCs *ex vivo* or *in vivo*. These mechanisms underlying tissue injury and therapeutic strategies can be applied to most tissue injury associated with MSCs.

The involvement of the SDF-1/CXCR4 axis of MSCs in bone repair has recently been elucidated. Otsuru et al. showed (using an ectopic bone formation model induced by implantation of a bone morphogenic protein-2-containing collagen pellet in mouse muscle tissues) that circulating bone marrow-derived osteoblast progenitor cells migrated to the region of bone formation through chemoattraction by SDF-1 expressed on vascular endothelial cells and on the *de novo* osteoblasts of the region (Otsuru et al. 2008). Granero-Molto et al. published an interesting article in which they showed that implanted MSCs migrated to a fracture site in an exclusively CXCR4-dependent manner (Granero-Molto et al. 2009). We also showed that SDF-1 was induced in the periosteum in bone injury and promoted endochondral bone repair by recruiting MSCs to the injury site. In mouse models of structural femoral live and dead bone grafts, bone formation was decreased in SDF-1$^{+/-}$ and CXCR4$^{+/-}$ mice. This was rescued by grafted bones from CXCR4$^{+/-}$ mice transplanted into the SDF-1$^{+/-}$ femur, but not *vice versa*. Our study demonstrated that after bone injury, SDF-1 is expressed on the periosteum of the bone graft and recruits CXCR4-expressing MSCs to bone repair sites in the acute phase of bone repair (Kitaori et al. 2009). In terms of association with another factor, Shinohara et al. showed SDF-1 and MCP-3 cooperatively improved recruitment of MSCs into sites of fracture repair. Chondrogenesis and even ligament repair have been documented to work through SDF-1 delivery on MSC migration (Shinohara et al. 2011). Accumulating evidence collectively and strongly support the notion that SDF-1 potently recruits MSCs from surrounding or peripheral circulation, or both, for bone repair. The remaining big questions are whether the same mechanism of MSC migration would be applicable in humans, and whether the similar delivery of SDF-1 to induce bone formation would be effective in humans. The expression pattern of chemokines and their receptors in MSCs appears to be substantially different between rodents and humans. Can SDF-1 induce the migration of MSCs and promote fracture repair in humans? The answer and the modalities to make it effective in humans are much anticipated.

Other possible roles of SDF-1 in bone healing

A number of studies have so far shown the effects of SDF-1 on MSC migration. Recently, however, growing evidence has suggested that SDF-1 functions not only as a chemoattractant, but also an important survival and differentiation factor for bone-forming cells, including osteoclasts, osteoblasts, and chondrocytes.

Osteoclasts are a unique and highly specialized class of HSC that share a common circulating CD14+ precursor with monocytes and macrophages and that are derived from pluripotent bone marrow granulocyte-macrophage colony-forming units. In addition to the well-characterized cytokines associated with osteoclast development and function, recent studies imply that chemokines have a role in the process of physiological and pathological osteoclast formation and activation, and that SDF-1 exerts its actions at the early stages of osteoclastogenesis. The high production of SDF-1 by immature stromal cells suggests that this chemokine is a chemoattractant and perhaps a survival factor for circulating osteoclast precursor cells. Several studies show that SDF-1 increases the survival, migration, and activation of osteoclasts (Gronthos and Zannettino 2007). However, unlike other chemokines, such as MIP-1α, that predominantly affect committed osteoclasts, SDF-1 seems to be crucial at the early phase of osteoclastogenesis, when human CD14+ osteoclast precursor cells express high levels of CXCR4, which gradually declines during osteoclast maturation. Collectively these findings indicate a potential role for SDF-1 during the early phase of osteoclast precursor cell recruitment and development, as a positive regulator of bone resorption. Then, what does SDF-1 do in fracture repair in terms of osteoclast development? A large number of studies show that SDF-1 has certain, distinct roles in pathological bone remodeling such as metastatic bone resorption, but the contribution of SDF-1 on osteoclastogenesis is not well examined in fracture repair. Elucidation of the role of SDF-1 in osteoclastogenesis for bone healing would be fascinating, but remains to be investigated.

Previous studies have suggested the expression of SDF-1 and CXCR4 in bone, and the intimate association of the SDF-1/CXCR4 pathway with progenitor cells that have the potential to become bone-producing osteoblasts or form bone. Expression of SDF-1 and CXCR4 is found in bone sections with greater levels in less differentiated cells or immature osteoblasts than in mature osteoblasts and osteocytes. Several studies reveal the direct involvement of the SDF-1/CXCR4 pathway in osteogenic differentiation from MSCs and osteogenic progenitor cells *in vitro*. Very recently a report demonstrated that CXCR4 functions in postnatal bone development by regulating osteoblast development in cooperation with BMP signaling and that CXCR4 acts as an endogenous signaling component necessary for bone formation (Zhu et al. 2011). Indeed, exogenously-released SDF-1 cooperatively enhanced the recruitment of osteogenic cells and angiogenesis with BMP-2, resulting in a synergistic effect on bone regeneration (Ratanavaraporn et al. 2011). There is still a lack of studies on SDF-1-related osteogenic bone formation in normal fracture healing, but this opens a novel therapeutic application of SDF-1, at least in conjunction with BMP-2, for bone regeneration.

Another key cell type in fracture repair is the chondrocyte. There are only a few reports on the expression and function of the SDF-1/CXCR4 axis in chondrocyte differentiation. During chondrogenesis from MSCs,

SDF-1, and CXCR4 expression were down modulated *in vitro* (Cristino et al. 2008) as they were during osteoclastogenesis and osteogenesis. Elevation of SDF-1 concentration in rabbit growth plates *in vivo* leads to increased type X collagen gene expression, degradation of the cartilage matrix, potentially from MMP-13, and premature closure of the growth plate. A recent report demonstrated that SDF-1 is expressed in the bone marrow adjacent to hypertrophic chondrocytes, while CXCR4 is predominantly expressed in hypertrophic chondrocytes (Wei et al. 2010). Thus, these two are expressed in a complementary pattern in the chondro–osseous junction of the growth plate. The report also showed that interaction of SDF-1 and CXCR4 is required for RUNX2 expression and that a positive feedback loop of stimulation of chondrocyte hypertrophy by SDF-1/CXCR4 is mediated by RUNX2. There seems to be a vast area for investigation of the roles of SDF-1 in chondrogenesis.

HSC migration for tissue healing

HSCs were identified as early as 1961 as "colony-forming units" and defined as cells able to give rise to hematopoietic nodules in the spleen following transplantation of the bone marrow into irradiated animals. HSCs are self-renewing cells located in the bone marrow that are responsible for the maintenance of homeostasis through continual replenishment of all of the cellular components of the blood such as leukocytes, erythrocytes, lymphocytes, and platelets. Recently some studies have proposed an interesting hypothesis that endogenous HSCs are recruited to injury sites, one type of which is possibly fracture sites (Kavanagh and Kalia 2011).

First, HSCs are mobilized from the bone marrow before being recruited to the injury sites. A number of factors involved in bone marrow retention of HSCs are those secreted by injured tissues to recruit HSCs. Soluble factors are released from injured sites and subsequently generate a concentration gradient that facilitates the egress of HSCs from the bone marrow. The SDF-1/CXCR4 axis is the most important candidate for the critical role of HSC retention within the bone marrow and also promotes HSC recruitment to injury sites. One of the major regulators of SDF-1 expression is oxygen tension via the activities of the transcription factor HIF-1α. As the bone marrow is hypoxic, the expression level of SDF-1 is high in the stromal cells that line the bone marrow where HSCs reside. This high concentration of SDF-1 retains HSCs within the bone marrow, and once SDF-1 expression is reduced, HSCs are mobilized into circulation. In case of tissue injury, the levels of circulating SDF-1 are increased, and the concentration gradient promotes emigration of HSCs from the bone marrow. While the SDF-1/CXCR4 axis undeniably plays a key role in HSC mobilization, other pathways and factors have also been implicated in HSC mobilization after tissue injury. For example, vascular endothelial growth factor and MMPs such as MMP9 can also induce HSC mobilization from the bone marrow.

Second, circulating HSCs are recruited from the circulation to injury sites. A well characterized multistep adhesion and activation cascade describes the processes by which mature leukocytes adhere to the vessel endothelium and subsequently transmigrate into underlining tissues. Recent studies demonstrated that the kinetics of stem and progenitor cell homing to the bone marrow compartment involves a similar rolling, firm adhesion, and transmigration cascade. Emerging data suggests that similar adhesive events govern HSC recruitment to the injured extramedullary tissue endothelium. However, it is unclear which secreted factor(s) and its concentration gradient actually recruits HSCs to injury sites. The SDF-1/CXCR4 axis would play a crucial role as it does in MSC migration to injured tissues, but this remains to be proven. Moreover, while the therapeutic benefit of HSCs has been demonstrated in numerous clinical and experimental studies for a broad range of conditions, whether HSCs facilitate bone repair remains unclear. A recent study revealed that HSCs and MSCs form a unique niche in the bone marrow made of heterotypic stem-cell pairs (Mendez-Ferrer et al. 2010), which leads to a hypothesis that this partnership between two distinct somatic stem-cell types works cooperatively for tissue repair, including bone healing.

Clinical application of SDF-1 for bone healing

As described in the section regarding osteogenesis, several reports provide promising results of clinical application of SDF-1 for bone regeneration. Whether SDF-1 only recruits osteogenic stem/progenitor cells from surrounding tissues or the circulation, or both, or exerts direct effects of cell differentiation, remains to be investigated. It seems to have certain positive effects, at least with BMP-2. There remains a lack of human studies of the clinical application of SDF-1 for bone healing. The induction of MSC migration may be less efficient and less applicable than cell transplantation for stimulating bone healing. However, the method has certain clinical advantages, such as fewer ethical issues and less possibility of infection, and it could be used to augment cell transplantation to enhance cell targeting. More studies must be conducted to elucidate the mechanisms so that practical therapeutic modalities can be developed in the near future.

Conclusion

Skeletal injuries remain among the most prevalent clinical problems, especially in an aging society. MSCs are, without doubt, the most attractive candidate for cell-based bone regeneration, but current results present several notable shortcomings such as vulnerability to infection, the uncertainty of the capability of MSCs for differentiation in specific *in vivo* situations, the high cost of *ex vivo* cell handling, the limited numbers of cells actually obtainable, and even possible malignant transformation of the cells during *ex vivo* cell expansion. The induction of MSC migration is a promising way to overcome these issues. Cell sources may be the bone marrow, periosteum, vessel walls, muscle, circulation, and elsewhere, but differences among them still cause controversy and require more detailed investigation. The SDF-1/CXCR4 axis is one of the most potent and promising molecules that has been shown to be involved in the migration of MSCs. In certain skeletal injury models, the expression and function of SDF-1 has been clearly demonstrated. SDF-1 is even implicated to have direct effects on mesenchymal cell differentiation such as osteoclastogenesis, osteogenesis, and chondrogenesis. HSCs are already shown to be recruited to injury sites by SDF-1, but their actual potency for bone repair remains to be investigated. The induction of MSC or HSC migration, or both, by SDF-1 could emerge as an efficient method for treating difficult bone regeneration issues. Practical aspects of the use of SDF-1 including injection, coating, transduction, and even blocking of antagonized factors should be investigated. The elucidation of mechanisms in which SDF-1 may play a role, as well as seeking practical use of SDF-1 for bone healing are warranted.

Summary Points

- Stromal-derived factor-1 (SDF-1) was originally discovered as a unique growth factor and chemoattractant for lymphocytes expressed in bone marrow stromal cells.
- SDF-1 binds to its major receptor, CXC chemokine receptor-4 (CXCR4) and its minor receptor, CXCR7.
- SDF-1 works as a crucial chemoattractant for mesenchymal stem cells (MSCs) migrating from surrounding tissues and/or circulation to repair tissue injury including fracture.
- SDF-1 plays many other roles in mesenchymal cell proliferation and differentiation.
- SDF-1 recruits hematopoietic stem cells to sites of injury.
- Vigorous attempts are underway for clinical application of SDF-1.

List of Abbreviations

BMP-2	:	bone morphogenic protein-2
CXCR4	:	CXC chemokine receptor-4
HIF-1α	:	hypoxia inducible factor-1α

HSC   :   Hematpoietic stem cell
MMP   :   matrix metalloproteinase
MSC   :   mesenchymal stem cell
SDF-1   :   stromal-derived factor-1

References

Cristino, S., A. Piacentini, C. Manferdini, K. Codeluppi, F. Grassi, A. Facchini and G. Lisignoli. 2008. Expression of CXC chemokines and their receptors is modulated during chondrogenic differentiation of human mesenchymal stem cells grown in three-dimensional scaffold: evidence in native cartilage. Tissue Eng Part A. 14(1): 97–105.

Dar, A., P. Goichberg, V. Shinder, A. Kalinkovich, O. Kollet, N. Netzer et al. 2005. Chemokine receptor CXCR4-dependent internalization and resecretion of functional chemokine SDF-1 by bone marrow endothelial and stromal cells. Nat Immunol. 6: 1038–46.

Granero-Moltó, F., J.A. Weis, M.I. Miga, B. Landis, T.J. Myers, L. O'Rear, L. Longobardi, E.D. Jansen, D.P. Mortlock and A. Spagnoli. 2009. Regenerative effects of transplanted mesenchymal stem cells in fracture healing. Stem Cells. 27: 1887–98.

Gronthos, S. and A.C. Zannettino. 2007. The role of the chemokine CXCL12 in osteoclastogenesis. Trends Endocrinol Metab. 18(3): 108–13.

Ito, H. 2011. Chemokines in mesenchymal stem cell therapy for bone repair: a novel concept of recruiting mesenchymal stem cells and the possible cell sources. Mod Rheumatol. 21(2): 113–21.

Kavanagh, D.P. and N. Kalia. 2011. Hematopoietic stem cell homing to injured tissues. Stem Cell Rev. 7(3): 672–82.

Kitaori, T., H. Ito, E.M. Schwarz, R. Tsutsumi, H. Yoshitomi, S. Oishi, M. Nakano, N. Fujii, T. Nagasawa and T. Nakamura. 2009. Stromal cell-derived factor 1/CXCR4 signaling is critical for the recruitment of mesenchymal stem cells to the fracture site during skeletal repair in a mouse model. Arthritis Rheum. 60: 813–23.

Kuznetsov, S.A., M.H. Mankani, S. Gronthos, K. Satomura, P. Bianco and P.G. Robey. 2001. Circulating skeletal stem cells. J Cell Biol. 153: 1133–40.

Méndez-Ferrer, S., T.V. Michurina, F. Ferraro, A.R. Mazloom, B.D. Macarthur, S.A. Lira, D.T. Scadden, A. Ma'ayan, G.N. Enikolopov and P.S. Frenette. 2010. Mesenchymal and haematopoietic stem cells form a unique bone marrow niche. Nature. 466(7308): 829–34.

Nagasawa, T. 2006. Microenvironmental niches in the bone marrow required for B-cell development. Nat Rev Immunol. 6(2): 107–16.

Nagasawa, T., H. Kikutani and T. Kishimoto. 1994. Molecular cloning and structure of a pre-B-cell growth-stimulating factor. Proc Natl Acad Sci USA. 91(6): 2305–9.

Otsuru, S., K. Tamai, T. Yamazaki, H. Yoshikawa and Y. Kaneda. 2008. Circulating bone marrow-derived osteoblast progenitor cells are recruited to the bone-forming site by the CXCR4/stromal cell-derived factor-1 pathway. Stem Cells. 26: 223–34.

Ratanavaraporn, J., H. Furuya, H. Kohara and Y. Tabata. 2011. Synergistic effects of the dual release of stromal cell-derived factor-1 and bone morphogenetic protein-2 from hydrogels on bone regeneration. Biomaterials. 32(11): 2797–811.

Shinohara, K., S. Greenfield, H. Pan, A. Vasanji, K. Kumagai, R.J. Midura, M. Kiedrowski, M.S. Penn and G.F. Muschler. 2011. Stromal cell-derived factor-1 and monocyte chemotactic protein-3 improve recruitment of osteogenic cells into sites of musculoskeletal repair. J Orthop Res. (7): 1064–9.

Sun, X., G. Cheng, M. Hao, J. Zheng, X. Zhou, J. Zhang, R.S. Taichman, K.J. Pienta and J. Wang. 2010. CXCL12/CXCR4/CXCR7 chemokine axis and cancer progression. Cancer Metastasis Rev. 29(4): 709–22.

Takahashi, M. 2010. Role of the SDF-1/CXCR4 system in myocardial infarction. Circ J. 74(3): 418–23.

Tashiro, K., H. Tada, R. Heilker, M. Shirozu, T. Nakano and T. Honjo. 1993. Signal sequence trap: a cloning strategy for secreted proteins and type I membrane proteins. Science. 261(5121): 600–3.

Tavor, S. and I. Petit. 2010 Can inhibition of the SDF-1/CXCR4 axis eradicate acute leukemia? Semin Cancer Biol. 20(3): 178–85. Epub 2010 Jul 15. Review.

Wei, L., K. Kanbe, M. Lee, X. Wei, M. Pei, X. Sun, R. Terek and Q. Chen. 2010. Stimulation of chondrocyte hypertrophy by chemokine stromal cell-derived factor 1 in the chondro-osseous junction during endochondral bone formation. Dev Biol. 341(1): 236–45.

Wynn, R., C. Hart, C. Corradi-Perini, L. O'Neill, C. Evans, J. Wraith et al. 2004. A small proportion of mesenchymal stem cells strongly expresses functionally active CXCR4 receptor capable of promoting migration to bone marrow. Blood. 104: 2643–5.

Zhu, W., G. Liang, Z. Huang, S.B. Doty and A.L. Boskey. 2011. Conditional inactivation of the CXCR4 receptor in osteoprecursors reduces postnatal bone formation due to impaired osteoblast development. J Biol Chem. 2011 Jul 29; 286(30): 26794–805.

CHAPTER 12

Chromatin Remodeling and Transcriptional Control of Mesenchymal Cells Towards the Osteogenic Pathway

Marom R.[a] and Benayahu D.[b],*

ABSTRACT

Stem cell commitment and differentiation into specialized functional cells involves the coordinated activation of different sets of genes. Key players that control these processes in embryogenesis and adulthood are chromatin remodeling proteins and transcription factors. Tissue specific transcription factors direct the activation and regulation of lineage fate differentiation. Nevertheless these processes also depend on epigenetic regulation, namely-chromatin remodeling and structural changes that allow transcription factor-binding to functional elements at promoter sites. Understanding the mechanism of action of the different regulatory factors will allow resolving how chromatin remodeling and gene activation come together to regulate stem cell lineage fate decision. Consequently, one of the important challenges of today's research

Department of Cell and Developmental Biology, Sackler School of Medicine, Tel-Aviv University, Israel.
[a]E-mail: ronitm26@gmail.com
[b]E-mail: dafnab@post.tau.ac.il
*Corresponding author

List of abbreviations given at the end of the text.

is to characterize molecular pathways that coordinate the lineage specific function of such factors. In skeletal tissue, marrow stromal cells differentiate into osteoblasts under the well-coordinated action of bone specific regulators, such as Runx2 and Osterix, along with SWI/SNF (mating type switching/sucrose non fermenting) components, members of the CHD (chromodomain-helicase DNA binding) and HDAC (histone deacetylases) families of chromatin remodeling proteins. Recent studies demonstrate the functional interaction between chromatin remodeling and transcription factors during regulation of osteogenesis that was revealed using different cell culture and animal based models. Understanding the mechanisms of action of such factors will allow controlling cell differentiation, and utilizing stem cells for regenerative medicine of skeletal tissues.

Introduction

Adult stem cells possess self-renewal capacities which allow maintaining a reservoir of multipotent cells that differentiate to various types of cells or organs. Stem cells are essence for supporting tissue homeostasis and regeneration and here mesenchymal stem cells will be discussed. The mesenchymal stem cells function are supportive stroma in the bone marrow and also differentiate into lineage-specific pathways. The stroma component of bone marrow supports hematopoiesis and bone formation. During embryogenesis and post-natal development mesenchymal cells differentiate through unknown stages into several cell lineages, including osteoblasts, chondrocytes, hematopoietic-supportive fibroblasts, and adipocytes (Benayahu 2000; Benayahu et al. 1989, 2009; Friedenstein et al. 1978).

Control of Mesenchymal Cells' Differentiation

The differentiation of mesenchymal stem cells into the osteoblastic lineage is characterized by the expression of osteoblast-specific genes. The regulation of tissue specific gene expression or repression depends on changes in chromatin structure that control binding of master regulator genes to tissue specific promoters. Switching on and off distinct sets of genes to achieve lineage-specific differentiation is attained through functional interactions between transcription factors, chromatin remodelers and regulatory elements on target genes. Lately, growing attention is paid to the part that chromatin remodeling plays in lineage fate commitment (Benayahu et al. 2007, 2009; Flowers et al. 2009; Villagra et al. 2006; Young et al. 2005). The building block of chromatin is the nucleosome that is composed of DNA wrapped around eight proteins (i.e., the histone octamer). The degree of chromatin condensation has a regulatory role on gene expression. When regulatory elements of a gene are tightly compacted they are not accessible to the transcriptional machinery and this gene will remain silent. If the DNA is loosely wrapped, however, gene promoters are exposed to transcription factors and transcription may commence. Thus, stem cell fate is controlled by cues from their local niche that affect chromatin remodeling factors and epigenetic regulation of gene expression (Benayahu et al. 2007, 2009). However, it is yet unclear under what circumstances chromatin remodeling complexes are recruited to a specific promoter to allow tissue-selective gene transcription. Consequently, the challenge is to characterize the molecular pathways that coordinate lineage-specific transcription with chromatin-modifying factors, in order to achieve proper understanding of the specific pattern of gene expression. The commitment of adult mesenchymal stem cells in the bone marrow (BM-MSC) to differentiate into the osteogenic phenotype is regulated by chromatin remodeling and master transcription factors that orchestrate the expression of additional tissue-specific genes as detailed in this review.

In vitro Model Systems to Study Osteogenesis

Osteogenic potential was demonstrated *in vitro*, and bone formation *in vivo*, for primary stromal cells and cell lines (Benayahu et al. 1989; Friedenstein et al. 1978; Siggelkow et al. 1999). Two main *in vitro* models are used to study osteogenesis 1) Primary cell cultures, and 2) Immortalized cell lines (Table 12.1). Primary cultured marrow stromal cell derived from an individual subject make a good model for cell "healthy" development, but will undergo senescence following certain amount of population doublings in culture. The immortalized cell lines (such as U2-OS, SaOS2, MC3T3) have unlimited lifespan, which makes them a more convenient research tool,

Table 12.1 Cell culture models of osteoblast differentiation.

Cell type	Source
Primary bone cell cultures	Human (HOb cells, TBC, MSC), rat (rat calvaria cells), mouse (mouse calvaria cells)
hFOB 1.19	Immortalized human pre-osteoblasts
MC3T3, MBA-15, CH3T10/2	mouse pre-osteoblasts
ROS (rat); U2-OS, MG-63, SaOS2 (human)	Osteosarcoma cells

Different cell culture models for osteoblast development.

Hob: human osteoblasts; TBC: trabecular bone cells; MSC: marrow stromal cells; FOB: human fetal osteoblastic cells; MC3T3: mouse calvaria embryonic cells; MBA-15: mouse bone marrow-derived osteogenic cells; CH3T10/12: mouse embryonic fibroblasts; ROS: rat osteosarcoma cells; U2-OS, MG-63 and SaOS: human osteosarcoma cells.

but while going through numerous divisions they acquire mutations and modifications that might alter their original characteristics. We discuss the results of several studies that utilized both types of *in vitro* culture models, in addition to animal models (mice and rats), to analyze transcription regulation in osteogenesis.

Chromatin Remodeling Associated with Osteogenesis

Stem cells have the ability to self-renew and differentiate into lineage specific cells, depending on the changes in chromatin structure and the action of trasncription factors. The major transcription factors in osteogenesis are Runx2 and Osterix (Osx) (Ducy et al. 1997; Nakashima et al. 2002) that act in concert with chromatin remodeling factors to promote transcription activation or repression. The recruitment of these factors along with RNA polymerase II is regulated by cell signaling pathways, transcription factors availability, and chromatin structure. Epigenetic regulation comprise of dynamic chromatin remodeling from the compacted "heterochromatin" form which is inaccessible to transcription, to the activated "euchromatin" form which is open for gene expression, and vice versa. The remodeling of chromatin structure facilitates the access of transcription factors to DNA by repositioning nucleosomes at the promoter region (Benayahu et al. 2009; Narlikar et al. 2002; Orphanides and Reinberg 2002). Chromatin-remodeling proteins (Table 12.2) enable the access of transcription factors to promoters and regulate gene expression by providing promoter-specific targeting of hormone and growth factor receptors. The chromatin accessibility is achieved through two different mechanisms: (1) disruption of nucleosome stability in ATP-dependent manner, and (2) DNA methylation and covalent modification of histones.

Table 12.2 Chromatin remodeling activity associated with osteogenesis.

Example of regulated genes	Protein family	Remodeling factor
Osteocalcin, alkaline phosphatase, collagen type I	SWI/SNF ATPases	BRG1/BRM
Runx2, biglycan, osteocalcin	CHD (chromodomain/helicase/DNA binding) proteins	CHD9
Runx2	Histone deacetylases	HDAC1, HDAC3, HDAC4
Osteopontine, Dlx5, Osx	DNA methylases	Undefined

Examples of chromatin remodeling factors that participate in the regulation of osteoblast differentiation. BRG1: brahma-related gene 1; BRM: brahma gene, SWI/SNF: mating type switching/sucrose non fermenting family; Runx2: runt-related transcription factor 2.

ATP dependent chromatin remodeling

ATP dependent chromatin remodelers act as part of multi-protein complexes in a wide range of regulatory pathways. They belong to the SNF2 family of proteins, and present ATPase activity with helicase-like motifs. They are further classified into three groups by their functional motifs: (a) the SWI/SNF (mating type switching/sucrose non-fermenting) proteins contain bromodomains which bind acetylated histone tails; (b) the ISWI (imitation switch) proteins contain SANT domains which bind unmodified histone tails; and (c) the CHD proteins contain chromodomains which were reported to bind methylated histone tails (Benayahu et al. 2007, 2009; Marfella et al. 2007; Wang et al. 2007).

The mammalian SWI/SNF complex can contain either BRG1 or BRM as the catalytic ATPase component. In osteoblasts, knock-down of either BRG1 or BRM subunits correlated with changes in gene expression and differentiation pattern (Flowers et al. 2009). In the osteoblast precursor cell line MC3T3 SWI/SNF complexes were recruited to osteoblast specific gene promoters, including osteocalcin, alkaline phosphatase and collagen type I promoters. In ROS cells SWI/SNF complex was shown to be recruited to the osteocalcin promoter by C/EBPβ, and when bound to the promoter—to interact with Runx2 and activate transcription. SWI/SNF also mediated vitamin D—induced osteocalcin transcription, thereby enhancing osteoblast differentiation (Villagra et al. 2006). SWI/SNF requirement for osteoblast differentiation was also demonstrated in C2C12 cells treated with BMP-2 that exhibited up regulation of Runx2 and alkaline phosphatase, along with changes in SWI/SNF component expression (Young et al. 2005). Immunofluorescence analysis in primary mouse calvaria cells showed co-localization of BRG1 and Runx2, which is another example of the functional interaction between chromatin remodeling and transcription factors regulating osteogenesis.

Key Facts of CHD Family of Proteins

- The CHD protein family members all contain three motifs: chromodomain, helicase and DNA binding domain.
- CHD proteins were previously shown to be essential to proper development in drosophila and mice (Benayahu et al. 2009; Marfella and Imbalzano 2007).
- In humans, CHD7 mutations are associated with CHARGE syndrome (Lalani et al. 2006) causing various congenital anomalies in the CNS, retina, heart, inner ear and nasal regions. Mutations in CHD7 were also demonstrated in patients with immunedeficiency (Gennery et al. 2008).

- Human CHD3 was recently associated with Hodgkin's lymphoma, and human CHD5 with neuroblastoma (Marfella and Imbalzano 2007).
- Therefore, evidence exists that the CHD family is strongly associated with cell fate decisions during development.

The CReMM/CHD9, a member of the CHD protein family, was identified primarily in mesenchymal cells and is expressed differentially in skeletal tissue, in osteoprogenitors in mice embryos at 16.5 p.c. (Marom et al. 2006; Shur and Benayahu 2005; Shur et al. 2006a; Shur et al. 2006b). CReMM/CHD9 protein belongs to the third CHD subfamily, based on its additional SANT and BRK domains. The protein also contains other motifs such as A/T hook like DNA binding domain and nuclear receptor binding domain. The sequence of CReMM/CHD9 is highly conserved through drosophila (kismet), mice, rat and human. Studies done in our laboratory further elucidate its structure and function in mouse and human tissue. Immunoprecipitated protein from mesenchymal cells catalyzed ATP hydrolysis in a DNA dependent manner (Shur and Benayahu 2005). Recombinant CReMM/CHD9 fragment containing the A/T hook like DNA binding domain interacted with A/T rich DNA in electrophoresis mobility shift assay (Shur and Benayahu 2005). During development CReMM/CHD9 was identified primarily in multipotential MSC population and osteoprogenitors, such as MBA-15 cells (Benayahu et al. 1989), and is absent in mature bone cells population (Marom et al. 2006; Shur et al. 2006b). In human bone marrow stromal cells CReMM/CHD9 was found to bind osteogenesis-specific gene promoters, such as Runx2, biglycan, and osteocalcin—as demonstrated by chromatin immunoprecipitation (ChIP) analysis (Shur et al. 2006a).

Covalent modification of DNA and histones

Histone deacetylases (HDACs) affect chromatin structure and transcription factor activity by determining accessibility of transcription factors to the DNA. They are known to deacetylate histones and also non-histone proteins, including Runx2 and P53 (McGee-Lawrence et al. 2011). HDAC1, HDAC3 and HDAC4 interaction with Runx2 was shown to suppress osteoblast differentiation by repressing its transcriptional activity. On the contrary, inhibition of HDAC activity *in vivo*, by introducing HDAC inhibitors such as Valproic Acid—decreases bone mineral density and increases fracture risk (McGee-Lawrence et al. 2011). Epigenetic regulation at the DNA methylation level had been demonstrated in MSCs. In bone marrow derived MSCs mechanical stimulation induced hypomethylation of the osteopontine promoter region, and up-regulated osteopontine mRNA expression (Arnsdorf et al. 2010). The promoters of Dlx5 and Osx osteogenic transcription factors were shown to be differentially—methylated according to cell-specific fate, thus enabling the expression of these factors selectively in osteogenic precursors (Lee et al. 2006).

Transcriptional Control of Osteogenic Cell Differentiation

MSCs will differentiate to osteoblasts under proper stimuli. The potential of cells to differentiate to bone-forming cells relies upon molecular regulation. During osteoblast development the combined action of transcription factors, which bind to specific DNA sequences, determines gene expression pattern. Several classes of transcription factors have been identified in osteoblast development, including homeodomain proteins, steroid receptors, and Runx2 (Table 12.3; reviewed in Kobayashi and Kronenberg 2005). Expression of a specific set of genes, in a specific sequence, dictates the transition between osteoblast differentiation stages, from proliferation to matrix maturation and finally to mineralization.

Early osteoblast-committed MSCs express the homeodomain proteins Dlx3, Dlx5 and Msx2, which selectively regulate osteogenic gene expression, such as osteocalcin and bone sialoprotein (Singh et al. 2012). During osteoblast development they either up or down regulate the transcription of osteoactivin, an osteoblast-differentiation marker, depending on cell maturation state. They occupy the osteoactivin promoter in response to BMP-2 signaling, which in turn influences transcription of other osteoblast related genes. Members of the Sox family high mobility group transcription factors have also been associated with early osteoblastogenesis. Sox8-deficient mice showed premature maturation of osteoblasts, which resulted in osteopenic phenotype (Schmidt et al. 2005). Sox2 was shown to promote

Table 12.3 Transcription factors associated with osteogenesis.

Transcription factor	Protein family	Example of regulated genes
Runx2		Osx, osteocalcin, NELL-1
Osx		osteocalcin, alkaline phosphatase, NELL-1, VDR
Dlx3, Dlx5, Msx2	Homeodomain proteins	osteoclacin, osteoactivin, bone sialoprotein
Sox2, Sox8	Sox high mobility group transcription factors	cell cycle & mitosis-related factors
cFos		alkaline phosphatase, osteopontine, collagen type I
Vitamin D receptor	Nuclear hormone receptors	osteocalcin, osteopontine

Transcription factors that are active in the regulation of osteoblast differentiation. Runx2: runt-related transcription factor 2; Osx: osterix; NELL-1: Nel like molecule 1; VDR: vitamin D receptor; Dlx: homeodomain transcription factors related to the Dll (distal-less) gene in drosophila; Msx2: Msh–homeobox 2 protein; Sox: Sry-related HMG box protein.

osteoprogenitor self-renewal by activating transcription of "stemness" genes (such as cell cycle and mitosis—related factors), and by inhibiting the Wnt differentiation pathway (Seo et al. 2011). Thus, both Sox2 and Sox8 are associated with increase in the osteoblast-precursor pool that will later acquire a mature bone-forming cell phenotype.

Runx2 is a well-studied master regulator of osteogenesis. Its absence is associated with severe bone abnormalities (Ducy et al. 1997; Otto et al. 1997). The osteoblast specific transcription factor osterix (Osx) is a downstream target gene of Runx2. Osx is thought to be necessary to the final mineralization stage. Osx knock-out mice fail to express osteoblastic markers such as osteocalcin and alkaline phosphatase, and do not have mineralized bone (Nakashima et al. 2002). NELL-1 (Nel-like molecule-1) is a osteogenesis-promoting factor associated with craniosynostosis, and is a direct transcriptional target of both Runx2 and Osx (Chen et al. 2011). Recently, Osx was shown to enhance vitamin D receptor (VDR) expression through specific binding to its promoter (Zhang et al. 2011). VDR belongs to the family of steroid hormone receptor transcription factors, and regulates osteoblast specific genes osteocalcin and osteopontine, as well as genes associated with coagulation and immune functions in osteoblast cells (Tarroni et al. 2012). cFOS transcription factor ply a role in osteoblast development (Shur et al. 2005). Overexpression of cFos is associated with osteosarcoma, and its deficiency causes osteopetrosis (Matsuo et al. 1999). These factors regulate the transcription of genes through specific binding to promoters: Runx2 is known to regulate the transcription of osteocalcin, and cFos binds the promoters of alkaline phosphatase, osteopontine, and collagen I (Cowels et al. 2000). Molecular interaction between the leucine zipper domain of cFos and runt domain of Runx2 enables these factors to cooperate and regulate transcription of other genes related to bone remodeling, such as collagenase 3/MMP 13 (Hess et al. 2001).

Several transcription regulators that act in mesenchymal progenitors were recognized as *negative* regulators of osteoblast phenotype by directing MSC into different cell lineages. For example, PPARγ—a member of the nuclear receptor family, is associated with MSC differentiation towards adipogenesis vs. osteogenesis. Activation of PPARγ in bone marrow suppressed the expression of osteogenic transcription factors as Dlx5, Runx2 and Osx. A shift of MSC differentiation towards osteogenic phenotype was observed with PPARγ haploinsufficiency in mice (Kawai and Rosen 2010). Similarly, C/EBP (CCAAT/enhancer binding protein) β factors, which belong to the CREB/ATF family of transcription factors, play a role in adipogenesis. Recently, C/EBPβ was shown to reduce osteoblast proliferation by down-regulation of Ric-8B gene (Grandy et al. 2011). Such differential activity emphasizes the well-controlled regulation of tissue specific gene expression.

Transcriptional Control of Abnormal Cell Differentiation in Osteosarcoma

Aberrant differentiation of osteoprogenitors may result in osteosarcoma, which is the most common primary malignant tumor of bone and predominates in adolescence. The neoplastic cells are derived from mesenchymal bone marrow cells, have high proliferation capacity and produce an abnormal osteoid substance or sclerotic bone that result in bone lesions. The understanding of osteosarcoma pathogenesis can serve as a model for improper cell differentiation, and thus shed light on the normal development. Indeed, osteosarcoma cell lines (including U2-OS, MG-63, and SAOS-2) have been used as experimental models for osteoblast differentiation. However, malignant growth dictates multiple cellular changes that need to be thought of when implicating the data based upon these cells on the phenotype and characteristics of normal osteoblasts. In a study that compared the expression of transcription factors (cFos, cJun, cMyc), cytokines (IL-6, IL-11) and extracellular matrix proteins (osteocalcin, osteonectin, biglycan) between osteosarcoma cells and normal osteoblasts from marrow stroma or trasbecular bone (Benayahu et al. 2001). Similar levels of expression were observed for all mentioned genes, except for biglycan which was not detected in either U2OS, SAOS-2 cells, or in osteosarcoma biopsies. The absence of biglycan expression in osteosarcoma may affect matrix composition and organization in a way that will compromise osteoblast maturation and impair mineralization (Benayahu et al. 2001). Another study demonstrated disruption of Runx2 function in osteosarcoma cell lines (Thomas et al. 2004). The factors contributing to osteosarcoma pathogenesis are poorly understood. Cytogenetic and molecular studies of osteosarcoma reveal extremely complex karyotypes, including numerical chromosomal gains and losses, deletions and translocations. Needless to say that such change of DNA sequence is expected to affect transcription regulation at both the chromatin and transcription factor levels. One of the frequently aberrated chromosomal regions is 8q24 that contains the RECQL4 gene locus. Patients with Rothmand-Thomson syndrome who carry the RECQL4 mutations are highly prone to develop osteosarcoma (Maire et al. 2009). Curiously, RECQL4 encodes a DNA helicase involved in chromatin remodeling, and was also linked to the retinoblastoma pathway. Mutations of the retinoblastoma gene are associated with osteosarcoma. pRB was found to co-activate Runx2 and associate with osteocalcin, alkaline phosphatase and osteopotine promoters. In MC3T3-E1 cells pRB was shown to be essential for SWI/SNF recruitment to the alkaline phosphatase promoter, thereby shifting cells towards differentiation (Flowers et al. 2010). In osteoblasts the inverse relation between proliferation and differentiation, which involves cell-cycle arrest, may explain why pRB-deficient cells are more prone to osteosarcoma progression.

Summary

Bone formation requires the well-orchestrated action of transcription factors, signaling pathways and epigenetic regulators that control mesenchymal cell differentiation. The osteogenic cell fate is mandated by several master regulatory genes that are typical of each stage of differentiation. Bone specific transcription factors are further regulated by chromatin remodeling proteins which determine DNA accessibility. The signaling milieu which allows cells differentiation may change in aging or under de-regulated metabolic state and may alter stem cells fate by diverting them from entering towards osteogensis. The availability of MSCs and their potential to differentiate into the fibroblasts, adipocytes, myocytes, chondrocytes and osteoblasts cell types—make them attractive for future use in tissue engineering, in treating degenerative and age-related diseases. Thus, revealing the pathways of osteogenic development at the molecular level will enable further understanding of skeletal formation in health and disease, and future development of diagnostic approaches and therapeutic strategies.

Summary Points

- Bone specific gene expression or repression depends on changes in chromatin structure that control binding of master regulator genes to tissue specific promoters.
- Understanding the functional interactions between transcription factors, chromatin remodelers and regulatory elements on bone specific genes is a great challenge in the research of osteoblast development.
- *In vitro* models used to study osteogenesis include adherent cells originating from culturing of marrow stromal cells, and immortalized cell lines.
- Chromatin-remodeling proteins enable the access of transcription factors to promoters through two different mechanisms: (1) disruption of nucleosome stability in ATP-dependent manner, and (2) DNA methylation and covalent modification of histones.
- CReMM/CHD9 belongs to the CHD (chromdomain-helicase-DNA binding) family of ATPases, is expressed in osteoprogenitors and was

found to bind osteogenesis-specific gene promoters, such as Runx2, biglycan, and osteocalcin.
- Other chromatin remodelers known to play a role in osteoblast development include BRG1/BRM components of SWI/SNF complexes, histone deacetylases and DNA methylases.
- The regulation of osteoblast differentiation is well-controlled under the action of tissue specific master genes, including Runx2, Osx, homeodomain proteins and Sox transcription factors.
- Abnormal cell differentiation in osteosarcoma can lead to better understanding of the proper regulation of osteoblast biology.

Dictionary

- *SWI/SNF*: mating type switching/sucrose non fermenting family of chromatin remodeling proteins containing ATPase catalytic subunit.
- *CHD*: chromodomain/helicase/DNA binding chromatin remodeling proteins containing ATPase catalytic subunit.
- *HDAC*: histone deacetylase family of chromatin remodeling proteins which deactylate histone as well as non-histone proteins to regulate gene transcription.
- *BM-MSC/MSCs*: stromal cells derived from bone marrow and function as mulipotent stem cells that are able to differentiate to multiple lineages.
- *Osx*: Osterix, a master-regulator gene essential in osteoblast development.
- *C/EBPβ*: CCAAT/enhancing binding protein β, a key transcription factor in mesenchymal cell differentiation to the osteoblast and adipocyte lineages.
- *CHARGE syndrome*: an association of multiple congenital anomalies, namely coloboma of the eye, heart defects, choanal atresia, growth retardation, genitourinary anomalies, ear anomalies and deafness. Results from CHD7 gene mutations.
- *CReMM/CHD9*: chromatin remodeling mesenchymal modulator, CHD9 is a member of the CHD family of remodeling proteins that is expressed in differentiating pre-osteoblasts.
- *ChIP*: chromatin immunoprecipitation, a laboratory method designed to study DNA-protein interactions at promoter regions.
- *NELL1*: Nel like molecule 1 a transcription factor that is a downstream target of Runx2 and Osx in osteoblast differentiation.
- *VDR*: vitamin D receptor, a transcription factor belong to the nuclear (steroid) receptor family and mediates vitamin D transcriptional regulation.

- *PPARγ*: peroxisome proliferator activated receptor γ is a transcription factor that belongs to the nuclear receptor family of proteins, and is a key regulator of adipocyte development. pRB—the retinoblastoma protein encoded by retinoblastoma gene, a tumor suppressor gene. The pRB recruits chromatin remodeling factors to promoter regions on the DNA to regulate transcription.

List of Abbreviations

BM-MSC/ MSCs	:	bone marrow stromal cells/marrow stromal cells
C/EBPβ	:	CCAAT/enhancing binding protein β
CHD	:	chromodomain/helicase/DNA binding
CHARGE syndrome	:	Coloboma of the eye, Heart defects, choanal Atresia, growth Retardation, Genitourinary anomalies, Ear anomalies
ChIP	:	Chromatin Immuno Precipitation
CReMM/ CHD9	:	chromatin remodeling mesenchymal modulator, CHD9
HDAC	:	histone deacetylase
NELL1	:	Nel like molecule 1
Osx	:	Osterix
PPARγ	:	peroxisome proliferator activated receptor γ
pRB	:	the retinoblastoma protein
Runx2	:	runt related transcription factor 2
SWI/SNF	:	mating type switching/sucrose non fermenting
VDR	:	vitamin D receptor

References

Arnsdorf, E.J., P. Tummala, A.B. Castillo, F. Zhang and C.R. Jacobs. 2010. The epigenetic mechanism of mechanically induced osteogenic differentiation. J Biomech. 43(15): 2881–6.

Benayahu, D. 2000. The hematopoietic microenvironment: the osteogenic compartment of bone marrow: cell biology and clinical application. Hematology. 4(5): 427–35.

Benayahu, D., Y. Kletter, D. Zipori and S. Wientroub. 1989. Bone marrow-derived stromal cell line expressing osteoblastic phenotype *in vitro* and osteogenic capacity *in vivo*. J Cell Physiol. 140(1): 1–7.

Benayahu, D., I. Shur, R. Marom, I. Meller and J. Issakov. 2001. Cellular and molecular properties associated with osteosarcoma cells. J Cell Biochem. 84(1): 108–14.

Benayahu, D., N. Shacham and I. Shur. 2007. Insights into the functional role of chromatin remodelers in osteogenic cells. Crit Rev Eukaryot Gene Expr. 17(2): 103–13.

Benayahu, D., G. Shefer and I. Shur. 2009. Insights into the transcriptional and chromatin regulation of mesenchymal stem cells in musculo-skeletal tissues. Ann Anat. 191(1): 2–12.

Chen, F., X. Zhang, S. Sun, J.N. Zara, X. Zou, R. Chiu, C.T. Culiat, K. Ting and C. Soo. 2011. NELL-1, an osteoinductive factor, is a direct transcriptional target of osterix. PLOS One. 6(9): e24638.

Cowles, E.A., L.L. Brailey and G.A. Gronowicz. 2000. Integrin-mediated signaling regulates AP-1 transcription factors and proliferation in osteoblasts. J Biomed Mater Res. 52(4): 725–37.

Ducy, P., R. Zhang, V. Geoffroy, A. Ridall and G. Karsenty. 1997. Osf2/Cbfa1: a transcriptional activator of osteoblast differentiation. Cell. 89(5): 747–54.

Flowers, S., N.G. Nagl, G.R. Beck and E. Moran. 2009. Antagonistic roles for BRM and BRG1 SWI/SNF complexes in differentiation. J Biol Chem. 284(15): 10067–75.

Flowers, S., G.R. Beck and E. Moran. 2010. Transcriptional activation by pRB and its coordination with SWI/SNF recruitment. Cancer Res. 70(21): 8282–7.

Friedenstein, A., A. Ivanov-Smolenski, R. Chajlakjan, U. Gorskaya, A. Kuralesova, N. Latzinik and U. Gerasimow. 1978. Origin of bone marrow stromal mechanocytes in radiochimeras and heterotopic transplants. Exp Hematol. 6(5): 440–4.

Gennery, A.R., M.A. Slatter, J. Rice, L.H. Hoefsloot, D. Barge, A. McLean-Tooke, T. Montgomery, J.A. Goodship, A.D. Burt, T.J. Flood, M. Abinun, A.J. Cant and D. Johnson. 2008. Mutations in CHD7 in patients with CHARGE syndrome cause T-B+natural killer + severe combined immune deficiency and may cause Omenn-like syndrome. Clin Exp Immunol. 153(1): 75–80.

Grandy, R., H. Sepulveda, R. Aguilar, P. Pihan, B. Henriquez, J. Olate and M. Montecino. 2011. The Ric-8B gene is highly expressed in proliferating preosteoblastic cells and downregulated during osteoblast differentiation in a SWI/SNF- and C/EBPbeta-mediated manner. Mol Cell Biol. 31(14): 2997–3008.

Hess, J., D. Porte, C. Munz and P. Angel. 2001. AP-1 and Cbfa/Runt physically interact and regulate parathyroid hormone-dependent MMP13 expression in osteoblasts through a new osteoblast-specific element 2/AP-1 composite element. J Biol Chem. 276(23): 20029–20038.

Kawai, M. and C.J. Rosen. 2010. PPARγ: a circadian transcription factor in adipogenesis and osteogenesis. Nat Rev Endocrinol. 6(11): 629–36.

Kobayashi, T. and H. Kronenberg. 2005. Minireview: Transcriptional regulation in development of bone. Endocrinology. 146(3): 1012–17.

Lalani, S.R., A.M. Safiullah, S.D. Fernbach, K.G. Harutyunyan, C. Thaller, L.E. Peterson, J.D.McPherson, R.A.Gibbs, L.D.White, M. Hefner, S.L.Davenport, J.M. Graham, C.A. Bacino, N.L. Glass, J.A. Towbin, W.J. Craigen, S.R. Neish, A.E. Lin and J.W. Belmont. 2006. Spectrum of CHD7 mutations in 110 individuals with CHARGE syndrome and genotype-phenotype correlation. Am J Hum Genet. 78(2): 303–14.

Lee, J.Y.,Y.M. Lee, M.J. Kim, J.Y. Choi, E.K. Park, S.Y. Kim, S.P. Lee, J.S. Yang and D.S. Kim. 2006. Methylation of the mouse Dlx5 and Osx gene promoters regulates cell-type specific gene expression. Mol Cells. 22(2): 182–8.

Marfella, C.G. and A.N. Imbalzano. 2007. The Chd family of chromatin remodelers. Mutat Res. 618(1-2): 30–40.

Maire, G., M. Yoshimoto, S. Chilton-MacNeill, P.S. Thorner, M. Zielenska. and J.A. Squire. 2009. Recurrent RECQL4 imbalance and increased gene expression levels are associated with structural chromosomal instability in sporadic osteosarcoma. Neoplasia. 11(3): 260–8.

Marom, R., I. Shur, G.l. Hager and D. Benayahu. 2006. Expression and regulation of CReMM/CHD9, a chromodomain helicase-DNA-binding (CHD), in marrow stroma derived osteoprogenitors. J Cell Physiol. 207(3): 628–35.

Matsuo, K., W. Jochum, J. Owens, T. Chambers and E. Wagner. 1999. Function of Fos protein in bone cell differentiation. Bone 25: 141.

McGee-Lawrence, M.E. and J.J. Westendorf. 2011. Histone deacetylases in skeletal development and bone mass maintenance. Gene. 474(1-2): 1–11.

Nakashima, K., X. Zhou, G. Kunkel, Z. Zhang, J.M. Deng, R.R. Behringer and B.de Crombrugghe. 2002. The novel zinc finger-containing transcription factor osterix is required for osteoblast differentiation and bone formation. Cell. 108(1): 17–29.

Narlikar, G., H. Fan and R. Kingston. 2002. Cooperation between complexes that regulate chromatin structure and transcription. Cell. 108(4): 475–87.

Orphanides, G. and D. Reinberg. 2002. A unified theory of gene expression. Cell. 108(4): 439–51.

Otto, F., A. Thornell, T. Crompton, A. Denzel, K. Gilmour, I. Rosewell, G. Stamp, R. Beddington, S. Mundlos, B.R. Olsen, P.B. Selby and M.J. Owen. 1997. Cbfa1, a candidate gene for cleidocranial dysplasia syndrome, is essential for osteoblast differentiation and bone development. Cell. 89(5): 765–71.

Schmidt, K., T. Schinke, M. Haberland, M. Priemel, A. Schilling, C. Mueldner, J. Rueger, E. Sock, M. Wegner and M. Amling. 2005. The high mobility group transcription factor Sox8 is a negative regulator of osteoblast differentiation. J Cell Biol. 168(6): 899–910.

Seo, E., U. Basu-Roy J. Zavadil, C. Basilico and A. Mansukhani. 2011. Distinct Functions of Sox2 control self-renewal and differentiation in the osteoblast lineage. Mol Cell Biol. 31(22): 4593–608.

Shur, I. and D. Benayahu. 2005. Characterization and functional analysis of CReMM/CHD9, a novel chromodomain helicase DNA-binding protein. J Mol Biol. 352(3): 646–55.

Shur, I., R. Socher and D. Benayahu. 2005. Dexamethasone regulation of cFos mRNA in osteoprogenitors. J Cell Physiol. 202(1): 240–5.

Shur, I., R. Solomon and D. Benayahu. 2006a. Dynamic interactions of chromatin-related mesenchymal modulator, a chromodomain helicase-DNA-binding protein, with promoters in osteoprogenitos. Stem Cells. 24(5): 1288–93.

Shur, I., R. Socher and D. Benayahu. 2006b. *In vivo* association of CReMM/CHD9 with promoters in osteogenic cells. J Cell Physiol. 207(2): 374–8.

Siggelkow, H., K. Rebenstorff, W. Kurre, C. Niedhart, I. Engel, H. Schulz, M. Atkinson and M. Hufner. 1999. Development of the osteoblast phenotype in primary human osteoblasts in culture: comparison with rat calvarial cells in osteoblast differentiation. J Cell Biochem. 75(1): 22–35.

Singh, M., F.E. Del Carpio-Cano, M.A. Monroy, S.N. Popoff and F.F. Safadi. 2012. Homeodomain transcription factors regulate BMP-2-induced osteoactivin transcription in osteoblasts. J Cell Physiol. 227(1): 390–9.

Tarroni, P., I. Villa, E. Mrak, F. Zolezzi, M. Mattioli, C. Gattuso and A. Rubinacci. 2012. Microarray analysis of 1,25(OH)2D3 regulated gene expression in human primary osteoblasts. J Cell Biochem. 113: 640–9.

Thomas, D.M., S.A. Johnson, N.A. Sims, M.K. Trivett, J.L. Slavin, B.P. Rubin, P. Waring, G.A. McArthur, C.R. Walkley, A.J. Holloway, D. Diyagama, J.E. Grim, B.E. Clurman, D.D. Bowtell, J.S. Lee, G.M. Gutierrez, D.M. Piscopo, S.A. Carty and P.W. Hinds. 2004. Terminal osteoblast differentiation, mediated by runx2 and p27KIP1, is disrupted in osteosarcoma. J Cell Biol. 167(5): 925–34.

Villagra, A., F. Cruzat, L. Carvallo, R. Paredes, J. Olate, A.J. van Wijnen, G.S. Stein, J.G. Lian, J.L. Stein, A.N. Imbalzano. and M. Montecino. 2006. Chromatin remodeling and transcriptional activity of the bone-specific osteocalcin gene require CCAAT/enhancer-binding protein beta-dependent recruitment of SWI/SNF activity J Biol Chem. 281(32): 22695–706.

Wang, G.G., C.D. Allis and P. Chi. 2007. Chromatin remodeling and cancer, Part 1: covalent histone modification; Part 2: ATP–dependent chromatin remodeling. Trends Mol Med. 13(9): 363–80.

Young, D.W., J. Pratap, A. Javed, B. Weiner, Y. Ohkawa, A. van Wijnen, M. Montecino, G.S. Stein, J.L. Stein, A.N. Imbalzano and J.B. Lian. 2005. SWI/SNF chromatin remodeling complex is obligatory for BMP2-induced, Runx2-dependent skeletal gene expression that controls osteoblast differentiation. J Cell Biochem. 94(4): 720–30.

Zhang, C., W. Tang, Y. Li, F. Yang, D.R. Dowd and P.N. MacDonald. 2011. Osteoblast-specific transcription factor Osterix increases vitamin D receptor gene expression in osteoblasts. PLoS One. 6(10): e26504.

Section 3
Conditions, Applications, Treatments and Repairs

CHAPTER 13

Use of Chondrogenic Progenitor Cells in Osteoarthritis

Boris Schminke, Nicolai Miosge[a],***** and
Hayat Muhammad

ABSTRACT

It is thought that the general increase in life expectancy will make osteoarthritis the fourth leading cause of disability by the year 2020. Even though the pathogenesis of idiopathic osteoarthritis has not been fully elucidated, the main features of the disease process are the altered interactions between the chondrocytes and their surrounding extracellular matrix. In the course of these disturbances fibroblast-like chondrocytes take part in tissue regeneration especially in advanced stages of osteoarthritis. However, only fibrocartilaginous or scar tissue, since only collagen type I, and not collagen type II, typical for healthy cartilage, is synthesized. It remains a great challenge to enhance the regeneration potential of hyaline cartilage tissue. Tissue degeneration overrides the generally limited self-renewal capacity of this tissue.

Tissue regeneration work group, Medical Faculty, Department of Prosthodontics, Georg August University, Goettingen, D-37075, Germany.
[a]E-mail: nmiosge@gwdg.de
*Corresponding author

List of abbreviations given at the end of the text.

Adult mesenchymal stem cells, which are thought to be capable of repairing injured tissue can be differentiated into chondrocyte-like cells *in vitro*. During embryonic development, some cells of the inner cell mass will develop into the mesoderm. This will be the founder of the mesenchymal cells in connective tissues of adult life, such as bone, tendon, muscle, and cartilage. Some of these embryonic mesenchymal cells are believed not to differentiate, but reside in each of the tissues. These are now collectively described as adult mesenchymal stem cells, which are thought to be capable of repairing injured tissue. We found that repair tissue from human articular cartilage during the late stages of osteoarthritis harbors a unique progenitor cell population, termed chondrogenic progenitor cells (CPC). These exhibit stem cell characteristics together with a high chondrogenic potential. They will be relevant in the development of novel therapeutic regenerative approach for a progenitor cell-based therapy of late stages of OA.

Introduction

Osteoarthritis (OA) is a chronic and mainly degenerative joint disease. Degeneration is progressive and the loss of articular cartilage finally leads to the eburnation of the subchondral bone (Fig. 13.1). The process is accompanied by an inflammatory synovial reaction (Poole et al. 1993). OA is the most common musculoskeletal disease in the elderly, according to Reginster (2002) up to 1.75 million people alone in England and Wales suffer from symptomatic OA. However, the number of asymptomatic cases is estimated to be much higher. There is a strong association between its prevalence and increasing age, since up to 20 percent of the population over 60 yr of age show signs of OA (Haq et al. 2003). The severity of OA also increases indefinitely with age and up to now the condition is not reversible (Woolf and Pfleger 2003). As OA often remains asymptomatic until late in the disease progress and early markers as reliable tools of diagnosis are still lacking up to now, therefore, total knee replacement is the ultimate therapeutic intervention. This means that important parts of health care resources have to be spent on coping with this disease (Reginster 2002). The general increase in life expectancy and the resulting aging populations are expected to make OA the fourth leading cause of disability by the year 2020 (Woolf and Pfleger 2003). This warrants the further elucidation of the pathogenesis of OA with the final goal of gaining insight into the disease processes to render a cell biological therapy possible and within reach. Regenerative medicine and tissue engineering approaches are being investigated and developed further. Here, we review the current knowledge on progenitor cells and their possible usage in future therapies of osteoarthritis.

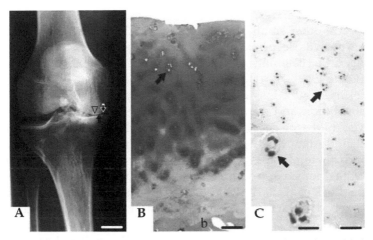

Figure 13.1 Adapted from Bock et al. 2001 (with permission from the publisher). (A) Radiograph of a patient with grade IV OA. Areas from which the cartilage samples were taken are marked (black arrow=main defect, open triangle=adjoining area and black-on-white arrow=macroscopically intact), bar=2 cm. (B) Alcian blue staining of a cartilage sample from the macroscopically intact area; note that chondrocytes are organized in clusters (black arrow), bar=50 μm. (C) Light microscopic in situ hybridization in a sample similar to (B), intracellular staining for biglycan mRNA (Black arrows), bar=50 μm, inset: higher magnification of two cell clusters.

Articular Cartilage: Chondrocytes and Matrix Composition in Health and Disease

Normal, healthy articular cartilage is a highly specialized and uniquely designed tissue, which covers the articulating ends of long bones (Kuettner 1992). It is an avascular, aneural and alymphatic tissue. The resilience, integrity and function of articular cartilage all depend on the composition of the abundant extracellular matrix (ECM) synthesized by the single cell type found in this tissue, the chondrocytes. These are responsible for the secretion and maintenance of the ECM of articular cartilage (Kuettner 1992). The abundant extracellular matrix of articular cartilage is composed of two major elements: the collagens and the proteoglycans. Normal articular cartilage contains types II, III, VI, IX, X, XI, XII and XIV collagens, the most abundant being collagen type II (Mayne and Brewton 1993). Collagens type II, IX and XI form fibrillar alloys with type XI collagen as core and type IX collagen on the outside possibly limiting the fiber diameter (Kuettner 1992). In addition, the proteoglycans, a heterogeneous group of proteins, consisting of a central core protein substituted with one or more glycosaminoglycan side chains constitute the other major extracellular matrix components. A few good examples would be, first of all, aggrecan, the large cartilage matrix proteoglycan responsible for cartilage tissue maintenance together with several other small proteoglycans, which are also important for its function, such as decorin, biglycan and fibromodulin.

A disturbed cell-matrix relationship lies at the center of the pathogenesis of OA (Poole 1999). The degradation of the tissue by matrix metalloproteases is underlined by a loss of the main proteoglycan, aggrecan, collagen fiber fibrillation and surface splits (Poole 1999; Martel-Pelletier 1999). However, this tissue degeneration is intermingled with regeneration efforts (Sandell and Aigner 2001), which might possibly be seen in the occurrence of chondrocyte clusters and are certainly seen in the appearance of fibrocartilaginous tissue with a more fibrillar matrix and a newly emerging cell type (Bock et al. 2001; Poole 1999). These cells were initially identified and described at the ultrastructural level and named elongated secretory type 2 cells (Fig. 13.2) and had an irregular shape with a prominent rough endoplasmic reticulum (Kouri et al. 1996). We called them fibroblast-like chondrocytes, which build the fibrocartilagenous tissue at the more advanced stages of OA (Sandell and Aigner 2001; Miosge et al. 1998). This repair tissue is mainly composed of collagen type I, whereas physiological articular cartilage reveals only collagen type II. Collagen type I protein has been detected at the light microscopic level in osteoarthritic cartilage with the help of immunohistochemistry. Recently, we identified a subpopulation of these cells as chondrogenic progenitor cells (Koelling et al. 2009; Koelling and Miosge 2010).

From Embryonic Stem Cells to Mesenchymal Cells and Adult Progenitor Cells

The original stem cells reside in the inner cell mass of the embryo proper at the stage of the blastocyst. Before this developmental stage, one could argue, that every cell of the morula is a stem cell. Obviously, cells of the morula are pluripotent and capable of developing into each of the three germ layers that will later, during the course of embryogenesis, develop into the specific tissues to form the organs (O'Rahilly et al. 1981). During early embryonic development, some of the embryonic stem cells of the inner cell mass of the blastocyst will turn into ectoderm. Derivatives of this germ layer will, for example, develop into skin and brain-tissues devoted to connecting organisms to the exterior world. Some will turn into endoderm. This germ layer will mainly form internal organs, for example, the gut and the liver. The third germ layer, the mesoderm, will develop from the ectoderm, known as the first ectoderm-mesenchyme transition. This mesoderm will be the founder of the mesenchymal cells later found in the connective tissues. In adult life, these tissues, such as bone, tendon,

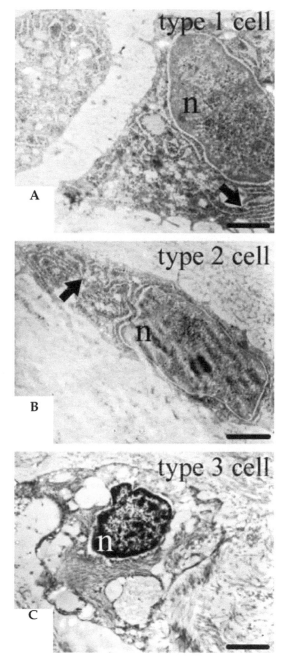

Figure 13.2 Adapted from Bock et al. 2001 (with permission from the publisher), Electron micrographs of the three chondrocyte types: (A) normal, (B) fibroblast-like, (C) degenerating, n = nucleus, black arrows = endoplasmic reticulum, bars = 0,7 μm.

muscle, and cartilage, are the building blocks of the skeletal system. Adult mesenchymal cells exhibit a profound plasticity. Adult differentiated cells like chondrocytes, which were kept in culture on plastic dishes for prolonged periods of time, dedifferentiate to an extent that allows for the reoccurrence of characteristics of stem cells (Dell'Accio et al. 2003). *In vitro*, adult skin fibroblasts have also been turned into stem cell-like cells with the help of the transfection of the transcription factors oct3/4, sox2, c-myc and klf4 (Takahashi and Yamanaka 2006). Mesenchymal stem cells are believed to be capable to regenerate diseased tissues. We would prefer to call them mesenchymal *progenitor* cells, as all those cells described *in vivo* and *in vitro* are migratory cells that have left their original stem cell niche and, therefore, belong to the transient amplifying pool of cells. *Per definitionem*, the stem cell resides in its niche composed of adjacent, more differentiated, cells and extracellular matrix molecules (Fuchs et al. 2004).

Friedenstein et al. (1970) was the first to describe fibroblast colonies derived from guinea-pig bone marrow and this paved the way to the further characterization of bone marrow stroma cells. To date, various populations of bone marrow stroma cells have been described as stem cells and differentiated into cartilage tissue *in vivo* and *in vitro*. Due to space limitation, we can only highlight a few of these studies here. Johnstone et al. (1998) described rabbit mesenchymal cells differentiated *in vitro* into a tissue staining positive for collagen type II protein. Mackay et al. (1998) also described the chondrogenic differentiation of mesenchymal stem cells derived from bone marrow to produce a chondrocyte-like extracellular matrix in pellet culture. Micro-mass culture of human bone marrow stromal cells with the addition of the chondrogenic mediators BMP-6 and TGFß$_3$ resulted in the formation of cartilage-like tissue *in vitro*. Gronthos et al. (2003) have also applied bone marrow derived mesenchymal stem cells partially characterized by their expression of STRO-1. The other marker related to stem cells that this group applied is CD106 or VCAM-1, which is an adhesion molecule, is also found on endothelial cells. Chondrogenesis of these STRO-1$^+$/CD106$^+$ cells was proven by the detection of mRNA for collagen type II, type X and aggrecan.

Stem cells from tissue sources other than bone marrow have also been described. Synovia-derived stem cells and cells isolated from the synovial fluid can be differentiated into cartilage-like tissue. The Hoffa fat pad in the knee joint has been described as the origin of stem cells driven into the chondrocyte lineage *in vitro* (English et al. 2007).

Up until now, only one study in a goat animal model has described the use of mesenchymal stem cells to treat an osteoarthritic defect. In this case, intra-articular injections of mesenchymal stem cells resulted in minor improvement of the disease process. However, the cells migrated to all of the tissues of the knee, except the cartilage tissue itself (Murphy et al. 2003).

Chondrogenic Progenitor Cells for Cartilage Repair

Especially in connective tissues, adult stem cell-like cells have been long known to be responsible for tissue repair after injury. In muscle, stellate cells are found and the broken bone heals via activation of mesenchymal cells derived from the inner layers of the periost, the connective tissue surrounding each bone. There is evidence that mesenchymal cells characterized by their surface antigens are found in osteoarthritic cartilage tissue. Alsalameh et al. (2004) isolated CD105$^+$ and CD166$^+$ cells from osteoarthritic cartilage tissue by enzymatic digestion and drove them into cartilage-like tissue with the help of micro-mass culture *in vitro*. In addition Fickert et al. (2004) isolated cells, this time positive for CD9, CD90 and CD166 and were able to demonstrate their differentiation into such a tissue. Moreover, microfracture and Pridie drilling to open the bone marrow underneath the cartilage defect are still used as a therapeutic option and result in a fibrocartilaginous repair tissue. This repair tissue is thought to originate from migrating mesenchymal cells (Fig. 13.3). However this regeneration tissue exhibits

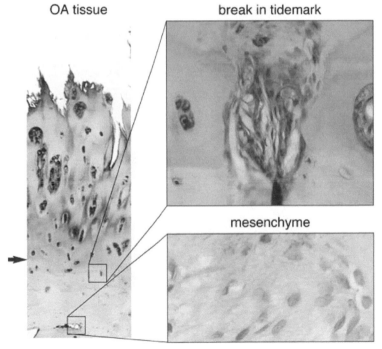

Figure 13.3 The diseased cartilage tissue exhibits deep surface fissures, chondrocytes in clusters and breaks in the tidemark (arrow), taken from Koelling et al. 2009 (with permission from the publisher).

less resistance to mechanical stress and is composed mainly of collagen type I, which is not typically present in healthy articular cartilage tissue (Miosge et al. 1998). Finally, postnatal stem cells have been identified in the superficial zone of healthy bovine cartilage believed to be responsible for the appositional growth of the joint surface (Dowthwaite et al. 2004).

Progenitor Cells in Repair Tissue of Late-stage Osteoarthritis

Physiological repair mechanisms of diseased hyaline cartilage tissues are sparse and overridden by matrix destruction resulting in less functional fibrocartilaginous, collagen type I-rich scar tissue (Koelling and Miosge 2009). Despite the evidence that stem cells might be involved in regeneration activities seen in osteoarthritis, no studies to date have identified an already committed chondrogenic progenitor cell population in late-stage osteoarthritis. We identified migratory cells derived from repair tissue of late-stage osteoarthritis (Fig. 13.4) which possess a high chondrogenic potential and progenitor cell characteristics. We called them

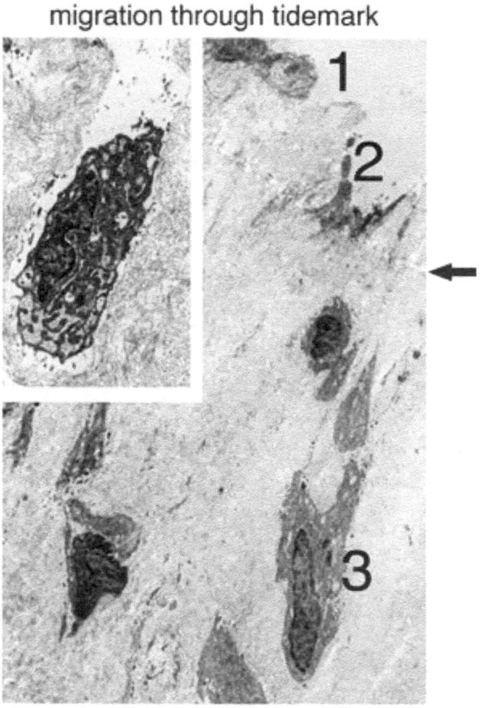

Figure 13.4 A brake in the tidemark with mesenchymal tissue and blood vessels entering the cartilage tissue. Via this route, progenitor cells from the bone marrow enter the diseased tissue, taken from Koelling et al. 2009 (with permission of the publisher).

chondrogenic progenitor cells (Koelling et al. 2009). These cells possess a multipotent differentiation capacity especially towards the chondrogenic lineage, as well as a migratory potential (Fig. 13.5) and furthermore they

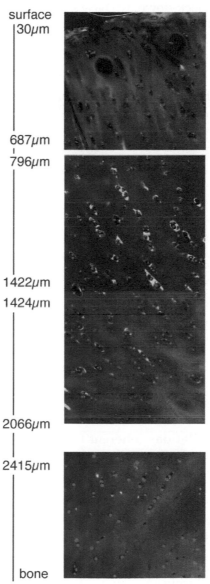

Figure 13.5 Chondrogenic progenitor cells transfected with green fluorescent protein migrate into osteoarthritic tissue *ex vivo*. These cells are found approx. 1400µm deep in the tissue after 2 d, taken from Koelling et al. 2009 (with permission of the publisher).

also populate diseased tissue *in vivo* (Fig. 13.6). Because these cells show heterogeneity in these properties and because of their migratory potential, we prefer to call them chondrogenic progenitor cells. Furthermore, with the help of RNA knock-down, we have shown that sox-9 and runt-related transcription factor 2 (runx-2) play a central role in the chondrogenic differentiation process of these cells that are also influenced by mediators from the extracellular matrix. These CPCs are an ideal starting point for a cell biological regenerative therapy of osteoarthritis.

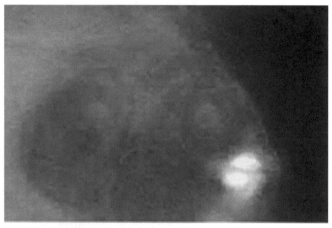

Figure 13.6 Cells with marker proteins related to stem cells are also found *in vivo* in osteoarthritic cartilage tissue. Here two cells in a cluster are positive for CD29 and CD73, taken from Koelling et al. 2009 (with permission of the publisher).

Future Perspectives of a Regenerative Therapy of Osteoarthritis

There are key limitations inherent to any cell biological therapy of osteoarthritic defects that have to be overcome before a regenerative therapy with progenitor cells will be applicable. First, it has to be shown that these cells can be manipulated to enhance their chondrogenic potential and that it remains present over a long time. The cells present in osteoarthritic tissue are not able to alter the disease process. Their physiological repair capacity is not sufficient. Therefore, it remains to be seen if these cells will produce an extracellular matrix that results in a repair tissue with a higher physical resistance to mechanical stress than the fibrocartilaginous tissue developed during the course of the disease. Finding the optimal conditions to manipulate such cells will be crucial for the development of a cell biological therapy for the treatment of osteoarthritis. Resident cells that are already used as a physiological response to the cell biological stimuli of the cartilage tissue, the tissue they are supposed to repair, may be more sufficient than cells derived from a totally different source such as adipose tissue stem cells. Furthermore, stem cells have been shown to gradually lose their differentiation potential with age. Mesenchymal stem cells from patients with osteoarthritis exhibited a reduced potential for differentiation towards a cartilage-like tissue (Murphy et al. 2002). Thus, a new concept sees mesenchymal stem cells as a therapeutic means to positively influence the microenvironment of the stem cells already present in the diseased tissue and to direct those cells in their regeneration activities (Caplan 2007). This is derived from results demonstrating a positive immunomodulatory effect of mesenchymal stem cells, even in allogenic transplantations (Wolbank et al. 2007). However, there is also the possibility that stem cells found in osteoarthritic tissue are responsible for the disease process. Before a cell biological therapy of osteoarthritis becomes a clinical reality, numerous scientific questions remain to be addressed *in vitro*. However, the crosstalk of these transcription factors mediating pluripotency remains to be elucidated, to gain insight into molecular mechanisms of reversibility of commitment and consecutive plasticity in the cells of the transient amplifying pool. It is well possible, that the transcription factors named above are also important for CPCs and might enhance their multipotency. One of the shortfalls of stem cell therapy is that the stemness of true stem cells is altered, as soon, as it is removed from its niche (Fuchs et al. 2004) and that these cells are the transient amplifying pool already committed towards terminal differentiation (Fuchs 2009). Furthermore, the *in vitro*-cultivation of these cells further alters their cell biological properties. Therefore, understanding of the underlying mechanisms that govern stemness, multipotency and differentiation is essential for all future stem cell therapy approaches.

Dictionary of Key Terms

- *Osteoarthritis (OA)*: OA is mainly degenerative joint disease. Cartilage plays an important role in the joints mechanics for painless and frictionless motion. Loss of the cartilage in the joints permits direct bone to bone exposure which leads to pain, inflammation and ultimately in the loss of the joint.
- *Sox9*: Sox9 is a transcription factor, DNA binding protein and essential for chondrogenesis. Sox9 is directly involved in the regulation of type II collagen gene, which is the major components in extra cellular matrix (ECM) of hyaline cartilage. Sox9 is considered as master regulator and as a key player during chondrocyte differentiation.
- *Runx2*: Runx2 is mainly involved in osteogenesis and cartilage development. Runx2 is a transcription factor and acts as "main/master

switch" in regulating many other genes that are involved in promoting the osteogenic characters in cells as in osteoblasts.

- *Mesenchymal stem cells (MSCs)*: MSCs mostly reside in bone marrow, basically mesenchyme is embroyonic connective tissue that derived from the mesoderm. MSCs differentiates into multiple cell types including adipocytes, chondrocytes, and cell of the osteoblastic lineage. In short, MSCs are capable to regenerate the diseased tissue.
- *Progenitor stem cells*: Progenitor stem cells behave like stem cells and have the ability to differentiate into specific types of target cell. The most prominent and clear difference between progenitor cells and stem cells is that the stem cells can divide indefinitely and are open to choose their differentiation direction while progenitor cells are already determined to a more narrow cell fate decision for example cells of the skeletal system and have a limit number of replication cycles.

Summary Points

- Osteoarthritis (OA) is such a wide spread complication associated with aging that it is expected to become the fourth most frequent cause of disability by the year 2020 and cartilage regeneration becomes a challenge for the regenerative and bio-engineering community.
- Osteoarthitis is a mainly degenerative joint disease. Eventual loss of cartilage in the joint uncovers the bone and allows direct bone to bone contact. The frictional moment of joint damaging the bone and ultimately results in pain and inflammation. This mechanical disability arises due to destruction and vanishing of extracellular matrix proteins particularly collagen type II and proteoglycans in cartilaginous tissue.
- Chondrogeinc progenitor cells (CPCs) are present in the late stages of human OA. They migrate from the fibrocartilaginous tissue *in vitro*.
- CPCs exhibit stem cell characteristics such clonogencity, multipotency and migratory activity.
- CPCs have tremendous ability towards chondrogensis and this chondrogenic potential can be enhanced by knocking down (using siRNA) Runx2 (an osteogenic protein) or by over expression of sox9 (a chondrogenic protein).

Acknowledgements

We would like to apologize to all the colleagues whose work we could not mention due to space limitations. This work was supported by the German Research Council (DFG) and the Medical Faculty in Goettingen.

List of Abbreviations

OA	:	osteoarthritis
ECM	:	extracellular matrix
CPC	:	chondrogenic progenitor cells
MSC	:	mesenchymal stem cells

Transcription factors

oct3/4	:	octamer-binding transcription factor ¾
sox2	:	SRY (sex determining region Y)-box 2
c-myc	:	cellular myelocytomatosis oncogene
klf4	:	Krüppel like factor 4
STRO-1	:	antigen identified by Simmons and Torok-Storb for stem cells
CD106, CD9, CD90 and CD166, CD105	:	cluster of differentiation, so called stem cell markers
VCAM-1	:	vascular cell adhesion molecule 1
BMP-6	:	bone morphogenetic protein 6
TGFß$_3$	:	transforming growth factor beta 3
Runx-2	:	runt-related transcription factor 2
Sox9	:	SRY (sex determining region Y)-box 9

References

Alsalameh, S., R. Amin, T. Gemba and M. Lotz. 2004. Identification of mesenchymal progenitor cells in normal and osteoarthritic human articular cartilage. Arthritis Rheum. 50(5): 1522–1532.

Bock, H.C., P. Micheali, C. Bode, W. Schultz, H. Kresse, R. Herken and N. Miosge. 2001. The small proteoglycans decorin and biglycan in human articular cartilage of late-stage osteoarthritis. Osteoarthritis and Cartilage. 9: 654–663.

Caplan, A.I. 2007. Adult mesenchymal stem cells for tissue engineering versus regenerative medicine. J Cell Physiol. 213(2): 341–347.

Dell'Accio, F., J. Vanlauwe, J. Bellemans, J. Neys, C. De Bari and F.P. Luyten. 2003. Expanded phenotypically stable chondrocytes persist in the repair tissue and contribute to cartilage matrix formation and structural integration in a goat model of autologous chondrocyte implantation. J Orthop Res. 21(1): 123–131.

Dowthwaite, G.P., J.C. Bishop, S.N. Redman, I.M. Khan, P. Rooney, D.J. Evans, L. Haughton, Z. Bayram, S. Boyer, B. Thomson et al. 2004. The surface of articular cartilage contains a progenitor cell population. J Cell Sci. 117(Pt 6): 889–897.

English, A., E.A. Jones, D. Corscadden, K. Henshaw, T. Chapman, P. Emery and D. McGonagle. 2007. A comparative assessment of cartilage and joint fat pad as a potential source of cells for autologous therapy development in knee osteoarthritis. Rheumatology (Oxford). 46(11): 1676–1683.

Fickert, S., J. Fiedler and R.E. Brenner. 2004. Identification of subpopulations with characteristics of mesenchymal progenitor cells from human osteoarthritic cartilage using triple staining for cell surface markers. Arthritis Res Ther. 6(5): R422–432.

Friedenstein, A.J., R.K. Chailakhjan and K.S. Lalykina. 1970. The development of fibroblast colonies in monolayer cultures of guinea-pig bone marrow and spleen cells. Cell Tissue Kinet. 3(4): 393–403.

Fuchs, E., T. Tumbar and G. Guasch. 2004. Socializing with the neighbors: stem cells and their niche. Cell. 116(6): 769–778.

Fuchs, E. 2009. Finding one's nice in the skin. Cell Stem Cell. 4: 499–502.

Gronthos, S., A.C. Zannettino, S.J. Hay, S. Shi, S.E. Graves, A. Kortesidis and P.J. Simmons. 2003. Molecular and cellular characterisation of highly purified stromal stem cells derived from human bone marrow. J Cell Sci. 116(Pt 9): 1827–1835.

Haq, I., E. Murphy and J. Dacre. 2003. Osteoarthritis. Postgrad Med. J. 79: 377–383.

Johnstone, B., T.M. Hering, A.I. Caplan, V.M. Goldberg and J.U. Yoo. 1998. *In vitro* chondrogenesis of bone marrow-derived mesenchymal progenitor cells. Exp Cell Res. 238(1): 265–272.

Koelling, S. and N. Miosge. 2009. Stem cell therapy for cartilage regeneration in osteoarthritis. Expert Opinion on Biological Therapy. 9: 1–7.

Koelling, S., J. Kruegel, M. Irmer, J. Path, B. Sadowski, X. Miró and N. Miosge. 2009. Migratory chondrogenic progenitor cells from repair tissue during the late stages of human osteoarthritis. Cell Stem Cell. 4: 324–335.

Koelling, S. and N. Miosge. 2010. Sex differences of chondrogenic progenitor cells in late stages of osteoarthritis. Arthritis Rheum. 62: 1077–1087.

Kouri, J.B., S.A. Jimenez, M. Quintero and A. Chico. 1996. Ultrastructural study of chondrocytes from fibrillated and non-fibrillated human osteoarthritic cartilage. Osteoarthritis Cartilage. 4: 111–125.

Kuettner, K.E. 1992. Biochemistry of articular cartilage in health and disease. Clin Biochem. 25: 155–163.

Mackay, A.M., S.C. Beck, J.M. Murphy, F.P. Barry, C.O. Chichester and M.F. Pittenger. 1998. Chondrogenic differentiation of cultured human mesenchymal stem cells from marrow. Tissue Eng. 4(4): 415–428.

Martel-Pelletier, J. 1999. Pathophysiology of osteoarthritis. Osteoarthritis Cartilage. 6: 371–783.

Mayne, R. and R.G. Brewton. 1993. New members of the collagen superfamily. Curr Opin Cell Biol. 5: 883–890.

Miosge, N., K. Waletzko, C. Bode, F. Quondamatteo, W. Schultz and R. Herken. 1998. Light and electron microscopic *in situ* hybridization of collagen type I and type II mRNA in the fibrocartilaginous tissue of late-stage osteoarthritis. Osteoarthritis Cartilage. 6: 278–285.

Murphy, J.M., D.J. Fink, E.B. Hunziker and F.P. Barry. 2003. Stem cell therapy in a caprine model of osteoarthritis. Arthritis Rheum. 48(12): 3464–3474.

Murphy, L.M., K. Dixon, S. Beck, D. Fabian, A. Feldman and F. Barry. 2002. Reduced chondrogenic and adipogenic activity of mesenchymal stem cells from patients with advanced osteoarthritis. Arthritis Rheum. 46: 704–713.

O'Rahilly, R., J. Bossy and F. Muller. 1981. Introduction to the study of embryonic stages in man. Bull Assoc Anat (Nancy). 65(189): 141–236.

Poole, A.R., G. Rizkalla, M. Ionescu, A. Reiner, E. Brooks, C. Rorabeck, R. Bourne and E. Bogoch. 1993. Osteoarthritis in the human knee: a dynamic process of cartilage matrix degradation, synthesis and reorganization. Agents Actions Suppl. 39: 3–13.

Poole, A.R. 1999. An introduction to the pathophysiology of osteoarthritis. Front Biosci. 4: D662–670.

Reginster, J.Y. 2002. The prevalence and burden of arthritis. Rheumatology 41 Suppl1: 3–6.

Sandell, L.J. and T. Aigner. 2001. Articular cartilage and changes in arthritis. An Introduction: Cell biology of osteoarthritis. Arthritis Res. 3: 107–113.

Takahashi, K. and S. Yamanaka. 2006. Induction of pluripotent stem cells from mouse embryonic and adult fibroblast cultures by defined factors. Cell. 126(4): 663–676.

Wolbank, S., A. Peterbauer, M. Fahrner, S. Hennerbichler, M. van Griensven, G. Stadler, H. Redl and C. Gabriel. 2007. Dose-dependent immunomodulatory effect of human stem cells from amniotic membrane: a comparison with human mesenchymal stem cells from adipose tissue. Tissue Eng. 13(6): 1173–1183.

Woolf, A.D. and B. Pfleger. 2003. Burden of major musculoskeletal conditions. Bull. World Health Organ. 81: 646–656.

Allogeneic Mesenchymal Stem Cells for Bone Tissue Engineering: Clinical and Basic Research on Perinatal Hypophosphatasia

Hiroe Ohnishi,[a] Yoshihiro Katsube,[b] Mika Tadokoro,[c] Shunsuke Yuba[d] and Hajime Ohgushi[e,*]

ABSTRACT

Bone marrow mesenchymal stem cells (MSCs) have high proliferation and differentiation capabilities, especially towards the osteogenic lineage. Since 2001, we have used MSCs derived from patients with osteoarthritis, bone tumor, and osteonecrosis for their own treatment. However, if the patients have a genetic disorder resulting in impairment of osteogenic differentiation capability of their MSCs, the patient's stem

Tissue Engineering Research Group, Health Research Institute, National Institute of Advanced Industrial Science and Technology (AIST), 3-11-46 Nakouji, Amagasaki, Hyogo, Japan.
[a]E-mail: hiroe-oonishi@aist.go.jp
[b]E-mail: yo-katsube@aist.go.jp
[c]E-mail: mika-tadokoro@aist.go.jp
[d]E-mail: yuba-sns@aist.go.jp
[e]E-mail: hajime-ohgushi@aist.go.jp
*Corresponding author

List of abbreviations given at the end of the text.

cells cannot be used for bone tissue repair and regeneration. However there are solutions: utilization of allogeneic MSCs having normal osteogenic function, and manipulation of the affected gene to fight or prevent the disease. We have detailed our experience of a patient with a severe type of hypophosphatasia (HPP), having short stature, brittle bones, and deformity of the long bones. The disease is caused by mutations in the tissue-nonspecific alkaline phosphatase (*TNSALP*) gene. The patient's serum alkaline phosphatase (ALP) activity was extremely low and the MSCs did not show bone forming capability. Therefore, the patient was treated with allogeneic MSCs instead of autologous MSCs. The patient responded to the treatment; though bone deformity did not improve, and the ALP activity remained at a low level, indicating the limitation of the treatment. The impaired bone forming capability of the patient's MSCs was experimentally salvaged by normal human TNSALP gene transduction into the MSCs, suggesting the possibility of gene therapy using the patient's MSCs. This chapter focuses on the *in vivo* transplantation of autologous and allogeneic MSCs for bone tissue regeneration, based on our basic research and clinical experience.

Introduction

MSCs are somatic stem cells that are able to differentiate toward various cell lineages including osteocytes, chondrocytes, adipocytes, hepatocytes, neurocytes, and cardiomyocytes. MSCs are found in various tissues, such as bone marrow, adipose tissue, tooth germ, and synovial membranes. They have been suggested to be an effective tool for regenerative medicine (Caplan and Bruder 2001). MSCs can be culture-expanded using small amounts (3 to 12ml) of aspirated bone marrow, even from aged people (Kotobuki et al. 2004; Ohgushi et al. 2005). We developed new tissue engineering techniques that use osteogenic constructs consisting of *in vitro* culture-expanded MSCs followed by osteogenic differentiation on porous hydroxyapatite (HA) ceramics (Ohgushi et al. 2005). The cultured osteoblastic cells on ceramics enable immediate new bone formation after *in vivo* transplantation. Since 2001, we have used the osteogenic constructs derived from osteoarthritis, bone tumor, and osteonecrosis patients' own bone marrow for their treatment (Ohgushi et al. 2005; Morishita et al. 2006; Kawate et al. 2006). As the donor cells and recipients are the same individual, the transplantation is an autologous transplantation, with no risk of immunorejection. However, there are some reports of immunomodulatory effects of MSCs and the possible survival of transplanted allogeneic MSCs, which circumvent immunorejection.

Concerning immunorejection, we previously investigated the survival of transplanted fresh allogeneic marrow cells, which contain about

0.01–0.001 percent of MSCs. The results showed that the administration of the immunosuppressant known as FK506 (tacrolimus hydrate) was required for successful allogeneic transplantation (Akahane et al. 1999). The importance of an immunosuppressant for allogeneic MSCs transplantation has been reported by others (Eliopoulos et al. 2005; Nauta et al. 2006). We also reported a unique experimental rat model to examine the survival and differentiation capability of allogeneic MSCs, which clearly demonstrated the necessity of an immunosuppressant for successful allogeneic transplantation (Kotobuki et al. 2008).

These reports confirmed that allogeneic MSCs find it difficult to survive in the recipients. However, when patients have mutations in genes essential for bone metabolism and show decreased bone forming capability because of the mutation, the patient's MSCs cannot be used for bone tissue regeneration. HPP is a genetic disorder characterized by impaired bone mineralization and extremely low ALP activity in the serum. The disease is caused by mutations in the *TNSALP* gene (Weiss et al. 1988). The activity of ALP is required for bone formation, thus the patients show impaired bone architecture, such as short stature and skeletal deformities (Fig. 14.1) (Whyte et al. 1996; Zurutuza et al.1999). Various therapies, such as cortisone, plasma, and traditional enzyme (ALP) replacement therapy have been attempted, but the results were inconsistent and did not lead to significant clinical improvement. A recent report indicated a possible therapeutic approach using bone-targeted human recombinant TNSALP fusion protein, which has affinity to HA crystals (McKee et al. 2011). Although there are no established therapies for this disease, clinical studies have suggested that conventional bone marrow transplantation (BMT) improved skeletal mineralization in the genetic disorder of osteogenesis imperfecta (OI) (Horwitz et al. 1999; Horwitz et al. 2002). HPP is also reported to respond to BMT and additional infusions of cultured MSCs, furthermore bone grafting was used to improve the patients' clinical symptoms and skeletal mineralization (Whyte et al. 2003; Cahill et al. 2004; Cahill et al. 2007). Based these results, and our own findings, after the administration of immunosuppressants, we transplanted allogeneic MSCs, osteoblastic cells, and osteogenic constructs in our HPP patient.

Our clinical experience of allogeneic transplantation under minimum immunosuppression showed some therapeutic effects, but these effects were limited and as described, simple transplantation of allogeneic MSCs resulted in their rejection (Kotobuki et al. 2008). Therefore, a new strategy, using autologous (patient's own) MSCs for the treatment of HPP is required. HPP is a genetic disorder; therefore, transduction of the normal gene into the patient's MSCs is required for this therapeutic approach. Thus, we also discuss the restoration of the mineralization capability of normal *TNSALP* gene-transduced HPP patient's MSCs in

Normal   Patient

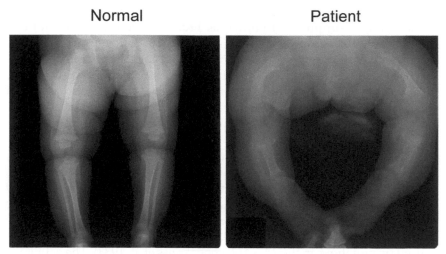

Figure 14.1 Radiographs of the patient and a normal infant at 6 mon. The patient presented with the severe skeletal deformity with poor mineralization (from Katsube et al. 2010; with permission from Nature Publishing Group).

response to osteogenic differentiation. In this chapter, we first describe the allogeneic MSCs transplantation in a rat experimental model to address MSC immunorejection, and our clinical experience of an HPP patient using allogeneic MSCs. Finally, we discuss perspectives on the therapeutic approach using gene-transduced autologous MSCs for HPP patients.

Autologous/Allogeneic MSCs Transplantation Model Using Rat Cells

Transplantation design and in vitro osteogenic potential of MSCs

For the purpose of the transplantation experiments using small animals, harvesting of the donor cells/tissues is usually associated with sacrifice of the donor animals, which makes autologous transplantation impossible. However, Fischer 344 (F344) inbred rat strains are genetically identical (syngeneic); therefore, we used bone marrow cells from F344 rats as donors and used other F344 rats as recipients. Although the donor and recipient rats are not the same animal, the transplantation experiments can be considered an autologous transplantation model. We used F344 rats as recipient rats and used donor MSCs from either F344 rats as the syngeneic (autologous) transplantation model or other rat strains (ACI and Lewis (LEW) rats) as allogeneic transplantation models. The ACI rat is a major antigen mismatch to the F344 rat, which means their histocompatibility antigens differ greatly.

The LEW rat is a minor antigen mismatch to the F344 rat, which means their histocompatibility antigens are only slightly different. Therefore, we used donor marrow cells from ACI and LEW rats as major and minor mismatch cells against recipient F344 rats, respectively. A schematic diagram of the experimental design is shown in Fig. 14.2 (Kotobuki et al. 2008). The experiment is simple and unique, because one recipient animal (F344 rat) could receive allogeneic as well as autologous (syngeneic) donor cells.

For the preparation of donor cells, fresh marrow cells were obtained from femora of three rat strains (F344, ACI and LEW). Dermis was also obtained from the F344 rat as a negative control for MSCs. Two days after the primary culture of fresh marrow cells, floating hematopoietic cells could be removed and adherent cells grew well on the surface of the culture dish. We labeled these cells as MSCs, based on the proliferation and differentiation capability of the cells (Kotobuki et al. 2004). The fibroblasts from dermis also grew well on the dish surface. These primary cultured cells were then subcultured in the presence of β–glycerophosphate, ascorbic acid, and dexamethasone (Dex). All MSCs from the three strains showed high ALP activity after 14 d of subculture. In contrast, the activity of the fibroblasts from the dermis was very low. These data indicate that the MSCs from the three strains, but not fibroblasts, show *in vitro* osteogenic differentiation capability; therefore we could confirm that these MSCs from three strains have differentiation capability toward osteogenic lineage. These primary

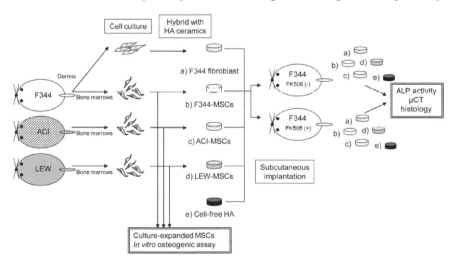

Figure 14.2 Schematic for the cell preparation and *in vivo/-in vitro* osteogenic assay. Hydroxyapatite (HA) ceramics with syngeneic (Fischer (F344)) MSCs, allogeneic (ACI and Lewis (LEW)) MSCs, and syngeneic fibroblasts were subcutaneously transplanted into recipient F344 rats. The F344 rats were treated with FK506 or saline. After 4 wk, the transplants were examined by an alkaline phosphatase (ALP) activity assay, μ-CT analysis, and by histology (from Kotobuki et al. 2008; with permission from Cognizant Communication Corporation).

cultured MSCs and fibroblasts were then combined (hybridized) with porous HA ceramics disks (HA disks) and transduction at subcutaneous sites of recipient F344 rats to investigate engraftments and the *in vivo* osteogenic potential of the donor MSCs.

Comparison of in vivo osteogenic potential of syngeneic and allogeneic MSCs

As seen in Fig. 14.2, hybrids of primary cultured allogeneic as well as syngeneic MSCs and HA disks were transplanted for 4 wk. HA disks without cells (cell free HA) were also transplanted. Half of the recipient F344 rats were treated with immunosuppressant FK506. In the FK506-treated group (FK506(+)), not only syngeneic but also allogeneic MSCs formed osteoblasts lining the surface of newly formed bone matrix in the pore regions of HA disks (Fig. 14.3). These histological findings indicated bone formation by osteogenic differentiation of both syngeneic and allogeneic MSCs (Fig. 14.3A, B, C). The fibroblast from dermis and HA disks without cells did not show bone formation. However, without FK506 treatment (FK506 (–)), only syngeneic MSCs showed bone formation (Fig. 14.3D). Furthermore, significant inflammation was seen in the allogeneic MSCs (Fig. 14.3E, F).

We also performed micro-computed tomography (μ-CT) analysis, which showed that all MSCs in FK506(+) recipient F344 rats produced new bone formation (Fig. 14.4A, gray areas) in the porous regions (Fig. 14.4A, black

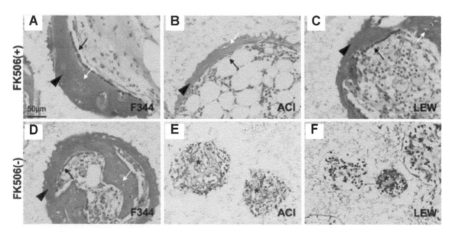

Figure 14.3 Histological analyses of transplants 4 wk after transplantation by H&E staining. Ceramic sections of transplants from FK506 treated rats (A-C) or non-treated rat (D-F). Scale bar: 50μm. Black arrows: osteoblasts; white arrows: osteocytes; arrowheads: bone matrix; asterisk: inflammation (from Kotobuki et al. 2008; with permission from Cognizant Communication Corporation).

Color image of this figure appears in the color plate section at the end of the book.

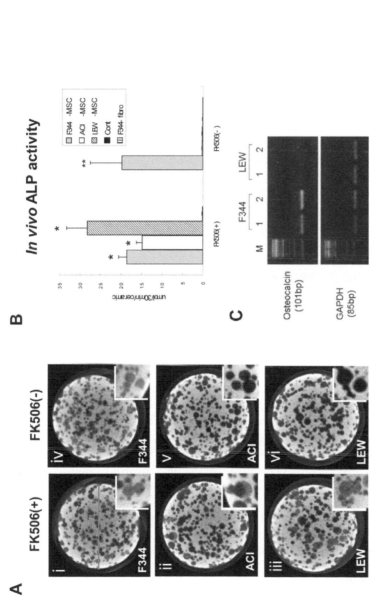

Figure 14.4 *In vivo* analysis of syngeneic and allogeneic MSCs 4 wk after transplantation. (A) Two dimensional μ-CT image of transplants. The rat origin (F344, ACI, and LEW) of donor MSCs is indicated in each photograph. Typical areas are enlarged at the lower right. (B) ALP activity of the transplants. ALP activity was calculated per transplants. *p>0.05 versus F344-fibro (n=3). **p> 0.05 versus MSCs derived from ACI and LEW rats, cell-free HA, and F344-fibro (n=3). (C) Gene expression of bone-specific osteocalcin (OC) and the housekeeping GAPDH gene in the transplants (from Kotobuki et al. 2008; with permission from Cognizant Communication Corporation).

areas) of the HA ceramics (Fig. 14.4A, white areas). In contrast, none of the allogeneic MSCs showed bone formation in the HA ceramics in FK506 (–) recipient F344 rats. The harvested HA disks were also analyzed by measuring ALP activity and bone specific osteocalcin mRNA. Syngeneic and allogeneic MSCs showed high ALP activity in transplants from the FK506 (+) recipient F344 rats. However, of the transplants from the FK506 (–) recipient F344 rats, only syngeneic MSCs showed high ALP activity (Fig. 14.4B). Furthermore, expression of osteocalcin was detected only from syngeneic MSCs in the FK506 (–) recipient F344 rats (Fig. 14.4C). These data indicate that only FK506 (+) recipient rats allowed osteogenic differentiation of allogeneic MSCs. Taken together, administration of an immunosuppressant is necessary for engraftments of allogeneic MSCs, which ultimately show osteogenic differentiation.

Treatment of Perinatal HPP with Allogeneic MSCs Transplantation

Case reports of BMT/MSCs transplantation in HPP

Depending on the age of onset and clinical symptoms of HPP, five clinical forms are categorized: perinatal, infantile, childhood, adult, and odont forms. Almost all infants with severe forms of the disease (perinatal form) die *in utero* or shortly after birth: the disease has no established therapy. Whyte et al. (Whyte et al. 2003) first reported the treatment of infantile HPP with BMT and an infusion of MSCs and Cahill et al. (Cahill et al. 2004; Cahill et al. 2007) reported the treatment of an 8-mon-old girl with infantile HPP using BMT, infusion of cultured osteoblastic cells, and transplantation of bone fragments from her father.

   Our patient underwent two trial treatments consisting of a 1st trial using BMT, MSCs, osteoblastic cells, and osteogenic constructs, and a 2nd trial using only MSCs (Fig. 14.5). The 1st trial was based on previous reports by Cahill et al. (Cahill et al. 2004; Cahill et al. 2007). However, in their reports, the osteoblastic cells used were cultured cells that grew from harvested bone fragments. They also transplanted the bone fragments into the peritoneal and subcutaneous sites of the patient. Their approach thus required the use of many bone fragments from the donor. In this regard, our treatment required only bone marrow and ceramics. In particular, we reported that the osteogenic constructs showed evidence of new bone formation even after *in vivo* transplant (Ohgushi and Caplan 1999; Ohgushi et al. 2005). The constructs were covered with abundant donor osteoblastic cells having high ALP activity. Therefore, we used the osteogenic constructs instead of bone fragments. The details are described in the following section.

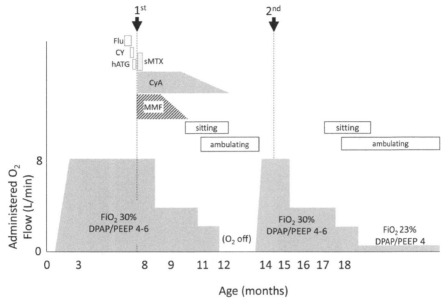

Figure 14.5 History of the patient. At age 8 mon, the 1st trial consisting of bone marrow transplantation (BMT) and a MSC-based treatment was performed. O_2 volume of ventilation was gradually decreased and finally could be shut off at 12 mon. At age 14 mon, artificial ventilation was administered (FiO$_2$ 30%, flow 8L/min, n DPAP/PEEP 4 6 cmH$_2$O) because of a relapse of the symptoms. At age 15 mon, the patient received the 2nd trial. One month later, the O_2 volume was reduced and finally kept at 0.5 L/min, FiO$_2$ 23% during the day and n-DPAP (PEEP 4 cmH$_2$O) was used at night (from Tadokoro et al. 2009; with permission from Elsevier Inc.).

Patient under study and preparation of donor cells

As described in our report (Tadokoro et al. 2009), radiographic examination of the patient revealed severe bowing of the long bones and widened metaphyses with poor mineralization, which formed radiolucent "tongues" projecting into the diaphyses from the proximal metaphyses (cupping) (Figs. 14.1 and 14.6A). Serum biochemical data at birth showed that ALP was 77 IU/l (normal: 327–774 IU/l). The ALP level in cultured MSCs from the patient's bone marrow was very low, even under the osteogenic culture conditions. However, the MSCs from the father (aged 33, HLA typing 3/6 matches, normal ALP activity in serum) showed high ALP activity and mineralization ability comparable to normal human MSCs. We diagnosed our patient with the perinatal form of HPP by clinical findings and the mutation sites of the *TNSALP* gene. Dyspnea with cyanosis and tachypnea, and poor weight gain appeared at age 3 mon, and respiratory care was initiated by controlled nasal directional positive airway pressure

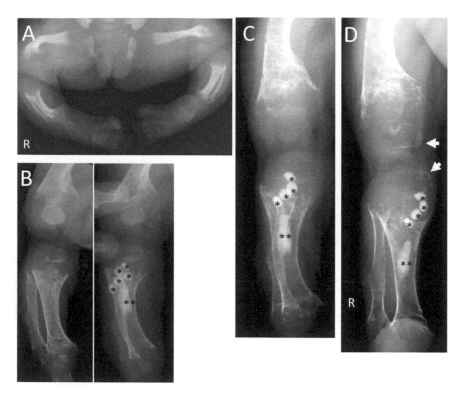

Figure 14.6 Sequential radiographs. (A) Immediate postnatal radiographs of both lower limbs, showing severe skeletal deformation with bowing of the long bones, and cupping and fraying in the epiphyses. (B) At age 8 mon, the patient underwent the 1st trial. The left and right radiographs were taken just before and after transplantation of the osteogenic constructs, respectively. Osteogenic constructs were transplanted subcutaneously (*) and intraosseously (**). (C) At age 24 mon (9 mon post 2nd trial), skeletal abnormalities remain. (D) At age 31 mon (16 mon post 2nd trial), the epiphysis appeared faintly in the femur and tibia (white arrows) (from Tadokoro et al. 2009; with permission from Elsevier Inc.).

(DPAP). As respiratory and radiographic deteriorations progressed, the patient underwent trial treatments consisting of BMT and an MSC-based treatment (Fig. 14.5).

First trial treatment

At age 8 mon, the patient received flutarabine phosphate (Flu), cyclophosphamide (CY), and human antithymus globulin (hATG). On day 0, the patient underwent BMT. Methotrexate (MTX), cyclosporine A (CyA), and mycophenolate mofetil (MMF) were dosed for graft-versus host disease (GVHD) prophylaxis (Fig. 15.5). On day 15 post BMT, MSCs were infused intravenously. The osteogenic constructs, using granular ceramics, were injected intramedullarily into both tibiae. The disk-shaped constructs were placed subcutaneously through the skin incisions (Fig. 14.6B). Post-treatment care was performed under controlled prophylaxis using CyA and MMF.

After the 1st trial, the patient's clinical progress was similar to the findings of Cahill et al. (Cahill et al. 2004; Cahill et al. 2007) in aspects such as clinical improvement, radiographic progression despite lack of donor chimerism, and unchanged biochemical data. After the MSC-based treatment, the patient's condition gradually improved (Fig. 14.5). Nasal-DPAP was no longer necessary, which was reflected in improved chest X-rays (Fig. 14.7A). Four months after the 1st trial, the patient was able to sit and ambulate using a wheelchair during the day. The bone mineral density (BMD) of the total body and right humerus showed gradual improvement

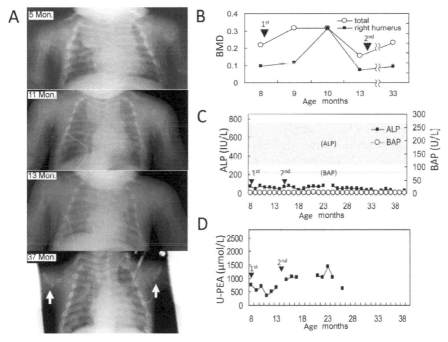

Figure 14.7 Sequential chest radiographs, physical and biochemical data. (A) Chronological chest radiographs at age 5, 11, 13, and 37 mon. At age 5 mon, just before treatment, the patient's ribs are deformed and hypomineralized. At age 11 mon (3 mon post 1st trial), the mineralization of the ribs and humeri seems to be improving. At age 13 mon (5 mon post 1st trial), radiolucency increases overall. At age 37 mon (22 mon post 2nd trial), irregular mineralization around the proximal humeral metaphyses (white arrows) can be seen. (B) Bone mineral density (BMD) at age 8, 9, 10, 13, and 33 mon. (C) Serum alkaline phosphatase (ALP) and bone alkaline phosphatase (BAP). The shaded areas indicate the normal ranges of ALP and BAP. (D) Chronically high levels of urinal PEA (normally, urinal PEA cannot be detected) (from Tadokoro et al. 2009; with permission from Elsevier Inc.).

(Fig. 14.7B). However, the biochemical parameters were not changed (Figs. 14.7C and D) and STR analysis showed no donor chimerism in the blood.

Five months after the 1st trial, radiographic findings showed increasing radiolucency and decreased BMD (Figs. 14.7A and B). The patient's respiratory condition gradually worsened. The patient again required respiratory care using artificial ventilation.

Second trial treatment

At age 15 mon, the patient received a booster shot of donor MSCs because of her deterioration (Fig. 14.5). This 2nd trial included neither myeloablation nor immunosuppression. The procedure for the expansion of the MSCs was the same as that for the 1st trial. MSCs were used for intravenous infusion and intraosseous injections into both tibiae, respectively. During the 2nd trial, one of the osteogenic constructs transplanted during the 1st trial was retrieved from its subcutaneous site to ascertain any osteogenic properties. The patient's clinical condition improved after one month, mimicking the results of the 1st trial (Fig. 14.5). The patient's respiratory condition stabilized with nasal-DPAP flowing at 0.5 L per min.

At age 24 mon (9 mon post-2nd trial), due to craniostenosis the patient underwent a craniotomy, in which small skull fragments were removed and used for histological analysis, as well as DNA extraction.

Chest radiographs showed notable mineralization around the metaphyses of both humeri at age 37 mon (Fig. 14.7A at 37 mon). Total and right humerus BMD also gradually increased after the 2nd trial (Fig. 14.7B). The biochemical parameters were not improved (Figs. 14.7C and D). Her physical development improved. Significant mineralization in the lower limbs was not seen until age 24 mon (Fig. 14.6C). At age 31 mon, a radiograph showed that the epiphysis could be seen faintly in the femora and tibiae and the bowing of the tibiae had improved (Fig. 14.6D).

Detection of allogeneic donor cells in the patient

A histological section of the construct retrieved from subcutaneous transplantation sites showed *de novo* bone formation. PCR analysis and fluorescence *in situ* hybridization (FISH) suggest that the donor osteoblastic cells on the ceramic pore surfaces survived and fabricated new bone (Figs. 14.8A, B and C).

At age 24 mon, the patient underwent a craniotomy and small skull fragments were obtained. The histological section showed a woven bone formation with osteoblasts in a decalcified paraffin section. Interestingly, the DNA sample showed male DYZ1 sequences (Fig. 14.8D). Thus, the donor MSCs delivered via blood flow could have ultimately engrafted onto the

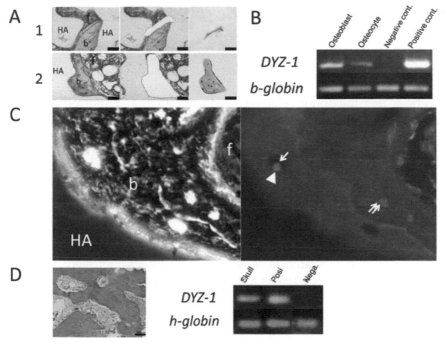

Figure 14.8 Analyses of the retrieved constructs and skull fragment. (A) Areas of the osteoblast lining (1) and osteocytes with bone matrix (2) were dissected from the original section using laser-assisted microdissection. The images show the original sections (left), after dissection (middle), and retrieved samples (right). The retrieved samples were used for detection of the Y chromosome. New bone (b) as well as fibrous tissue (f) can be seen within the pores of the ceramic (HA). Bar = 50 μm. (B) Detection of male-specific DYZ1 on the Y chromosome using PCR analysis. *β-globin* was used as an internal control. Osteoblast: DNA extracted from the retrieved osteoblast sample, as shown in A1; Osteocyte: from the retrieved osteocyte sample, as shown in A2; Negative cont.: negative control (female DNA); Positive cont.: positive control (male DNA). (C) Phase contrast (left) and fluorescence (right) images of FISH analysis of the retrieved construct. The osteocytes have the donor's (XY) and the patient's (XX) chromosomes. Nuclei are stained with DAPI (blue). The green and red dots indicate the Y chromosome (arrowhead) and the X chromosome (arrows), respectively. New bone (b) and fibrous tissue (f) were observed within the ceramic (HA). Original magnification ×400. (D) Decalcified section of the skull fragments (left). H-E staining, Bar = 100 μm. DNA samples from the other skull fragments were used for detection of the donor DYZ1 sequences using PCR analysis (right). Skull: extracted DNA sample from the skull fragment; Posi: positive control (male DNA); Nega: negative control (female DNA) (from Tadokoro et al. 2009, partially modification; with permission from Elsevier Inc.).

Color image of this figure appears in the color plate section at the end of the book.

skull. There have been similar observations in previous cases (Cahill et al. 2004; Cahill et al. 2007), which detected bone formation from systemically transplanted progenitor cells.

Implications of treatment by allogeneic MSCs transplantation

Although donor chimerism in peripheral blood was not confirmed, we detected new bone formation with donor signals, not only in transplanted local areas, but also in the distant areas of skull lesion. These findings indicate that the donor cells survived (engrafted) and showed continual osteogenic function, resulting in new bone formation. The second MSCs transplantation trial, without immunosuppressants, showed improvement of clinical symptoms, suggesting that the MSCs used in the 2nd trial avoided the normal alloresponses because of the immunomodulatory effects of the MSCs (Nauta and Fibbe 2007). It is also possible that the 1st trial contributed to survival of the MSCs through their acquisition of immunotolerance (Cahill et al. 2004; Cahill et al. 2007).

We could not definitively demonstrate clinical efficiency, but the trials show marginal effects and confirm the engraftment and bone formation of the donor cells. Elevation of serum ALP activity was not seen; therefore, alternative therapies, such as autologous transplantation of normal *ALP* gene-transduced MSCs, might be necessary for future curative therapy of HPP patients, as described in the following section.

Towards Autologous Transplantation for Curative Treatment of HPP Patients

Characterization of MSCs from the perinatal form HPP patient and TNSALP-transduction

As described in the previous sections, bone marrow MSCs, as well as osteogenic constructs, can be used for bone tissue regeneration. However, allogeneic MSCs are hard to survive after transplantation and show limited effects in HPP patients. The patient cells have mutations in the *TNSALP* gene; therefore, we attempted to transduce the normal *TNSALP* gene into the patient's MSCs, aiming for future curative treatment by autologous transplantation.

Before TNSALP transduction, we examined the characteristics of cells from the patient's bone marrow. The cultured cells from the patient's bone marrow had the same shape as normal MSCs (without gene mutation) and had a mesenchymal immunophenotype (Katsube et al. 2010). We could detect chondrogenesis and adipogenesis of the patient's MSCs under chondro and adipogenic differentiation conditions, respectively (Katsube et al. 2010). However, the MSCs derived from the patient did not produce a mineralized extracellular matrix, and showed very low ALP activity, even under osteogenic culture conditions. We therefore concluded that these cultured cells from patient's bone marrow are MSCs, but lack the mineralization capability because of low ALP activity.

To gain high ALP activity in the patient MSCs, we transduced the human *TNSALP* gene, under the control of its own promoter, into the MSCs using a retrovirus system. The cell shapes of the *TNSALP*-transduced MSCs were similar to non- and mock-transduced cells, and the cell surface expression of TNSALP was strongly enhanced in the *TNSALP*-transduced MSCs.

Examination of cellular functions of TNSALP-transduced MSCs from HPP patient

To investigate *in vitro* their osteogenic differentiation capability, *TNSALP*-transduced MSCs were cultured under osteogenic culture conditions for 4 wk. The culture showed many ALP-stainable cells and produced mineralized extracellular matrices, as seen by calcium staining using alizarin red S (Fig. 14.9A). In contrast, mock-transduced MSCs showed neither ALP-stainable cells nor mineralized matrix formation. Biochemical analysis confirmed that the *TNSALP*-transduced MSCs had approximately seven-fold higher ALP activity than did the mock-transduced MSCs. Calcium deposition and bone-specific osteocalcin (OC) were only detected in *TNSALP*-transduced MSCs (Fig. 14.9B).

To further analyze osteogenic differentiation, expressions of other bone-related genes were investigated by RT-PCR analyses. As seen in Fig. 14.9C, expressions of osteopontin (OPN), bone sialoprotein (BSP), OC, and Matrix Gla protein (MGP) were increased in *TNSALP*-transduced MSCs under osteogenic culture conditions. Mock-transduced MSCs also showed similar expression patterns (Fig. 14.9C). These data showed that the patient's MSCs expressed these bone related genes in osteogenic culture, although they hardly showed any mineralization because of their low ALP activity.

TNSALP-transduced MSCs were combined with porous HA ceramic disks and subcutaneously transplanted in athymic nude rats for 6 wk. Histological sections showed new bone formation in retrieved MSCs/HA hybrids (Fig. 14.10A). In contrast, bone formation was not detected in any mock-transduced MSCs/HA hybrids and or HA without cells. To confirm the origin of the newly formed bone, the bone areas of the sections were cut with LMD and used for PCR analysis. The analysis demonstrated the existence of the transgene in the dissected samples, indicating that the bone formation was caused by transplanted donor MSCs (Fig. 14.10B).

These results indicate that *TNSALP*-transduced MSCs have capability of *in vitro* and *in vivo* bone formation. Therefore, the patient's own MSCs, instead of allogeneic MSCs, can be used for transplantation therapy, and might facilitate bone mineralization in the HPP patient. Further studies are

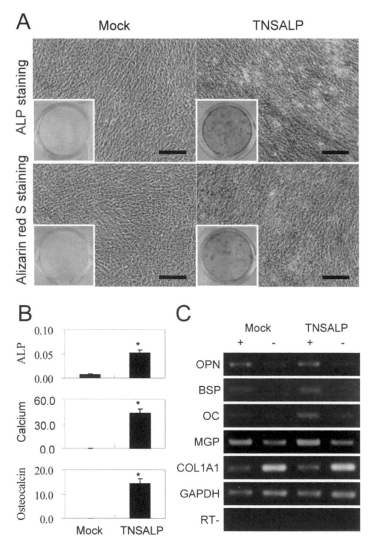

Figure 14.9 *In vitro* **osteogenic differentiation potential of** *TNSALP*-**transduced MSCs.**
(A) ALP activity and mineralization. Cells were stained by ALP and alizarin red S staining.
100× magnification, bar: 100 μm. Inset: Observation of whole areas of stained culture plate.
(B) Biochemical analysis. ALP activity (μmol/30 min/μg DNA), calcium content (μg/well),
and osteocalcin (OC) content (ng/well) are shown. Values are expressed as averages and
standard deviations (n = 4). Asterisks indicate statistically significant differences between
mock- and *TNSALP*-transduced MSCs (*t*-test: *P* < 0.01). (C) Bone-related gene expression
of MSCs. Mock- and *TNSALP*-transduced MSCs were cultured in osteogenic differentiation
medium (+) or control medium (–) (from Katsube et al. 2010; with permission from Nature
Publishing Group).

Color image of this figure appears in the color plate section at the end of the book.

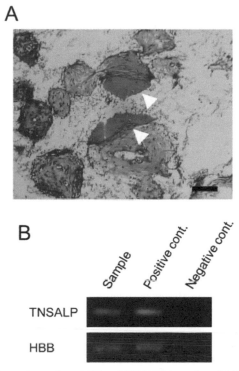

Figure 14.10 *In vivo* **bone formation ability of** *TNSALP*-**transduced MSCs**. (A) Histological appearance of osteogenic constructs using *TNSALP*-transduced MSCs. The harvested sample was decalcified, sectioned, and stained with hematoxylin and eosin. Arrowheads indicate newly formed bone. The newly formed bone areas in the section were harvested by laser-assisted microdissection. (B) PCR analyses of DNA samples of newly formed bone harvested by microdissection. The dissected samples (sample), human MSCs with *TNSALP* expression vector (positive control), and rat MSCs (negative control) were used to detect the transduced *TNSALP* gene (TNSALP) and human β-globin (HBB) (from Katsube et al. 2010; with permission from Nature Publishing Group).

Color image of this figure appears in the color plate section at the end of the book.

required to elucidate the efficacy of normal gene transduction into MSCs for clinical applications.

Conclusion and Perspective

Our basic and clinical data indicate that MSCs are useful stem cells for bone tissue engineering. Although allogeneic MSCs are hard to show osteogenic potential after *in vivo* transplantation, gain of the potential could be achieved using various approaches. In fact, we transplanted allogeneic MSCs into a severe type of HPP patient and found the osteogenic potential of the transplanted MSCs. We also confirmed the *in vitro* and *in vivo* osteogenic potential of *TNSALP*-transduced MSCs from the HPP patient. MSCs are capable of differentiating into various cell types; therefore, our strategy using genetically modified autologous MSCs may also be effective for treating other types of genetic diseases. One point to note is that MSCs are adult stem cells and their proliferation and the differentiation capability decrease after serial culture passages. In this regard, we have reported that MSCs are useful cells for generation of induced pluripotent stem cells (iPS cells), which have almost unlimited proliferation capability as well as pluripotency (Aoki et al. 2010; Oda et al. 2010; Ohnishi et al. 2011). The iPS cells derived from MSCs might be available for various tissue engineering purposes.

Table 14.1 Key features of Hypophosphatasia.

1. Hypophosphatasia is a genetic disorder characterized by impaired bone mineralization and low alkaline phosphatase (ALP) activity in serum.
2. Hypophosphatasia is caused by mutations in the tissue-nonspecific alkaline phosphatase gene that is required for bone formation.
3. Five clinical forms of hypophosphatasia are categorized: perinatal, infantile, childhood, adult, and odont forms. The perinatal form is a severe form of the disease.
4. Perinatal hypophosphatasia patients have short stature, brittle bones, and deformity of the long bones.
5. Various therapies, such as cortisone, plasma, and traditional enzyme (ALP) replacement therapy, have been attempted, but the results were inconsistent and there are no established therapies for this disease.

Key facts concerning hypophosphatasia.

Dictionary

- *HA (hydroxyapatite) ceramics*: Ceramics made from hydroxyapatite. Hydroxyapatite is naturally occurring crystal of calcium phosphate compositions. It is referred to as $Ca_5 (PO_4)_3(OH)$. In clinical situations, synthetic hydroxyapatite ceramics are used as bone graft substitutes.
- *ALP*: Alkaline phosphatase (ALP) is a hydrolase enzyme that removes phosphate groups from various types of molecules, including inorganic pyrophosphate (PPi). Its activity is high in an alkaline environment.
- *Osteoblast*: Bone tissue is composed of cellular and extracellular organic/inorganic matrix components. The main cellular components are osteocytes and bone forming osteoblasts. The main organic matrix is collagen fibers and the inorganic matrix comprises HA crystals. The crystals are produced by osteoblasts and laid down around the collagen fibers. The cell membrane of the osteoblasts shows high ALP activity, which is essential for HA crystal formation.

- *Syngeneic*: Syngeneic means genetically identical. It is also referred to as "Syngenic".
- *Allogeneic*: Allogeneic means genetically different. In allogeneic transplantation experiments, the donor cells or tissues are transplanted into genetically different recipients of the same species. It is also referred to as "Allogenic".
- *TNSALP (tissue-nonspecific alkaline phosphatase)*: One of the alkaline phosphatase isozymes. It is a ubiquitously expressed in the human body, mainly in the liver, bone, and kidney. Hypophosphatasia (HPP) is caused by mutations in this TNSALP gene.

Summary Points

- Using a rat MSCs transplantation model, we found that not only major, but also minor, mismatched allogeneic MSCs were hard to show engraftments and osteogenic differentiation.
- However, using the rat model, these allogeneic as well as syngeneic MSCs could show osteogenic differentiation followed by *in vivo* new bone formation, when the recipient rats were treated with immunosuppressant of FK506 (tacrolimus hydrate).
- HPP is caused by mutations in the *TNSALP* gene and there are no established therapies for this disease.
- We experienced the severe type (perinatal) of HPP. The MSCs from the patient showed neither high ALP activity nor mineralization in culture condition. However, the patient father's MSCs showed the high ALP activity as well as mineralization.
- We treated the patient by conventional BMT followed by transplantation of MSCs and osteogenic constructs derived from father's bone marrow.
- The treatments improved some features of the clinical symptoms and retrieved bone fragments showed the signal of donor (father); however, the serum ALP level of the patient remained very low.
- We transduced human *TNSALP* gene under the control of its own promoter into the patient's MSCs using a retrovirus system. The gene transduced MSCs showed high ALP activity and showed *in vivo* new bone formation after their subcutaneous transplantation into athymic nude rats.
- Based on our clinical and experimental experiences, genetically modified autologous MSCs might be available as a future therapeutic tool in various fields of regenerative medicine. Further data are needed to address this issue.

Acknowledgements

The MSCs were cultured in a cell processing center (CPC) at the National Institute of Advanced Industrial Science and Technology (AIST) and transplantation into the HPP patient was done at the Department of pediatrics, Shimane University, Faculty of Medicine. We thank all our colleagues at the AIST and Professor S. Yamaguchi and Dr. T. Taketani at Shimane University. This work was partially supported by Health and Labour Sciences Research Grants and the Project for Realization of Regenerative Medicine from the Ministry of Education, Culture, Sports, Science and Technology of Japan.

List of Abbreviations

ALP	:	alkaline phosphatase
BAP	:	bone alkaline phosphatase
BMD	:	bone mineral density
BMT	:	bone marrow transplantation
BSP	:	bone sialoprotein
CD	:	cluster of differentiation
CY	:	cyclophosphamide
CyA	:	cyclosporine A
Dex	:	dexamethasone
DPAP	:	directional positive airway pressure
F344	:	Fischer 344
FACS	:	fluorescence-activated cell sorter
FISH	:	fluorescence *in situ* hybridization
Flu	:	fludarabine phosphate
GVHD	:	graft-versus host disease
HA	:	hydroxyapatite
hATG	:	human antithymus globulin
HBB	:	human β-globin
HLA	:	human leukocyte antigen
HPP	:	hypophosphatasia
iPS cells	:	induced pluripotent stem cells
LEW	:	Lewis
LMD	:	laser-assisted microdissection
μ-CT	:	micro-computed tomography
MGP	:	matrix Gla protein
MMF	:	mycophenolate mofetil
MSCs	:	mesenchymal stem cells, also referred to as mesenchymal stromal cells
MTX	:	methotrexate

OC	:	osteocalcin
OI	:	osteogenesis imperfecta
OPN	:	osteopontin
PEA	:	phosphoethanolamine
PPi	:	pyrophosphate
STR	:	short tandem repeat
TNSALP	:	tissue-nonspecific alkaline phosphatase

References

Akahane, M., H. Ohgushi, T. Yoshikawa, T. Sempuku, S. Tamai, S. Tabata and Y. Dohi. 1999. Osteogenic phenotype expression of allogeneic rat marrow cells in porous hydroxyapatite ceramics. J Bone Miner Res. 14: 561–568.

Aoki, T., H. Ohnishi, Y. Oda, M. Tadokoro, M. Sasao, H. Kato, K. Hattori and H. Ohgushi. 2010. Generation of induced pluripotent stem cells from human adipose-derived stem cells without c-MYC. Tissue Eng Part A. 16: 2197–2206.

Cahill, R.A., O.Y. Jones, M. Klemperer, A. Steele, T.O. Mueller, N. el-Badri , Y. Chang and R.A. Good. 2004. Replacement of recipient stromal/mesenchymal cells after bone marrow transplantation using bone fragments and cultured osteoblast-like cells. Biol. Blood Marrow Transplant. 10: 709–717.

Cahill, R.A., D. Wenkert, S.A. Perlman, A. Steele, S.P. Coburn, W.H. McAlister, S. Mumm and M.P. Whyte. 2007. Infantile hypophosphatasia: transplantation therapy trial using bone fragments and cultured osteoblasts. J. Clin. Endocrinol. Metab. 92: 2923–2930.

Caplan, A.I. and S.P. Bruder. 2001. Mesenchymal stem cells: building blocks for molecular medicine in the 21st century. Trends Mol Med. 7: 259–264.

Eliopoulos, N., J. Stagg, L. Lejeune, S. Pommey and J. Galipeau. 2005. Allogeneic marrow stromal cells are immune rejected by MHC class I- and class II mismatched recipient mice. Blood. 106: 4057–4065.

Horwitz, E.M., D.J. Prockop, L.A. Fitzpatrick, W.W. Koo, P.L. Gordon, M. Neel, M. Sussman, P. Orchard, J.C. Marx, R.E. Pyeritz and M.K. Brenner. 1999. Transplantability and therapeutic effects of bone marrow-derived mesenchymal cells in children with osteogenesis imperfecta. Nat Med. 5: 309–313.

Horwitz, E.M., P.L. Gordon, W.K. Koo, J.C. Marx, M.D. Neel, R.Y. McNall, L. Muul and T. Hofmann. 2002. Isolated allogeneic bone marrow-derived mesenchymal cells engraft and stimulate growth in children with osteogenesis imperfecta: Implications for cell therapy of bone. Proc Natl Acad Sci USA. 99: 8932–8937.

Katsube, Y., N. Kotobuki, M. Tadokoro, R. Kanai, T. Taketani, S. Yamaguchi and H. Ohgushi. 2010. Restoration of cellular function of mesenchymal stem cells from a hypophosphatasia patient. Gene Ther. 17: 494–502.

Kawate, K., H. Yajima, H. Ohgushi, N. Kotobuki, K. Sugimoto, T. Ohmura, Y. Kobata, K. Shigematsu, K. Kawamura, K. Tamai and Y. Takakura. 2006. Tissue-engineered approach for the treatment of steroid-induced osteonecrosis of the femoral head: transplantation of autologous mesenchymal stem cells cultured with beta-tricalcium phosphate ceramics and free vascularized fibula. Artif Organs. 30: 960–962.

Kotobuki, N., M. Hirose, Y. Takakura and H. Ohgushi. 2004. Cultured autologous human cells for hard tissue regeneration: preparation and characterization of mesenchymal stem cells from bone marrow. Artif Organs. 28: 33–39.

Kotobuki, N., Y. Katsube, Y. Katou, M. Tadokoro, M. Hirose and H. Ohgushi. 2008. *In vivo* survival and osteogenic differentiation of allogeneic rat bone marrow mesenchymal stem cells (MSCs). Cell Transplant. 17: 705–712.

McKee, M.D., Y. Nakano, D.L. Masica, J.J. Gray, I. Lemire, R. Heft, M.P. Whyte, P. Crine and J.L. Millán. 2011. Enzyme replacement therapy prevents dental defects in a model of hypophosphatasia. J Dent Res. 90: 470–447.

Morishita, T., K. Honoki, H. Ohgushi, N. Kotobuki, A. Matsushima and Y. Takakura. 2006. Tissue engineering approach to the treatment of bone tumors: three cases of cultured bonegrafts derived from patients' mesenchymal stem cells. Artif Organs. 30: 115–118.

Nauta, A.J. and W.E. Fibbe. 2007. Immunomodulatory properties of mesenchymal stromal cells. Blood. 110: 3499–3506.

Nauta, A.J., G. Westerhuis, A.B. Kruisselbrink, E.G. Lurvink, R. Willemze and W.E. Fibbe. 2006. Donor-derived mesenchymal stem cells are immunogenic in an allogeneic host and stimulate donor graft rejection in a nonmyeloablative setting. Blood. 108: 2114–2120.

Oda, Y., Y. Yoshimura, H. Ohnishi, M. Tadokoro, Y. Katsube, M. Sasao, Y. Kubo, K. Hattori, S. Saito, K. Horimoto, S. Yuba and H. Ohgushi. 2010. Induction of pluripotent stem cells from human third molar mesenchymal stromal cells. J Biol Chem. 285: 29270–29278.

Ohgushi, H. and A.I. Caplan. 1999. Stem cell technology and bioceramics: from cell to gene engineering. 1999. J Biomed Mater Res. 48: 913–927.

Ohgushi, H., N. Kotobuki, H. Funaoka, H. Machida, M. Hirose, Y. Tanaka and Y. Takakura. 2005. Tissue engineered ceramic artificial joint *ex vivo* osteogenic differentiation of patient mesenchymal cells on total ankle joints for treatment of osteoarthritis. Biomaterials. 26: 4654–4661.

Ohnishi, H., Y. Oda, T. Aoki, M. Tadokoro, Y. Katsube, H. Ohgushi, K. Hattori and S. Yuba. 2011. A comparative study of induced pluripotent stem cells generated from frozen, stocked bone marrow- and adipose tissue-derived mesenchymal stem cells. J Tissue Eng Regen Med (in press).

Tadokoro, M., R. Kanai, T. Taketani, Y. Uchio, S. Yamaguchi and H. Ohgushi. 2009. New bone formation by allogeneic mesenchymal stem cell transplantation in a patient with perinatal hypophosphatasia. J Pediatr. 154: 924–930.

Weiss, M.J., D.E. Cole, K. Ray, M.P. Whyte, M.A. Lafferty, R.A. Mulivor and H. Harris. 1988. A missense mutation in the human liver/bone/kidney alkaline phosphatase gene causing a lethal form of hypophosphatasia. Proc Natl Acad Sci USA. 85: 7666–7669.

Whyte, M.P., D.A. Walkenhorst, K.N. Fedde, P.S. Henthorn and C.S. Hill. 1996. Hypophosphatasia: levels of bone alkaline phosphatase immunoreactivity in serum reflect disease severity. J Clin Endocrinol Metab. 81: 2142–2148.

Whyte, M.P., J. Kurtzberg, W.H. McAlister, S. Mumm, M.N. Podgornik, S.P. Coburn, L.M. Ryan, C.R. Miller, G.S. Gottesman, A.K. Smith, J. Douville, B. Waters-Pick, R.D. Armstrong and P.L. Martin. 2003. Marrow cell transplantation for infantile hypophosphatasia. J Bone Miner Res. 18: 624–636.

Zurutuza, L., F. Muller, J.F. Gibrat, A. Taillandier, B. Simon-Bouy, J.L. Serre and E. Mornet. 1999. Correlations of genotype and phenotype in hypophosphatasia. Hum Mol Genet. 8: 1039–1046.

CHAPTER 15

Haematopoietic Stem Cell Transplantation in Autosomal Recessive Osteopetrosis

Anna Teti[1,*] and Ansgar S. Schulz[2]

ABSTRACT

Autosomal recessive osteopetrosis is a life-threatening infantile genetic bone disease affecting approximately one in every 100,000 born. The skeleton is primarily affected and presents with dense and brittle bones prone to fracture, and with severe haematological failure due to the collapse of the bone marrow that does not have enough space to develop properly. The nervous system is affected, and in most cases because of nerve compression syndromes impairing especially vision and hearing. The cells implicated in this skeletal dysplasia are the osteoclasts. They are multinucleated giant cells that resorb bone, contributing to the modelling and remodelling processes indispensable for the harmonic bone accrual and homeostasis. Due to the impairment of the bone resorbing function of otherwise normally formed osteoclasts

[1]University of L'Aquila, Department of Biotechnological and Applied Clinical Sciences, via Vetoio, Coppito 2, 67100 L'Aquila, Italy.
E-mail: teti@univaq.it; anna.maria.teti@alice.it
[2]University Medical Center Ulm, Department of Pediatrics and Adolescent Medicine, Eythstr. 24, D-89075 Ulm, Germany.
E-mail Ansgar.Schulz@uniklinik-ulm.de
*Corresponding author

List of abbreviations given at the end of the text.

(osteoclast-rich osteopetrosis), or to their absence due to the disturbance of the process of osteoclastogenesis (osteoclast-poor osteopetrosis), bone resorption is precluded causing the persistence of primary bone and the compression of the inner cavities lodging vital organs, such as the bone marrow and the nervous system. Since the discovery of the osteoclast haematogenous origin, osteopetrosis is cured by haematopoietic stem cell transplantation, while pharmacological treatments remain only palliative. The haematopoietic stem cell transplantation outcome has greatly improved over the last decade, especially for human leukocyte antigen haplotype mismatched transplants. However, there are a number of problems and pitfalls that need to be addressed to make this treatment effective in a larger population of patients. In this chapter we will describe the pathogenesis of osteopetrosis and the haematopoietic stem cell transplantation options currently available. We will also discuss the weaknesses of this treatment and the potential areas of improvement to provide patients with better therapies and less debilitating life.

Introduction

Autosomal recessive OP is the most severe form of a disease characterized by fragile but dense bones (Fig. 15.1), bone marrow collapse and serious neurological symptoms (Del Fattore et al. 2011). The disorder is often lethal in infancy or childhood, and mortality rates are high because of anaemia, thrombocytopenia and recurrent infections (Del Fattore et al. 2011).

The cell implicated in OP is the osteoclast (Fig. 15.2). This is of haematogenous origin and fulfils the task of bone removal (Teti 2011). Failure of osteoclast bone resorption prevents the woven bone to be replaced by lamellar bone, and the internal bony cavities to be expanded, causing bone fragility, bone marrow failure and nerve compression syndromes (Del Fattore et al. 2011).

Osteoclasts

Osteoclasts are multinucleated cells that remove the bone matrix (Fig. 15.2). They arise from myeloid progenitors under the control of the osteoclastogenic cytokines M-CSF and RANKL (Teti 2011) (Fig. 15.3). M-CSF induces early monocytic precursors to express RANK (the receptor for RANKL) and become responsive to RANKL. The RANKL/RANK complex induces intracellular signalling leading to activation of NF-κB, AP-1 and NFATc1 transcription factors which trans-activate osteoclast-specific genes (Fig. 15.2) (Teti 2011).

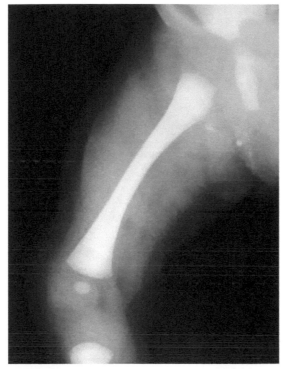

Figure 15.1 X-ray analysis of an osteopetrotic patient. Dense bone and lack of medullary cavity in femur (A. Teti, unpublished).

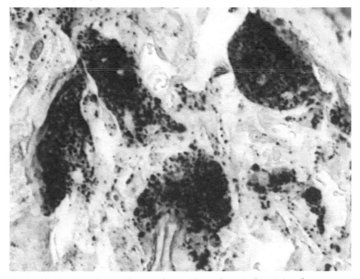

Figure 15.2 Histological sections of osteoclasts. Histochemical staining for tartrate-resistant acid phosphatase marking large osteoclasts in trabecular bone (A. Teti, unpublished).

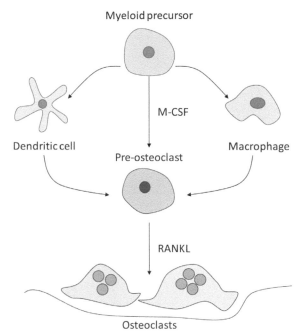

Figure 15.3 The process of osteoclastogenesis. Pathway of osteoclast formation from the myeloid precursor (A. Teti, unpublished).

Once formed, the osteoclast polarizes and arranges the plasma membrane in an area facing the bone matrix in which a peripheral adhesion zone is created by the assembly of a multitude of dynamic adhesions called podosomes (Teti 2011). These are composed by a core of actin filaments and actin-binding proteins, and a rosette of adhesion proteins, signalling proteins and integrins, especially the αVβ3 receptor. Due to the high content of actin, which runs all along the peripheral area of the cell, this microfilament assembly is called acting ring and the underneath membrane is called sealing membrane. The task of the latter is to ensure the dynamic segregation of the microenvironment between the osteoclast and the bone matrix, called resorbing lacuna, where the dissolution of the bone matrix takes place (Fig. 15.4) (Teti 2011).

Matrix dissolution is carried out by the highly convoluted plasma membrane, called ruffled border, entrapped internally to the sealing membrane, in which the molecular mechanisms implicated in bone resorption are located (Fig. 15.4). These consist mainly of the V-H$^+$ATPase which actively releases protons into the resorbing lacuna, the Cl$^-$/H$^+$ antiporter type 7, which charge balances the protons, and lysosomal and endosomal enzymes, especially cathepsin K and MMP9 respectively, which degrade the matrix proteins, particularly type I collagen. Of note, lysosomes

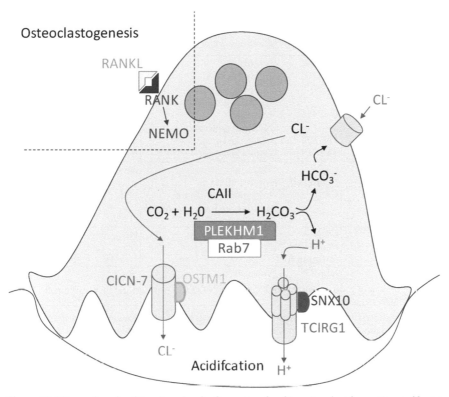

Osteoclastogenesis

RANKL

RANK

NEMO

CL^-

HCO_3^-

CAII

$CO_2 + H_2O \longrightarrow H_2CO_3$

PLEKHM1

Rab7

H^+

ClCN-7 OSTM1

SNX10

TCIRG1

CL^-

Acidifcation H^+

Figure 15.4 Genes involved in ostepetrosis. Genes involved in osteoclast formation and bone resorption that can be mutated in osteopetrotic patients (A. Teti, unpublished).

are secreted externally in the resorbing lacuna, a process requiring well controlled vesicular trafficking. This correctly localizes the molecules involved in the hydroxyapatite dissolution at the ruffled border plasma membrane and the lytic enzymes outside the cell.

Genetics of OP

The first gene recognized to cause OP is the *CAII* largely expressed by osteoclasts and renal tubular cells. This enzyme catalyzes the hydration of CO_2 in H_2CO_3, which spontaneously dissociates in H^+ and HCO_3^-. The reaction creates a source of protons (Fig. 15.4) and, since it is also spontaneous, inactivating mutations of the *CAII* cause a form of OP with intermediate severity (Table 15.1), uniquely characterized by tubular acidosis and, frequently, by cerebral calcifications (Bolt et al. 2005).

The *TCIRG1* gene encodes for a subunit (a3) of the V-H$^+$ ATPase expressed especially by osteoclasts (Fig. 15.4) and gastric parietal cells. Inactivating

Table 15.1 Classification of osteopetrosis.

Type	Gene	Characteristics	Haematopoietic Stem Cell Transplantation Indication
Autosomal Recessive Osteopetrosis	*TCIRG1*	Infantile	Yes (urgent)
	SNX10	Infantile	Yes (urgent)
	ClCN-7	Infantile +/– Neurodegeneration	Yes to no?
	OSTM1	Infantile + Neurodegeneration	No
	TNFRSF11A	Osteoclast-poor	Yes
	TNFSF11	Osteoclast-poor	No
Intermediate Recessive Osteopetrosis	*ClCN-7*	Variable	Severe cases only?
	PLEKHM1	Intermediate	
	CA II	Intermediate + Renal tubular acidosis + Cerebral calcifications	
Autosomal Dominant Osteopetrosis	*ClCN-7*	Mild (to variable)	No
	PLEKHM1 (mismatch)	Focal mild	
Ectodermal Dysplasia (X-linked)	*NEMO*	Variable + Immunodeficiency	Yes?

This table shows the genes implicated in the various forms of osteopetrosis, the characteristics of each form and the indication for hematopoietic stem cell transplantation.

mutations of this gene characterize about 50 percent of patients affected by autosomal recessive OP (Frattini et al. 2000). They are also characterized by increased gastric pH, impaired gastric calcium absorption and increased non-mineralized osteoid deposition (Schinke et al. 2009). This form is now called "osteopetro-rickets", being the increased osteoid typical of rickets. In osteoclasts, *TCIRG1* gene mutations cause impairment of proton transport into the resorbing lacuna, which prevents the hydroxyapatite dissolution. Recently, a honotygous missense mutation of the SNX10 gene has been found in patients with a similar osteopetrotic phenotype. This gene encodes a protein belonging to the sorting nexin family that interacts with the V-H$^+$ ATPase, probably implicated in the trafficking of this proton pump to the ruffled border membrane (Aker et al. 2012).

Coupled to the proton transport there is the release in the resorbing lacuna of negatively charged chloride ions (Fig. 15.4). This occurs through a dimeric protein encoded by the *ClCN-7* gene, which functions as a 2:1 chloride:proton antiporter. Patients with inactivating mutations in this gene may have OP associated with lysosomal storage disease and primary neurodegeneration (Kornak et al. 2001).

Heterozygous missense mutations of the *ClCN-7* gene can cause an autosomal dominant form of OP (Cleiren et al. 2001), the only one so far with a genetic diagnosis. This OP is phenotypically heterogeneous, characterized by focal sclerosis, especially of the skull base, pelvis and vertebral end plates. Patients have a generally normal life expectancy but may suffer from multiple fractures, haematological and neural malfunctions, dental problems and osteomyelitis (Del Fattore et al. 2008).

The *OSTM1* gene product is especially involved in the vesicular trafficking that delivers the ClCN-7 protein to the ruffled border (Fig. 15.4). The two proteins co-localize, and in the *OSTM1*-deficient patients the disruption of *OSTM1* prevents ClCN-7 to reach the plasma membrane of the ruffled border (Leisle et al. 2011). This tight association explains why *OSTM1*- and *ClCN-7*-dependent autosomal recessive OP share the lysosomal storage and neurodegenerative syndromes (Lange et al. 2006).

An inactivating mutation of the *PLEKHM1* gene has been found in an autosomal recessive OP with an intermediate course. This gene encodes a protein localized in the endosomal acidic vesicles of the osteoclast, in tight association with the small GTP-binding protein Rab7 (Fig. 15.4). Osteoclasts isolated from patients showed intracellular accumulation of TRAcP (Van Wesenbeeck et al. 2007). Interestingly, a heterozygous missense mutation of the *PLEKHM1* gene was associated with a peculiar syndrome characterized by focal sclerosis in an otherwise osteopenic phenotype (Del Fattore et al. 2008). The osteoclasts harbouring this Plekhm1 variant showed low intracellular TRAcP activity and increased TRAcP release in the supernatant, suggesting alteration of vesicular trafficking. These osteoclasts also showed reduced ability to acidify the endosomal compartment. The opposite results of these two reports suggest that different mutations in the *PLEKHM1* gene may lead to different phenotypes, as occurs in other bone diseases. For instance, inactivating mutations of the *LRP5* gene cause the osteoporosis pseudoglioma syndrome, while activating mutations are observed in the high bone mass syndrome (Van Wesenbeeck et al. 2003; Monroe et al. 2012). Incidentally, inactivating mutations of the *LRP5* gene are described in a mild form of autosomal dominant OP (Van Wesenbeeck et al. 2003) which, however, should be reclassified as high bone mass syndrome because the cells implicated are the osteoblasts.

A group of genes is responsible of OP forms which completely lack osteoclasts. These are classified as osteoclast-poor OP as opposed to the form described above, denominated osteoclast-rich OP, in which normal or even increased numbers of non-functional osteoclasts are identified in bone biopsies (Villa et al. 2009). Osteoclast-poor osteopetrosis may be due to mutations of the *TNFSF11* gene, encoding the RANKL (Sobacchi et al. 2007), or the *TNFRSF11A* gene, encoding its receptor RANK (Fig. 15.4) (Guerrini et al. 2008; Pangrazio et al. 2012). These forms are also characterized by immunological anomalies.

The gene called *NEMO* is associated with a peculiar form of OP. *NEMO* encodes for a regulatory subunit of IKK, a signal molecule involved in the RANKL/RANK pathway (Fig. 15.4) (Roberts et al. 2010).

Inactivating mutations of this gene cause an X-linked syndrome, in which OP is associated with lymphaedema, anhydrotic ectodermal dysplasia and immunodeficiency (OLEDAID). OLEDAID is lethal and, due to its extreme rarity and severity, there is little information regarding the osteoclast status and the therapeutic options (Permaul et al. 2009; Fish et al. 2009).

Rational of HSCT

Since osteoclasts are derived from the haematopoietic stem cell, HSCT in principle is a curative approach in OP. The proof of principle of HSCT in osteopetrosis was obtained for the first time in mice in 1975 (Walker 1975) and in man in 1980 (Coccia et al. 1980). However, in the last decade it was becoming evident that OP is a heterogeneous disease, which implicates some major considerations before a HSCT can be initiated in the individual patient.

- **Indication:** There are two "classical" and absolute indications to initiate HSCT in OP, a haematopoietic insufficiency requiring red cell and/ or platelet transfusions and the prevention of imminent blindness. In several OP patients these clinical indications are not, or not yet, existent and sometimes the clinical course over time may not be predictable. Furthermore, in milder variants of OP and in older patients other clinical symptoms may be prominent (pathological fractures, CNS problems, recurrent infections). In these non-classical OP variants the indication for HSCT may be relative or even not existent and therefore a detailed and interdisciplinary workup as well as a clarification of these facts with the parents (and patients) is mandatory.
- **Contra-indications:** There are two absolute contra-indications for HSCT, the osteoclast non-intrinsic forms and the neurodegenerative variants. A known osteoclast non-intrinsic variant is the OP due to RANKL mutations. Since RANKL is expressed in osteoblasts, patients with this form cannot be cured by HSCT (Sobacchi et al. 2007). In the neuronopathic form a progressive neurodegeneration with developmental delay and convulsions are prominent (Steward 2003). These symptoms due to a primary defect in nervous tissue must be carefully distinguished from neurological sequelae secondary to hyperostotic malformations such as blindness due to optical nerve compression and hydrocephalus. Since the neurodegeneration is not positively influenced by HSCT and affected patients have a very bad prognosis, there is usually no indication for HSCT.

Whereas the detection of mutations in the *OSTM1* and RANKL genes seem to be a clear contra-indication for HSCT, a clear black and white discrimination of other clinical variants by the genetic background is sometimes not possible so far (see Table 15.1). For instance, even though most patients with *TCIRG1* mutations show a "classical" infantile manifestation of OP, milder forms without haematological and visual impairments are possible (A.S. Schulz, unpublished observation). Furthermore, in patients with *ClCN-7* mutations, there seems to be a wide clinical spectrum ranging from severe neurodegenerative forms (as usually in *OSTM1* patients) over "classical" infantile forms without neurodegeneration to intermediate forms without haematopoietic insufficiency (Pangrazio et al. 2010). Though there is a wide spectrum ranging from an absolute and urgent indication in patients with "classical" infantile OP (harbouring bone marrow failure and impending visual loss, usually *TCIRG1, SNX10* and some *ClCN-7* mutations) over a relative indication in patients with intermediate forms (some *ClCN-7, RANK* and *CAII* variants) up to a clear contra-indication in neurodegenerative (*OSTM1* and some *ClCN-7* variants) and osteoclast non-intrinsic (RANKL) forms.

Stem Cell Donor and Stem Cell Source

As usual the best *donor* is an HLA-identical family donor (MFD) as an HLA-identical (and healthy!) sibling or another HLA-genoidentical relative, particularly in consanguineous families. If a matched family donor cannot be identified, a search for a matched unrelated donor (MUD) should be initiated. If neither an MFD nor an MUD can be found within a reasonable time frame, an HLA-haploidentical transplantation using a parent as the donor may be considered. HSCT from an HLA-haplotype mismatched donor is associated with additional risks; however, the probability of disease free survival in this setting improved in recent years from 24 percent (Driessen et al. 2003) to about 65 percent now (A.S. Schulz, unpublished observations). In a patient with classical infantile OP presenting with bone marrow failure but yet conserved vision without an MFD but with a rare HLA type, an HLA-haploidentical HSCT should be initiated without delay (Schulz et al. 2002). As an interesting alternative approach a cord blood progenitor transplantation may be considered. This approach may however result in unexpected severe GvHD, delayed engraftment and decelerated haematological recovery particularly in this disease because of the limited number of stem cells (Jaing et al. 2010).

With regard to the *stem cell sources,* in a HLA-identical setting, bone marrow HSCT should be preferred because peripheral blood HSCT is associated with a significant higher risk of extensive chronic GvHD in children and adults with malignant and non-malignant diseases (Stem Cell Trialists Collaborative Group 2005). Furthermore, theoretically committed osteoclast precursors in the bone marrow, probably not existent in peripheral blood stem cells, may facilitate and accelerate bone resorption and engraftment. In the HLA-haploidentical setting, however, an extensive T-cell depletion is necessary along with a "megadose" of stem cells to ensure engraftment and to avoid fatal GvHD (Schulz et al. 2002). *In vitro* removal of T lymphocytes can be performed by positive selection of CD34+ stem cells and/or by negative depletion of CD3+ T lymphocytes using paramagnetic beads coupled to monoclonal antibodies (Lang et al. 2008). In the future, modified techniques as a depletion of GvHD inducing α/β-T cells from peripheral blood or bone marrow may enhance and accelerate haematopoietic and immune reconstitution in HLA-haploidentical HSCT.

Conditioning Regimen and Prophylaxis of VOD and GvHD

The "classical" myeloablative conditioning regimen based on the myeloablative drug Busulfan, given orally, and the immunosuppressive agent Cyclophosphamid, adheres to an enhanced risk of severe toxicity, especially VOD/sinusoidal obstruction syndrome of the liver particularly in OP patients (Corbacioglu et al. 2006). For this reason, the conditioning regimens have been modified to reduce the toxicity by retaining the myeloablative potential in the last decade. For instance:

- *Busulfex* can be given i.v. avoiding the liver-toxic "first pass" effect of the p.o. formula. Furthermore, timely drug monitoring can, and should be done, to ensure a correct dosage of the drug in the individual patient.
- *Fludarabine*, a highly immunosuppressive but not alkylating agent, can replace cyclophosphamide significantly reducing the risk of VOD.
- *Mycophenolatmofetyl* can replace the liver toxic Methotrexate administered in addition to CSA for GvHD prophylaxis after HSCT (Bolwell et al. 2004).
- *Defibrotide*, a polydeoxyribonucleotide with aptameric activity on endothelium, can be administered to avoid and/or treat severe liver VOD (Corbacioglu et al. 2004, 2006).

These modifications have been included in the actual guidelines for myeloablative conditioning in HSCT for OP (Table 15.2, guidelines available at http://www.ebmt.org/Contents/Pages/Default.aspx).

Complications of HSCT

Engraftment problems

Variations to a normal haematological reconstitution are common and should be distinguished accurately:

Table 15.2 Suggested transplant procedures.

Donor	Stem Cell Source	Conditioning Regimen	Prophylaxis of Rejection and Graft-versus-Host-Disease
Matched Family Donor	Bone marrow Peripheral blood stem cells	Busulfex i.v. (myeloablative AUC) Fludarabine (160 mg/m²)	CSA MMF
Matched Unrelated Donor	Bone marrow Peripheral blood stem cells	Busulfex i.v. (myeloablative AUC) Fludarabine (160 mg/m²) Thiotepa (10 mg/kg)	ATG (or Campath-1H) CSA MMF
Haploidentical Donor	T-cell-depleted Peripheral blood stem cells	Busulfex i.v. (myeloablative AUC) Fludarabine (160 mg/m²) Thiotepa (15 mg/kg)	MMF (if > 5*10⁴ CD3+ T-cells in the graft)

This table shows the features of donors, stem cell sources, conditioning regimen and prophylaxis procedure to prevent the graft-versus-host disease according to the actual guidelines of the European Society of Immunodeficiencies (ESID) and European Group of Blood and Marrow Transplantation (EBMT). ATG: Anti-Thymocyte Globulin. AUC: Area Under Curve. CSA: Cyclosporin A. MMF: Mycophenolate MoFetil.

- A *delayed haematological recovery* is common in OP, possibly because of narrowed marrow space and/or hepatosplenomegaly particular in advanced disease and older patients. A large number of stem cells in the graft (> 10 x 10⁶ CD34+ cells/kg body weight recipient) and if necessary a stem cell boost about 4 wk after HSCT may prevent this complication (Schulz et al. 2002).

- An *active rejection* is often observed a few days after primary engraftment in particular when using HLA-haploidentical stem cells in patients >10 mon of age. In the case of an active acute rejection (rising recipient T-cells with CD8-phenotype, disappearing donor granulocytes and stem cells), a secondary highly immunosuppressive conditioning regimen prior to a stem cell boost or a second transplant from the other parent may be a successful rescue strategy.

- Stable *mixed chimerism* without recurrence of disease, sometimes characterized by "tolerant" T-cells of the recipient after transplantation, which do not reject the donor cells and disappear over time, may require no specific treatment.

- The *extinction of donor cells*, accompanied by worsening of haematopoiesis, can be successfully treated by a second HSCT.

Because of these common complications, a careful monitoring of haematological, immunological and chimerism parameters after HSCT is recommended. It is noteworthy, that these complications can be treated successfully, if diagnosed correctly and early (Schulz et al. 2002).

VOD

Patients with OP are at high risk of developing liver (and pulmonary) VOD (Corbacioglu et al. 2006). Data of a controlled prospective randomized trial in children demonstrated that administration of prophylactic Defibrotide efficiently prevents VOD (Corbacioglu et al. 2012). Therefore, the prophylactic use of Defibrotide or at least the start of this specific therapy early after diagnosis is indicated.

Respiratory problems

Respiratory problems are common during transplantation for several reasons:

- *Upper airway obstructions* (e.g., choanal stenosis).
- *Secondary ventilation problems:* These are due to fluid overload and hepatosplenomegaly (VOD, CLS) or CNS diseases (hydrocephalus, hypocalcaemic convulsions).
- *Infection:* In addition to the general risk of infections in bone marrow aplasia there seems to be an increased risk of pneumocystic jirovecii pneumonia, possibly because of the lack of prophylaxis before HSCT and the prolonged haematological and immunological recovery (A.S. Schulz, unpublished observation).
- *Primary pulmonary hypertension:* In the case of respiratory problems, particularly oxygen requirement for unknown reason, one should consider monitoring and treating pulmonary hypertension (Kasow 2008). This rare complication with an increased risk in OP patients has been successfully treated with Epoprostenol (Flolan) (Steward et al. 2004).

Serum calcium disturbance

Following engraftment of donor cells an elevation of serum calcium is commonly observed. Very high serum calcium levels, seen particularly in older patients with high bone mass following HLA-identical transplantation, can lead to life threatening symptoms (Martinez et al. 2010). Inhibition of overwhelming osteoclast function by bisphosphonates (Rawlinson et al. 1991) and/or by Denosumab (PROLIA, Amgen), a monoclonal RANKL antibody, was successful in very severe cases.

Outcome

In a retrospective analysis of 122 children who had received HSCT for OP between 1980 and 2001, the actuarial probabilities of 5 yr disease free survival were 73 percent for recipients of a genotype HLA-identical HSCT (n=40), 43 percent for recipients of a phenotype HLA-identical or one HLA-antigen mismatch graft from a related donor (n=21), 40 percent for recipients of a graft from a matched unrelated donor (n=20) and 24 percent for patients who received a graft from an HLA-haplotype-mismatch related donor (n=41). In a more recent survey on behalf of the ESID and the EBMT in 106 patients transplanted between 1993 and 2007, an improvement was achieved in all transplant groups but most markedly in the HLA-haploidentical HSCT group showing an actuarial probability of 5 yr disease free survival of 60 percent compared to 24 percent (Fig. 15.5) (A.S. Schulz, unpublished data, Schulz et al. 2002). Causes of death after HSCT were mainly graft failure and early transplant-related complications.

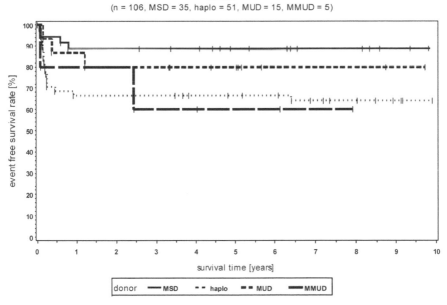

Figure 15.5 Survival of osteopetrotic patients after transplantation. Survival following haematopoietic stem cell transplantation according to donor (data of a survey of the European Society of Immunodeficiencies and the European Group of Blood and Marrow Transplantation. A.S. Schulz, unpublished).

Even after successful transplantation, patients with OP have a high risk of specific sequelae requiring a long-term interdisciplinary follow up:

- *Visual impairment:* In the study of Driessen et al. (2003) severe visual impairment was present in 42 percent of the children before HSCT. In the actual retrospective survey, about 1/3 of patients had normal vision, about 1/3 moderate visual impairment and 1/3 total optical nerve atrophy post HSCT. Conservation of vision was better in children transplanted before the age of 3 mon.
- *Dwarfism:* After HSCT, the height of the majority of patients is between 3rd and 10th percentile and about 20 percent of patients show dwarfism (< 3rd percentile) after transplantation. Final height was reported to be related to height at the time of HSCT and better preserved in children transplanted early (Driessen et al. 2003). There seems to be no primary growth hormone deficiency in OP patients (al Herbish et al. 1994) and treatment by growth hormone is not established but may be considered in severe cases after careful evaluation of pros and cons.
- *Mental retardation and CNS complications:* Developmental and mental retardation is reported in a small subgroup of patients. Undiscovered neurodegeneration due to the disease (in particular in patients with *CLCN-7* mutations), severe complications during the HSCT procedure, and handicaps because of visual impairment can contribute to several mental problems (Pangrazio et al. 2010). Increased cranial pressure secondary to craniosynostosis has been described in very few patients, which sometimes has to be treated surgically (Dowlati et al. 2007). In an also small subset of patients, symptoms of autism have been described, which may be due to visual or auditory compromise or to the particular genetic defect or both (Mazzolari et al. 2009).
- *Osteoporosis:* Very recent mouse data indicate the risk of impaired calcium homeostasis due to impaired gastric acidification leading to late hypocalcaemic osteoporosis (particularly in patients with *TCIRG1* mutations), which may be treated by calcium gluconate (Schinke et al. 2009). Therefore, bone marrow density measurements are recommended in long-term follow up after HSCT.

It is noteworthy that after successful HSCT most children attended regular school or education for the visually handicapped, and that about 2/3 of patients were judged by their parents to have a normal quality of life (A.S. Schulz, unpublished data, Driessen et al. 2003).

Problems and Pitfalls

Delayed Diagnosis: Unfortunately, OP is in many cases diagnosed with delay probably because of the rareness and of the sometimes confusing clinical symptoms in multiple organ systems. Some patients have been misdiagnosed (and treated) for weeks and months due to the leading symptom, for instance as neonatal leukaemia (because of haematological alterations), as Leber's hereditary optic neuropathy (because of visual impairment) or as isolated Arnold Chiary like malformation (due to CNS changes). This delay is disadvantageous particularly in rapid progressive infantile OP since sensorial defects like blindness are not reversible and the risks of transplantation increase with the progression of disease with age. Therefore it is imperative to improve the knowledge and awareness of this treatable rare disease by enforced actions of medical key players and political stakeholders.

Transplantation of patients missing an HLA-matched donor

Despite the improvement of HLA-haploidentical HSCT in recent years, particularly older patients with advanced disease often fail to respond to this approach. In addition to the improvement of early diagnosis (see above), steady amendments of this transplant procedure are necessary as well as visionary design and development of innovative treatment strategies and supportive methods, as for instance gene therapy approaches (under way for *TCIRG1*) and development of drugs enhancing osteoclast function (as for instance soluble RANKL). Therefore, concerted activities of treating physicians and scientific experts should be escalated in transnational networks. (IEWP/EBMT guidelines and surveys. see Table 15.2).

Treatment of patients with intermediate OP

There is uncertainty over the best management and treatment of intermediate forms of OP, for example for patients without haematological insufficiency. Conservative treatment strategies have to be improved by an enhanced interdisciplinary management of OP patients. Furthermore, prospective studies and registries are eligible clarifying not only the probability of survival with or without transplantation but also the quality of life and long-term sequelae of either treatment strategy (Prospective Registry of the ESID and the EBMT, see Table 15.2).

Treatment of OP patients with RANKL and OSTM1 mutations

HSCT is considered as non-effective in patients harbouring these gene defects. While patients with RANKL mutations usually have an intermediate clinical course and a life span of several years, patients with *OSTM1* mutation usually die as infants.

Table 15.3 Resources of information regarding the treatment of osteopetrosis.

Title	Available at	Description
Osteopetrosis-Consensus Guidelines for Diagnosis, Therapy and Follow Up	On request at ansgar.schulz@ uniklinik-ulm.de (will be published on ESID and EBMT homepages soon)	Consensus guidelines on behalf of the Inborn Error Working Party of the European Group of Blood and Marrow Transplantation and the European Society of Immunodeficiencies
European Society of Immunodeficiencies (ESID)	Homepage: http://www. esid.org	Guidelines, registries and expert forum, including Working Party Bone Marrow Transplantation and Gene Therapy
European Group of Blood and Marrow Transplantation (EBMT)	Homepage: http://www. ebmt.org	Guidelines, registries and expert forum, including Working Party Inborn Errors
Orphanet	Hompage: http://www. orpha.net	Expert Centres, diagnostic tests, research and trials, patient organizations, professionals and institutions for rare diseases
Genetic Home Reference	Homepage: http://ghr. nlm.nih.gov/condition/ osteopetrosis	Set of Links to References, Gene Reviews, Gene Tests, Clinical Trials

This table shows how useful information can be gained by patients and doctors.

Conclusions

When autosomal recessive OP is caused by osteoclast intrinsic defects and presents with no significant CNS alteration, it can be cured with HSCT. The procedure has improved over time and is now suitable for a wider OP patient group, which can be restored for their bone, bone marrow and sensorial functions. However, there are forms of OP that are still incurable or for which there are important contra-indications that require further investigations to identify effective therapies. For instance, the RANKL-dependent OP could theoretically be cured by administration of soluble RANKL or by mesenchymal stem cell transplantation, while new disease causative genes need to be identified for the large population of OP patients who still lack a genetic diagnosis. We expect major advances in the near future for the set up of new therapies and for the improvement of those already existent.

Key Features of Osteopetrosis

- Autosomal recessive osteopetrosis is a life-threatening infantile genetic bone disease affecting the skeleton and other organs.

- It is caused by impairment of osteoclast function or formation.
- It is cured by haematopoietic stem cell transplantation which improves the haematological failure but does not change the dwarfism and rarely reverts the vision impairment.
- Areas of improvement are the stem cell donor, the stem cell source, and the conditioning regimen.
- There are several contra-indications to be taken into account when treating the patients.

Dictionary

- *Autosomal recessive inheritance*: Transfer of a mutant gene located in autosomal chromosomes which requires biallelic alteration to cause the disease.
- *Bone marrow*: Tissue localized in the bone marrow cavity that provides the blood elements.
- *Donor*: The individual who donates a tissue or an organ for transplantation.
- *Engraftment*: The process whereby donor haematopoietic stem cells survive in the transplanted patients and replenish the bone marrow cells.
- *Haematopoiesis*: The process of formation of blood elements.

Summary Points

- Autosomal recessive osteopetrosis is a life-threatening infantile genetic bone disease affecting approximately one in every 250,000 born.
- The skeleton presents with dense and brittle bones, and with severe haematological and neural failures, due to osteoclast malfunction.
- Osteoclasts are multinucleated cells that resorb the bone.
- Lack of bone resorption causes the persistence of primary bone and the compression of the inner cavities lodging vital organs, such as the bone marrow and the nervous system.
- Osteopetrosis is cured by haematopoietic stem cells transplantation.
- The chapter describes the genetics of osteopetrosis, the rational of haematopoietic stem cells transplantation and the contra-indications.
- It also provides information on stem cell donors and sources, conditioning regimen and prophylaxis to prevent complications.
- The complications of haematopoietic stem cells transplantation are described in detail along with the outcome, with an emphasis on the improvements implemented over the last decade.

- The chapter describes problems, pitfalls and our vision for future improvements and alternative therapies.

Acknowledgments

We wish to thank Sandra Steinmann for excellent data management, Dr. Anna Villa, Dr. Despina Moshous and Dr. Colin Steward for many helpful discussions and Dr. Rita Di Massimo for excellent help in editing the manuscript. The original work has been supported by the E-Rare JTC 2007 OSTEOPETR grant to A.T. and A.S.S., and by the Telethon grant GGP09018 to A.T.

List of Abbreviations

AP-1	:	Activator Protein-1
ATG	:	Anti-Thymocyte Globulin
ATPase	:	Adenosine TriPhosphatase
AUC	:	Area Under Curve
BM	:	Bone Marrow
CAII	:	Carbonic Anhydrase type II
ClCN-7	:	Chloride ChaNnel 7
Cl⁻	:	Chloride ion
CLS	:	Capillary Leak Syndrome
CNS	:	Central Nervous System
CSA	:	Cyclosporin A
DC-STAMP	:	Dendritic Cell-Specific TrAnsMembrane Protein
EBMT	:	European Group for Blood and Marrow Transplantation
ESD	:	European Society for Immunodeficiencies
G-CSF	:	Granulocyte-Colony Stimulating Factor
GTP	:	Guanosine TriPhospahte
GvHD	:	Graft-versus-Host Disease
H⁺	:	Hydrogen ion (proton)
HLA	:	Human Leukocyte Antigen
HSC	:	Haematopoietic Stem Cell
HSCT	:	Haematopoietic Stem Cell Transplantation
IEWP	:	Inborn Errors Working Party
IKK	:	IκB Kinase
LRP5	:	Low-density Lipoprotein Receptor-related Protein 5
M-CSF	:	Macrophage-Colony Stimulating Factor
MFD	:	HLA-identical Family Donor
MMF	:	Mycophenolatmofetyl

MMP9	:	Matrix MetalloProteinase 9
MUD	:	Matched Unrelated Donor
NEMO	:	NF-κB Essential Modulator
NFATc1	:	Nuclear Factor of Activated T-cells c1
NF-κB	:	Nuclear Factor-κB
OLEDAID	:	Ectodermal Dysplasia, Anhidrotic, with Immunodeficiency, Osteopetrosis, and Lymphedema
OP	:	Osteopetrosis
OSTM1	:	Osteopetrosis-associated transmembrane protein 1
PBSC	:	Peripheral Blood Stem Cells
PLEKHM1	:	PLEcKstrin Homology domain containing, family M (with RUN domain) member 1
p.o.	:	per os (oral administration)
RANK	:	Receptor Activator of NF-κB
RANKL	:	Receptor Activator of NF-κB Ligand
TCIRG1	:	T-Cell Immune Regulator Gene 1
TNFRSF11A:		Tumuor Necrosis Factor Receptor SuperFamily, member 11a
TNFSF11	:	Tumuor Necrosis Factor (ligand) SuperFamily, member 11
TRAcP	:	Tartrate-Resistant Acid Phosphatase
VOD	:	Venous Occlusive Disease

References

al Herbish, A.S., A. al Jarallah, N.A. al Jurayyan, A.M. abo Bakr, S.A. al Rasheed and A.H. Mahdi. 1994. Growth hormone and IGF1 profile in short children with osteopetrosis. Clin Invest Med. 17: 26–30.

Aker, M., A. Rouvinski, S. Hashavia, A. ta-Shma, A. Shaag, S. Zenvirt, S. Israel, M. Weintraub, A. Taraboulos, Z. Bar-Shavit and O. Elpeleg. 2012. An SNX10 mutation causes malignant. Osteopetrosis of infancy. J Med Genet. 49: 221–226.

Bolt, R.J., J.M. Wennink, J.I. Verbeke, G.N. Shah, W.S. Sly and A. Bökenkamp. 2005. Carbonic anhydrase type II deficiency. Am J Kidney Dis. 46: A50, e71–3.

Bolwell, B., R. Sobecks, B. Pohlman, S. Andresen, L. Rybicki, E. Kuczkowski and M. Kalaycio. 2004. A prospective randomized trial comparing cyclosporine and short course methotrexate with cyclosporine and mycophenolate mofetil for GVHD prophylaxis in myeloablative allogeneic bone marrow transplantation. Bone Marrow Transplant. 34: 621–625.

Cleiren, E., O. Bénichou, E. Van Hul, J. Gram, J. Bollerslev, F.R. Singer, K. Beaverson, A. Aledo, M.P. Whyte, T. Yoneyama, M.C. deVernejoul and W. Van Hul. 2001. Albers-Schönberg disease (autosomal dominant osteopetrosis, type II) results from mutations in the ClCN7 chloride channel gene. Hum Mol Genet. 10: 2861–2867.

Coccia, P.F., W. Krivit, J. Cervenka, C. Clawson, J.H. Kersey, T.H. Kim, M.E. Nesbit, N.K. Ramsay, P.I. Warkentin, S.L. Teitelbaum, A.J. Kahn and D.M. Brown. 1980. Successful bone-marrow transplantation for infantile malignant osteopetrosis. N Engl J Med. 302: 701–708.

Corbacioglu, S., J. Greil, C. Peters, N. Wulffraat, H.J. Laws, D. Dilloo, B. Straham, U. Gross-Wieltsch, K.W. Sykora, A. Ridolfi-Lüthy, O. Basu, B. Gruhn, T. Güngör, W. Mihatsch and

A.S. Schulz. 2004. Defibrotide in the treatment of children with veno-occlusive disease (VOD): a retrospective multicentre study demonstrates therapeutic efficacy upon early intervention. Bone Marrow Transplant. 33: 189–195. Erratum in: 2004 Bone Marrow Transplant. 33: 673.

Corbacioglu, S., M. Hönig, G. Lahr, S. Stöhr, G. Berry, W. Friedrich and A.S. Schulz. 2006. Stem cell transplantation in children with infantile osteopetrosis is associated with a high incidence of VOD, which could be prevented with defibrotide. Bone Marrow Transplant. 38: 547–553.

Corbacioglu, S., N. Kernan, L. Lehmann, J. Brochstein, C. Revta, S.Grupp, P. Martin and P.G. Richardson. 2012. Defibrotide for the treatment of hepatic veno-occlusive disease in children after haematopoietic stem cell transplantation. Expert Rev Hematol. 5: 291–302.

Del Fattore, A. and A. Teti. Osteoclast Genetic Diseases. pp. 57–78 *In*: D. Plaseska-Karanfilska [ed.]. 2011. Human Genetic Diseases. Intech Open Access Publisher. Skopje, Republic of Macedonia.

Del Fattore, A., R. Fornari, L. Van Wesenbeeck, F. de Freitas, J.P. Timmermans, B. Peruzzi, A. Cappariello, N. Rucci, G. Spera, M.H. Helfrich, W. Van Hul, S. Migliaccio and A. Teti. 2008. A new heterozygous mutation (R714C) of the osteopetrosis gene, pleckstrin homolog domain containing family M (with run domain) member 1 (*PLEKHM1*), impairs vesicular acidification and increases TRACP secretion in osteoclasts. J Bone Miner Res. 23: 380–391.

Dowlati, D., K.R. Winston, L.L. Ketch, R. Quinones, R. Giller, A. Frattini and J. van Hove. 2007. Expansion cranioplasty with jackscrew distracters for craniosynostosis and intracranial hypertension in transplanted osteopetrosis. Pediatr Neurosurg. 43: 102–106.

Driessen, G.J., E.J. Gerritsen, A. Fischer, A. Fasth, W.C. Hop, P. Veys, F. Porta, A. Cant, C.G. Steward, J.M. Vossen, D. Uckan and W. Friedrich. 2003. Long-term outcome of haematopoietic stem cell transplantation in autosomal recessive osteopetrosis: an EBMT report. Bone Marrow Transplant. 32: 657–663.

Fish, J.D., R.E. Duerst, E.W. Gelfand, J.S. Orange and N. Bunin. 2009. Challenges in the use of allogeneic hematopoietic SCT for ectodermal dysplasia with immune deficiency. Bone Marrow Transplant. 43: 217–221.

Frattini, A., P.J. Orchard, C. Sobacchi, S. Giliani, M. Abinun, J.P. Mattsson, D.J. Keeling, A.K. Andersson, P. Wallbrandt, L. Zecca, L.D. Notarangelo, P. Vezzoni and A. Villa. 2000. Defects in TCIRG1 subunit of the vacuolar proton pump are responsible for a subset of human autosomal recessive osteopetrosis. Nat Genet. 25: 343–346.

Guerrini, M.M., C. Sobacchi, B. Cassani, M. Abinun, S.S. Kilic, A. Pangrazio, D. Moratto, E. Mazzolari, J. Clayton-Smith, P. Orchard, F.P. Coxon, M.H. Helfrich, J.C. Crockett, D. Mellis, A. Vellodi, I. Tezcan, L.D. Notarangelo, M.J. Rogers, P. Vezzoni, A. Villa and A. Frattini. 2008. Human osteoclast-poor osteopetrosis with hypogammaglobulinemia due to TNFRSF11A (RANK) mutations. Am J Hum Genet. 83: 64–76.

Jaing, T.H., S.H. Chen, M.H. Tsai, C.P. Yang, I.J. Hung and P.K. Tsay. 2010. Transplantation of unrelated donor umbilical cord blood for nonmalignant diseases: a single institution's experience with 45 patients. Biol Blood Marrow Transplant. 16: 102–107.

Kasow, K.A., R.M. Stocks, S.C. Kaste, S. Donepudi, D. Tottenham, R.A. Schoumacher and E.M. Horwitz. 2008. Airway evaluation and management in 7 children with malignant infantile osteopetrosis before hematopoietic stem cell transplantation. J Pediatr Hematol Oncol. 30: 225–229.

Kornak, U., D. Kasper, M.R. Bösl, E. Kaiser, M. Schweizer, A. Schulz, W. Friedrich, G. Delling and T.J. Jentsch. 2001. Loss of the ClC-7 chloride channel leads to osteopetrosis in mice and man. Cell. 104: 205–215.

Lang, P., I. Mueller, J. Greil, P. Bader, M. Schumm, M. Pfeiffer, W. Hoelle, T. Klingebiel, F. Heinzelmann, C. Belka, P.G. Schlegel, B. Kremens, W. Woessmann and R. Handgretinger. 2008. Retransplantation with stem cells from mismatched related donors after graft rejection in pediatric patients. Blood Cells Mol Dis. 40: 33–39.

Lange, P.F., L. Wartosch, T.J. Jentsch and J.C. Fuhrmann. 2006. ClC-7 requires Ostm1 as a beta-subunit to support bone resorption and lysosomal function. Nature. 440: 220–223.

Leisle, L., C.F. Ludwig, F.A. Wagner, T.J. Jentsch and T. Stauber. 2011. ClC-7 is a slowly voltage-gated 2Cl(–)/1H(+)-exchanger and requires Ostm1 for transport activity. EMBO J. 30: 2140–2152.

Martinez, C., L.E. Polgreen, T.E. DeFor, T. Kivisto, A. Petryk, J. Tolar and P.J. Orchard. 2010. Characterization and management of hypercalcemia following transplantation for osteopetrosis. Bone Marrow Transplant. 45: 939–944.

Mazzolari, E., C. Forino, A. Razza, F. Porta, A. Villa and L.D. Notarangelo. 2009. A single-center experience in 20 patients with infantile malignant osteopetrosis. Am J Hematol. 84: 473–479.

Monroe, D.G., M.E. McGee-Lawrence, M.J. Oursler and J.J. Westendorf. 2012. Update on Wnt signaling in bone cell biology and bone disease. Gene. 492: 1–18.

Pangrazio, A., M. Pusch, E. Caldana, A. Frattini, E. Lanino, P.M. Tamhankar, S. Phadke, A.G. Lopez, P. Orchard, E. Mihci, M. Abinun, M. Wright, K. Vettenranta, I. Bariae, D. Melis, I. Tezcan, C. Baumann, F. Locatelli, M. Zecca, F. Horwitz, L.S. Mansour, M. Van Roij, P. Vezzoni, A. Villa and C. Sobacchi. 2010. Molecular and clinical heterogeneity in CLCN7-dependent osteopetrosis: report of 20 novel mutations. Hum Mutat. 31: E1071–1080.

Pangrazio, A., E. Boudin, E. Piters, G. Damante, N. Lo Iacono, A.V. D'Elia, P. Vezzoni, W. Van Hul, A. Villa and C. Sobacchi. 2011. Identification of the first deletion in the LRP5 gene in a patient with autosomal dominant osteopetrosis type I. Bone. 49: 568–571.

Pangrazio, A., B. Cassani, M.M. Guerrini, J.C. Crockett, V. Marrella, L. Zammataro, D.Strina, A. Schulz, C. Schlack, U. Kornak, D.J. Mellis, A. Duthie, M.H. Helfrich, A. Durandy, D. Moshous, A. Vellodi, R. Chiesa, P. Veys, N.L. Iacono, P. Vezzoni, A. Fischer, A. Villa and C. Sobacchi. 2012. RANK-dependent autosomal recessive osteopetrosis: characterisation of 5 new cases with novel mutations. J Bone Miner Res. 27: 342–351.

Permaul, P., A. Narla, J.L. Hornick and S.Y. Pai. 2009. Allogeneic hematopoietic stem cell transplantation for X-linked ectodermal dysplasia and immunodeficiency: case report and review of outcomes. Immunol Res. 44: 89–98.

Rawlinson, P.S., R.H.Green, A.M.Coggins, I.T.Boyle and B.E.Gibson. 1991. Malignant osteopetrosis: hypercalcaemia after bone marrow transplantation. Arch Dis Child. 66: 638–639.

Roberts, C.M., J.E. Angus, I.H. Leach, E.M. McDermott, D.A. Walker and J.C. Ravenscroft. 2010. A novel NEMO gene mutation causing osteopetrosis, lymphoedema, hypohidrotic ectodermal dysplasia and immunodeficiency (OL-HED-ID). Eur J Pediatr. 169: 1403–1407.

Schinke, T., A.F. Schilling, A. Baranowsky, S. Seitz, R.P. Marshall, T. Linn, M. Blaeker, A.K. Huebner, A. Schulz, R. Simon, M. Gebauer, M. Priemel, U. Kornak, S. Perkovic, F. Barvencik, F.T. Beil, A. Del Fattore, A. Frattini, T. Streichert, K. Pueschel, A. Villa, K.M. Debatinm, M. Rueger, A. Teti, J. Zustin, G. Sauter and M. Amling. 2009. Impaired gastric acidification negatively affects calcium homeostasis and bone mass. Nat Med. 15: 674–681.

Schulz, A.S., C.F. Classen, W.A. Mihatsch, M. Sigl-Kraetzig, M. Wiesneth, K.M. Debatin, W. Friedrich and S.M. Müller. 2002. HLA-haploidentical blood progenitor cell transplantation in osteopetrosis. Blood. 99: 3458–3460.

Sobacchi, C., A. Frattini, M.M. Guerrini, M. Abinun, A. Pangrazio, L. Susani, R. Bredius, G. Mancini, A. Cant, N. Bishop, P. Grabowski, A. Del Fattore, C. Messina, G. Errigo, F.P. Coxon, D.I. Scott, A. Teti, M.J. Rogers, P. Vezzoni A. Villa and M.H. Helfrich. 2007. Osteoclast-poor human osteopetrosis due to mutations in the gene encoding RANKL. Nat Genet. 39: 960–962.

Stem Cell Trialists' Collaborative Group. 2005. Allogeneic peripheral blood stem-cell compared with bone marrow transplantation in the management of hematologic malignancies:

an individual patient data meta-analysis of nine randomized trials. J Clin Oncol. 23: 5074–5087.

Steward, C.G. 2003. Neurological aspects of osteopetrosis. Neuropathol. Appl Neurobiol. 29: 87–97.

Steward, C.G., I. Pellierm, A. Mahajanm, M.T. Ashworth, A.G. Stuart, A. Fasth, D. Lang, A. Fischer, W. Friedrich and A.S. Schulz. 2004. Severe pulmonary hypertension: a frequent complication of stem cell transplantation for malignant infantile osteopetrosis. Br. J Haematol. 124: 63–71.

Teti, A. 2011. Bone development: overview of bone cells and signaling. Curr Osteoporos Rep. 9: 264–273.

Van Wesenbeeck, L., E. Cleiren, J. Gram, R.K. Beals, O. Bénichou, D. Scopelliti, L. Key, T. Renton, C. Bartels, Y. Gong, M.L. Warman, M.C. De Vernejoul, J. Bollerslev and W. Van Hul. 2003. Six novel missense mutations in the LDL receptor-related protein 5 (LRP5) gene in different conditions with an increased bone density. Am J Hum Genet. 72: 763–771.

Van Wesenbeeck, L., P.R. Odgren, F.P. Coxon, A. Frattini, P. Moens, B. Perdu, C.A. MacKay, E. Van Hul, J.P. Timmermans, F. Vanhoenacker, R. Jacobs, B. Peruzzi, A. Teti, M.H. Helfrich, M.J. Rogers, A. Villa and W. Van Hul. 2007. Involvement of PLEKHM1 in osteoclastic vesicular transport and osteopetrosis in incisors absent rats and humans. J Clin Invest. 117: 919–930.

Villa, A., M.M. Guerrini, B. Cassani, A. Pangrazio and C. Sobacchim. 2009. Infantile malignant, autosomal recessive osteopetrosis: the rich and the poor. Calcif. Tissue Int. 84: 1–12.

Walker, D.G. 1975. Bone resorption restored in osteopetrotic mice by transplants of normal bone marrow and spleen cells. Science. 190: 784–785.

CHAPTER 16

Human Fetal Stem Cells and Osteogenesis Imperfecta: Rodent Studies

Gemma N. Jones[a] and Pascale V. Guillot[b],*

ABSTRACT

Mesenchymal stromal/stem cells have shown promise both in rodent studies and clinical trials to treat the genetic disease osteogenesis imperfecta (brittle bone disease). Cell therapy capitalizes on the innate ability of stem cells to migrate to target sites of injury, and then differentiate to the required cell lineages to repair the tissue. Paracrine effects from the stem cells may also play a role in tissue repair. Rodent models of osteogenesis imperfecta, such as the *oim* and *Brtl* mice, have allowed better understanding of the disease pathology and the mechanisms by which infused cells improve the disease phenotype.

Here the pathology of osteogenesis imperfecta in humans including the 11 identified types, current treatments and future therapies are discussed. The rationale behind cell therapy is explained as well as the different types of stem cells, including both fetal and adult mesenchymal stromal/stem cells. Finally, the various rodent models of osteogenesis imperfecta are described along with the previous and current work being carried out on them to bring cell therapy closer to the clinic.

Fetal Stem Cell Therapy Group, Institute of Reproductive and Developmental Biology, Imperial College London, Du Cane Road, London, W12 0NN, United Kingdom.
[a]E-mail: Gemma.Jones08@imperial.ac.uk.
[b]E-mail: Pascale.Guillot@imperial.ac.uk.
*Corresponding author

List of abbreviations given at the end of the text.

The *oim* model in particular has advanced our understanding of cell therapy for OI, with current research showing fetal mesenchymal stromal/stem cells migrate to sites of bone formation and injury when transplanted in *oim*, where they differentiate into functional osteoblasts that synthesize collagen type I alpha 2 chain, absent in non-transplanted mice. This ultimately improves bone plasticity, thereby reducing fracture incidence. Current research is looking into strategies to increase engraftment by improving donor cell homing to bones as well as identifying the most suitable cell source/fraction/preparation for the clinic.

Introduction

Osteogenesis Imperfecta (OI), or brittle bone disease, is caused by mutations in collagen type I or genes associated with its biosynthesis. OI is defined as a heterogeneous group of disorders that vary in severity and result in bones being more prone to fracture (Van Dijk et al. 2010). There is no cure for OI and current treatments merely manage the symptoms or offer temporary improvements in bone strength rather than treat the underlying cause. Rodent models of OI provide a tool to not only better understand the disorder, but also to investigate the effectiveness of novel therapies, before they are brought to the clinic.

OI and Current Treatments

Osteogenesis Imperfecta (OI) is a genetic disorder of connective tissue that occurs in one in 15,000–20,000 births. There are 11 types of OI (Table 16.1) of which the clinical manifestations range in severity from the perinatal lethal type II to the mild non-deforming type I. OI patients are more susceptible to fractures due to reduced bone mass, which increases the brittleness and fragility of their bones. Other symptoms of OI can include short stature, bone deformities including scoliosis and vertebral compressions, joint laxity, blue or grey sclerae, dentinogenesis imperfecta, hearing problems, triangular shaped face, cardiopulmonary problems and reduced life expectancy (Van Dijk et al. 2010).

Current clinical management of OI includes non-surgical methods that aim to treat fractures through bracing, splinting and physiotherapy as well as improving hearing, dental and psychological issues. Whilst surgery aims to prevent and correct deformities, such as using intramedullary telescopic rods to stabilize long bones, spine stabilization to correct scoliosis and basilar impression surgery to correct migration of the spine (Monti et al. 2010). Pharmacological therapies, namely bisphosphonates, have become the standard treatment for children with OI. Bisphosphonates are internalized

Table 16.1 Types of OI, inheritance and clinical symptoms.

OI	Severity	Diagnosis	Clinical Symptoms				Genetic Defect	Reference
			Fractures	Stature	Sclera Colour	Dental Problem		
			Autosomal Dominant Inheritance:					
I	Mild	Child-adulthood	A few	Normal/mildly Short	Blue	No	COL1A1	(Sillence et al. 1979)
II	Lethal	*In utero*	Multiple	Short	Dark	NA	COL1A1/COL1A2	(Sillence et al. 1979)
III	Severe	*In utero*/at birth	Multiple	Short	Grey	Yes	COL1A1/COL1A2	(Sillence et al. 1979)
IV	Mild-moderate	Birth-adulthood	Variable	Moderately short	Grey/white	Yes/No	COL1A1/COL1A2	(Sillence et al. 1979)
V	Moderate-severe	Birth-adulthood	Multiple with hypertrophic callus	Short/mildly short	Normal white	No	Unknown	(Glorieux et al. 2000)
			Autosomal Recessive Inheritance:					
VI	Moderate-severe	Birth-adulthood	Multiple	Moderately short	Normal white	No	SERPINF1	(Glorieux et al. 2002)
VII	Moderate-severe	Birth	Multiple	Mildly short	Normal white	No	CRTAP	(Morello et al. 2006)
VIII	Severe-lethal	*In utero*/at birth	Multiple	Short	Normal white	Yes	LEPRE1	(Cabral et al. 2007)
IX	Severe-lethal	Birth-childhood	Multiple	Moderately short	Normal white	Yes	PPIB	(van Dijk et al. 2009)
X	Severe	Birth-childhood	Multiple	Moderately short	Blue	Yes	SERPINH1	(Christiansen et al. 2010)
XI	Severe	Birth	Multiple	Short	Normal white	No	FKBP10	(Alanay et al. 2010)

This table lists the OI types I-XI and shows their inheritance pattern. Also shown for each type is the causative genetic defect, disease severity, age of diagnosis and clinical symptoms presented; including incidence of fractures, stature, eye sclera colour and whether dental problems are present. OI: Osteogenesis Imperfecta.

by osteoclasts and inhibit their bone resorption, thereby increasing bone mass, density and cortical width through accumulation of the defective bone matrix (Munns et al. 2005). However, controlled trials show the effectiveness of bisphosphonates reduces long term, with no significant improvements in growth, muscle strength, bone pain or lower extremity long bone fractures (Letocha et al. 2005). Furthermore neither surgery nor bisphosphonates address the genetic mutations that affect collagen biosynthesis (shown in Fig. 16.1) and subsequently lead to bone fragility.

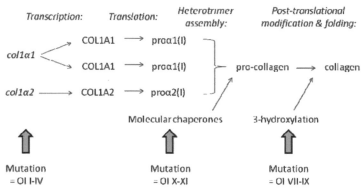

Figure 16.1 The biosynthesis of collagen type I. Diagram depicting the biosynthesis of collagen type I; *col1α1* and *col1α2* genes are transcribed and translated to two proα1(I) and one proα2(I), assembled by molecular chaperones, then modified with 3-hydroxylation and folded to form collagen type I. Mutations in *col1α1* and *col1α2* genes cause type I-IV OI, mutations in molecular chaperones cause type X-XI OI and mutations in the 3-hydroxylation process result in type VII-IX OI. Arrows represent mutations. OI: Osteogenesis imperfecta.

Future Treatments for OI

Some studies have shown growth hormone in combination with bisphosphonates may prove more effective than bisphosphonates alone, with greater bone mineral density and growth velocity measured in OI children, although fracture incidence was not improved (Antoniazzi et al. 2010). Alternatively antibodies, such as denosumab, anti-sclerostin and DKK1-IgG, have also been shown to increase bone formation over bone resorption, thereby increasing matrix deposition and bone mass (Forlino et al. 2011). However, the safety and long term efficacy of these methods is not yet known and the bone matrix deposited remains defective.

In contrast gene and cell therapy, improve bone fragility by acting during collagen formation to correct helical conformation and fibril assembly. Gene therapy aims to silence abnormal genes and/or increase normal gene expression. In autosomal recessive cases of OI (Table 16.1), where there is a deficiency in proteins required for post-translational modifications and folding (e.g., absence of 3-hydroxylation or deficiency in FKBP10 or Serpin H1 collagen chaperones), gene supplementation to over-express the normal collagen gene may be useful. Whereas in autosomal dominant OI types the defect is in the quantity or structure of type I procollagen; the presence of both mutant and normal procollagen chains results in matrix heterogeneity and the different severities of the OI phenotype. Therefore additional antisense gene approaches are required, such as using oligodeoxynucleotides, ribozymes, hammerhead ribozymes, RNAi or siRNA to promote degradation of mutant mRNA transcripts. However, the *col1α1* and *col1α2* genes span 51 and 52 exons respectively and over 800 different mutations have been reported within these genes to cause OI. Therefore the genetic heterogeneity of OI means vector design to specific mutations is both complicated and time consuming (Niyibizi et al. 2004; Forlino et al. 2011). Subsequently the reality of bringing gene therapy to the clinic as a standardized treatment for OI is unlikely in the near future.

Stem Cell Therapy

Cell therapy (Fig. 16.2) has the potential to treat OI, without the need to tailor treatments for individual mutations. Cell therapy relies on three main properties of stem cells. First, their ability to self-renew, either by symmetric (giving rise to two stem cells) or asymmetric replication (giving rise to one stem cell and one differentiated cell). Second, their ability to differentiate to specific cell lineages. Finally, their potential to be mobilized from a niche (e.g., bone marrow) and migrate to sites of injury where they regenerate and repair damaged tissues. The endogenous regenerative capacity of stem cells offers potential to treat diseases, by replacing damaged cells with normal ones from a healthy donor (Guillot et al. 2008a).

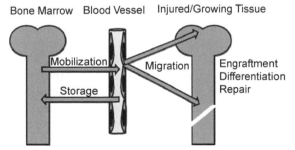

Figure 16.2 The rationale of cell therapy. This figure demonstrates the rationale of cell therapy; stem cells migrate through blood vessels to sites of tissue injury or growth where they engraft, differentiate and repair the tissue. The storage and mobilization of stem cells from niches such as the bone marrow is also shown.

Stem cells can be classified into two categories; embryonic stem (ES) cells, derived from the inner cell mass of embryos, and somatic stem cells, found in various adult (bone marrow, skin, liver), neonatal (cord) and mid-gestation fetal tissues (blood, liver, bone marrow). Embryonic stem cells can be almost indefinitely grown *in vitro* and are pluripotent (Table 16.2), but their use is restricted ethically and cells must be predifferentiated to prevent tumour formation when injected *in vivo*. In contrast adult mesenchymal stromal/stem cells (MSC) are multipotent, do not form tumours *in vivo*, can differentiate down the mesodermal lineages (bone, cartilidge, fat) and may also provide beneficial trophic effects through cytokine and growth factor production (Guillot et al. 2007). In addition, MSC do not express HLA class II and their expression of HLA class I is low, which means they are unlikely to elicit an immune reaction following allogeneic use (Le Blanc et al. 2003).

Table 16.2 Classification of stem cells according to their differentiation potential.

Potency classification	Description	Examples	Stem cells
Totipotent	Each cell can develop into any cell type	Cells in early embryos/zygotes (1–3 d post-fertilization or up to 8-cell stage of the morula)	
Pluripotent	Cells can form most cell types (e.g., all three germ layers)	Cells of the inner cell mass from blastocyst (5–14 d post-fertilization)	Embryonic and some fetal
Multipotent	Cells differentiate only into specific cell types	Cells from fetal tissue, extra-embryonic fetal tissue and adult tissues	Fetal and adult somatic cells
Unipotent	Cells form only one cell type	Progenitor cells in fetal and adult tissues	

This table defines the key potency terms used to describe the differentiation potential of different cell types, and shows which categories describe embryonic, fetal and adult stem cells.

Human fetal MSC (Table 16.3) share common characteristics with adult bone marrow MSC, such as spindle-shaped morphology, plastic adherence and gene expression profile (CD45$^-$/34$^-$/14$^-$/29$^+$/44$^+$/73$^+$/105$^+$), but present advantages over adult counterparts for cell-based therapies (LeBlanc et al. 2003; Guillot et al. 2008a). For example, human fetal bone marrow MSC grow faster, senesce later, are smaller in size, differentiate more readily and display basal expression of ES markers, suggesting a wider potency than adult MSC (Guillot et al. 2007). Other broadly multipotent, non-tumourigenic stem cells have subsequently been isolated from different fetal tissues; including blood and liver, as well as placenta and amniotic fluid.

Table 16.3 Key Facts of Human Fetal MSC.

1.	Human fetal MSC can be isolated from a variety of tissues following first trimester pregnancy termination; including fetal blood, bone marrow, liver and placenta.
2.	Human fetal MSC can also be safely isolated during ongoing pregnancy from placenta following first trimester chorionic villous sampling, or from amniotic fluid following second trimester amniocentesis.
3.	Finally human fetal MSC can be isolated from extra-embryonic tissues normally discarded at birth (e.g., placenta, umbilical cord, wharton's jelly, amniotic fluid).
4.	The phenotype, growth capacity and differentiation potential of human fetal MSC suggests they may be an intermediate cell type between early embryonic stem cells and adult MSC.
5.	Human fetal MSC can self renew in culture and can therefore be expanded for use in regenerative medicine.
6.	Human fetal MSC can differentiate to more specific tissue types and therefore have potential to be used for cell-based therapies or tissue engineering.
7.	Human fetal MSC can be repeatedly cryogenically frozen and thawed and therefore could be used to create a bank of cells for use in a wide variety of clinical applications.

This table lists the key facts of human fetal MSC, including tissues and gestational ages they can be isolated from, their cell phenotype and their applications for the clinic. MSC: Mesenchymal Stromal/Stem Cells. (Information from Abdulrazzak et al. 2010 and Guillot et al. 2007).

Clinical Trials

In contrast to gene therapy, there are currently clinical trials using cell therapy to treat OI patients; four published, two completed and one presently recruiting, with most studies carried out in the U.S.A and Sweden. One study showed transplantation of bone marrow MSC in OI children resulted in gains in body length and bone mineralization, although these effects were lost over time and donor cell engraftment was low at less than 2 percent (Horwitz et al. 2002). This group is currently recruiting participants for further clinical trials to evaluate the safety and effectiveness of repeated infusions of MSC in children with OI. In addition allogeneic fetal liver derived MSC were transplanted *in utero* in a human fetus and led to better than expected results with no visible immune rejection, engraftment of cells in bone nine months later and regularly arranged and configured bone trabecula. However, again engraftment was low at only up to 7 percent and results were confounded by concomitant use of bisphosphonates (Le Blanc et al. 2005).

Rodent Models of OI

To better understand OI at a cellular and molecular level, optimize new and current treatments and address issues raised by clinical trials; a number of rodent models of OI have been developed. A list of key OI rodent models is given in Table 16.4. The most popular rodent model, with over 57

Table 16.4 OI murine models.

Mouse Model	OI Type	Genetic Mutation	Inheritance	Reference
Mov13+/−	I	M-MuLV insertion	AD	(Bonadio et al. 1990)
Mov13−/−	II	M-MuLV insertion	AR	(Bonadio et al. 1990)
Oim−/−	III	Glycine deletion *col1α2*	AR	(Chipman et al. 1993)
Brtl+/−	IV	Knock-in *col1α1*	AD	(Forlino et al. 1999)
Brtl−/−	IV	Knock-in *col1α1*	AR	(Forlino et al. 1999)
OI-transgenic	II-IV	Human COL1A1 minigene	AD	(Khillan et al. 1991)

This table lists murine models of OI, the type of OI they most resemble, the causative genetic mutation and how the mutation is inherited. OI: Osteogenesis imperfecta. M-MuLV: Moloney murine leukaemia virus. AD: autosomal dominant. AR: autosomal recessive. $^{+/-}$: Heterozygous for mutation. $^{-/-}$: Homozygous for mutation.

publications to date, is a model of type III OI; named osteogenesis imperfecta murine (*oim*). These mice have reduced growth, multiple fractures, skeletal deformities, enlarged growth plates and decreased bone strength, length and volume. $Oim^{-/-}$ (i.e., *oim*) are homozygous for a naturally occurring Glycine deletion at nucleotide 3983 of the *col1a2* gene. As a result these mice cannot produce proα2(I) collagen, causing the replacement of normal heterotrimeric $COL1(\alpha1)_2(\alpha2)$ with homotrimeric $COL1(\alpha1)_3$, which accumulates in the extracellular matrix and causes bone fragility (Chipman et al. 1993).

Other key rodent models investigated include the *Mov13* mice, which were created by inhibition of *col1a1* allele transcription using the Moloney murine leukemia virus. Heterozygote *Mov13+/−* mice display mild type I OI phenotype, due to an insufficiency in the number of collagen fibrils synthesised (about 50 percent of the normal proα1(I) quantity) resulting in reduced bone strength and increased tissue porosity. In contrast homozygote *Mov13−/−* have embryonic lethality, this demonstrates the importance of COL1A1 for life (Bonadio et al. 1990).

There is also a knock-in murine model of type IV OI designated Brittle (*Brtl*) mouse, which was created by using the *Cre/lox* recombination system to generate a Glycine[349] to Cystine mutation in one *col1α1* allele. Heterozygote *Brtl+/−* mice have 30 percent chance of perinatal lethality, reduced body size, reduced bone mineral density, and long bone deformities and fractures. In comparison, homozygote *Brtl−/−* mice have no perinatal lethality, normal bone mineral density and only a moderately reduced body size. This less severe phenotype is due to homozygosity for the mutant allele resulting in a more homozygous extracellular matrix, which is subsequently better organized and therefore less brittle. Different mixtures of normal and mutant proα1(I) collagen chains may contribute to the range of severities seen in *Brtl+/−* mice as well as patients with type IV OI (Forlino et al. 1999).

Another important model of OI is the *OI-transgenic* strain created by insertion of a mini-gene version of human proα1(I) which had a large 41 exon deleted region. The synthesis of shortened proα1(I) chains that associate with each other and the normal proα1(I) chains results in their degradation in a process termed procollagen suicide. *OI-transgenic* mice have reduced growth, multiple fractures and reduced bone mineral and collagen content. The phenotype severity increases from moderate (type IV OI) to severe (type II OI) with increased expression of human mini-gene proα1(I) (Khillan et al. 1991). This murine model was the first used to assess the potential of cell therapy to treat OI (Pereira et al. 1998).

Cell Therapy in OI Rodent Models

A summary of the OI murine models that have been used for cell therapy and the findings are given in Table 16.5. Initial studies demonstrated that donor cells could engraft in a model OI. Pereira et al. infused wild type MSC isolated from bone marrow or whole bone marrow into *OI-transgenic* mice that expressed a human mini-gene for proα1(I) (Pereira et al. 1998). Recipient mice were injected when 3 wk old by an intraperitoneal infusion after irradiation to prevent the immune system attacking and destroying the transplanted cells. Similar results were seen between recipients of wild type MSC and wild type whole bone marrow. DNA from the donor cells were detected 1 and 2 mon after transplantation in the bone, bone marrow, cartilage and other non-hematopoietic tissues such as the lung, brain, skin and spleen. In addition a small, but significant increase in collagen content and mineral content was recorded in the transplanted mouse (Pereira et al. 1998). How the donor cells contributed to these changes, however, was unclear as analysis of donor cell osteogenic differentiation or changes at the bone matrix level were not carried out.

Subsequent experiments by Niyibizi's group using the *oim* mouse model were able to show both engraftment of donor cells and differentiation down the osteogenic lineage. In one study GFP tagged osteoprogenitors were prepared from wild type mouse bones and were transplanted into neonatal *oim* via the superficial temporal vein. Four weeks after transplantation GFP positive donor cells were distributed on the *oim* endosteal bone surfaces, the cortical bone, spongiosa, bone marrow and areas of active bone formation. Differentiation of donor cells to osteoblasts and osteoprogenitors following transplantation was confirmed by gene expression analysis; transplanted bone had high expression of osteoblast-specific genes (e.g., Osteocalcin and Osterix), whereas osteoprogenitor-specific genes were expressed more highly in the bone marrow (e.g., Osteopontin, Sox9 and Runx2) (Wang et al. 2006). Although protein analysis did not find the presence of the missing COL1A2 protein, a subsequent study by the same group

Table 16.5 The history of cell therapy with murine models of OI.

Mouse Model	Cells	Age	Irradiate	Route	Analysis age	Engraft	Differentiate	Results Bone	Phenotype	Reference
OI-transgenic	WT bone marrow (whole/MSC)	3 wk	Yes	i.p.	8–12 wk	Yes	Not measured	Not measured	Not measured	(Pereira et al. 1998)
Oim^-/-	WT osteoprogenitors	2 d	Yes	i.v.	2 & 4 wk	Yes	Yes	Not measured	Not measured	(Wang et al. 2006)
Oim^-/-	WT clonal bone marrow MSC	2 d	Yes	i.v.	2 & 4 wk	Yes	Yes	Not measured	Not measured	(Li et al. 2007)
Oim^-/-	Human fetal blood MSC	E13.5–15	No	i.p.	2,4,8,12 wk	Yes	Yes	Increase bone strength, thickness, plasticity & length	Decreased long bone fractures	(Guillot et al. 2008a)
Brtl^+/-	WT whole adult bone marrow	E13.5–14	No	IP	8 wk	Yes	Yes	Increase bone strength	No perinatal lethality	(Panaroni et al. 2009)
Oim^-/-	Human fetal blood MSC	E13.5–15	No	i.p.	8 wk	Yes	Yes	Improved plasticity, matrix stiffness & organization	Decreased long bone fractures	(Vanleene et al. 2011)
Oim^-/-	Human fetal blood MSC	3–5 days	No	i.p.	8 wk	Yes	Yes	Decreased bone brittleness	Decreased long bone fractures	(Jones et al. 2012)

This table lists in chronological order the cell therapy experiments carried out on murine models of OI; with details of the type of cells used, the mouse model, the age of transplantation and analysis, whether mice were irradiated before transplantation and the route of cell delivery. Results are also shown; including whether transplanted cells engrafted and differentiated in the recipient bones and any change to bone properties or mouse phenotype recorded. OI: Osteogenesis imperfecta. WT: wild type. MSC: mesenchymal stromal/stem cells. E: embryonic day. i.p: intraperitoneal. i.v: intravascular. +/-: Heterozygous for mutation. -/-: Homozygous for mutation.

did find COL1(α1)$_2$(α2) heterotrimers 4 wk after injection of single cell-expanded marrow progenitor cells in neonatal *oim* (Li et al. 2007). This data shows transplantation of normal cells into a mouse model of OI results in their migration to bone and subsequent differentiation to functional osteoblasts.

Fetal Cell Therapy in OI Rodent Models

Several factors influence the efficiency of cell therapy, such as cell source, time of transplantation, dosing regimen and pre-conditioning of donor cells. For osteogenesis imperfecta types that manifest *in utero* and can be diagnosed during pregnancy, prenatal stem cell transplantation has advantages over postnatal approaches by preventing fracture occurrence earlier in life and capitalizing on the small size of the fetus and immature fetal immune system. The fetal environment is also permissive to endogenous stem cell migration, which may help the movement of donor cells to sites of tissue injury (Guillot et al. 2008a). In addition a fetal-to-fetal approach to cell therapy may be optimal as fetal MSC are more primitive than adult MSC, showing greater expansion potential, smaller size, better osteogenic potential and greater immunogenicity (Guillot et al. 2007).

The potential of prenatal therapy to treat OI using human fetal MSC was assessed in the *oim* mouse model. *Oim* were transplanted prenatally at E13.5–15 with fetal MSC isolated from first trimester human fetal blood. At 8 wk transplanted *oim* had an overall two-thirds reduction in long bone fractures when compared to age-matched *oim* controls with significant reduction in fracture incidence for all long bones at all time points (4, 8 and 12 wk) analyzed. Furthermore there was normalization of growth plate height and increase in bone length, strength and cortical thickness (Guillot et al. 2008a). An independent study found that the dramatic reduction in fracture rate of these mice may be attributed to a reduction in bone brittleness, rather than increased bone strength (Vanleene et al. 2011).

Analysis of the mechanisms underlying these therapeutic effects revealed that the donor cells homed preferentially to bones, compared to other organs, where they were found in higher numbers at sites of fracture callus and active bone formation. Cells engrafted in bones, expressed late markers of osteoblasts (e.g., osteocalcin) and produced the Collagen type 1 alpha 2 (COL1A2) chain protein, which is absent in non-transplanted *oim* mice. This led to modification of the bone matrix composition, resulting in increased bone matrix stiffness and improved organization of lamellar bone (Guillot et al. 2008a; Vanleene et al. 2011).

However, engraftment levels of donor cells were low at only 3–5 percent and the long term effects of transplantation could not be measured due to strict severity limits on *oim* under British law (Guillot et al. 2008a). In addition fetal MSC isolated from first trimester blood are collected either by cardiocentesis during termination of pregnancy, which is ethically and technically challenging and prevents autologous application, or from the umbilical cord during ongoing pregnancy, which is highly invasive and carries a risk of miscarriage (Chan et al. 2008).

A similar study was carried out in the *Brtl* mouse, a model of autosomal dominant OI, and similar effects were shown. Donor cells were found engrafted in bone 8 wk after prenatal transplantation of wild type adult bone marrow into *Brtl* mice. These cells contributed to up to 20 percent of all collagen type I synthesized, which was associated with increased cortical thickness and improvements in bone stiffness, ultimate load and yield load. In addition transplantation *in utero* eliminated perinatal lethality seen in heterozygote *Brtl*$^{+/-}$ mice (Panaroni et al. 2009).

This study showed the potential of cell therapy to treat autosomal dominant types of OI in addition to recessive types. It also highlighted the possibility to treat perinatal lethal type II OI with a prenatal approach. The improvements observed occurred despite just 1–2 percent engraftment of donor cells (Panaroni et al. 2009). This engraftment rate was lower than that seen following fetal blood MSC transplantation in *oim* (3–5 percent) (Guillot et al. 2008a) or after fetal liver MSC transplantation in an OI fetus (7 percent) (Le Blanc et al. 2005). This suggests a fetal-to-fetal approach may have advantages over adult-to-fetal approach in terms of engraftment levels.

In addition human fetal MSC are more primative than adult stem cells with higher kinetics and later senescence, longer telomeres, greater telomerase activity and expression of pluripotency markers (e.g., Oct4, Nanog, SSEA-4, SSEA-3, Tra-1-60, Tra-1-81 and Rex1). Subsequently they have greater expansion potential. Second, fetal MSC are smaller in size, which is likely to aid their ability to pass through the endothelial wall of the blood vessels, allowing improved migration to sites of injury. Fetal MSC also have a unique immune-privileged status. For example, liver derived fetal MSC did not elicit alloreactive lymphocyte proliferation either when undifferentiated, differentiated or IFN-γ stimulated to upregulate HLA I and HLA II (Gotherstrom et al. 2003). Finally, fetal MSC more readily differentiate to osteoblasts, which is critical to produce normal type I collagen fibrils, improve bone matrix organization and subsequently improve the skeletal fragility caused by OI (Guillot et al. 2008b).

The Hurdles of Cell Therapy

Although cell therapy shows great potential, several pitfalls constrain their clinical application. First the clinical effectiveness of cell therapy is challenged by low level of engraftment in target organs. For example, engraftment levels in a given bone chip ranged from 0.3–28 percent following neonatal transplantation of adult murine MSC in *oim* mice, despite irradiation of the recipient mice (Li et al. 2007). Whilst *in utero* transplantation of human fetal blood MSC in *oim* mice that were not irradiated was associated with 3–5 percent engraftment levels in bone (Guillot et al. 2008a). Clinical experience is similar, with less than 2 percent engraftment of bone marrow MSC in transplanted children with OI (Horwitz et al. 2002), and up to 7 percent engraftment recorded following prenatal allogenic fetal liver MSC transplantation in a human fetus with OI (Le Blanc et al. 2005).

Donor cell engraftment levels are important to treat autosomal recessive OI types as they have a protein deficiency, but are more critical to treat autosomal dominant types of OI. This is because levels of differentiated donor cells must be high enough to significantly alter the ratio of normal to mutant proteins synthesized and counteract the structural instabilities caused by these abnormal pro-collagen proteins (Forlino et al. 2011). The proof of this principle has been demonstrated in patients that are mosaic carriers of autosomal dominant OI and subsequently have both affected and normal somatic and germline cells. Despite 40–75 percent of their osteoblasts presenting a collagen I mutation the patients have normal skeletal growth, density and histology and often are only identified as a carrier when they have more than one child with OI (Cabral and Marini 2004).

Second, the cell sources currently used are not applicable to the clinic as the collection procedures to aspirate bone marrow and isolate adult MSC are often painful and highly invasive and cell numbers are low with only one estimated MSC per 250,000 bone marrow cells. The limited expansion potential of adult MSC is also a practical disadvantage in terms of clinical application. Whereas stem cells from fetal organs and blood isolated after termination of pregnancy are difficult to procure, the small size of the fetus limits total number of cells that can be collected, they have a high rate of bacterial and fungal infections and their clinical use is restricted by ethical legislations (Guillot et al. 2007).

The Focus of the Future

To address the hurdles of cell therapy, strategies to increase donor cell engraftment are critical. However, simply increasing the number of cells transplanted has been shown to have little effect on engraftment rate, with one study showing maximal engraftment levels in lethally irradiated mice at just 15 percent (Marino et al. 2008). The importance of the CXCR4-SDF1 pathway in homing of MSC to fractures was shown in a mouse model of a stabilized tibia fracture, whereby only CXCR4+ MSC and not CXCR4- MSC migrated to the site of injury (Granero-Molto et al. 2009). Fetal MSC express high levels of intracellular CXCR4, but only a small number of cells present

CXCR4 on the cell surface. In recent work priming human fetal blood MSC with SDF1 externalized CXCR4, and was associated with a threefold upregulation of engraftment in *oim* and wild type bone and bone marrow. Furthermore higher engraftment in *oim* mice was associated with reduced bone brittleness, supporting the hypothesis that increasing engraftment improves the therapeutic effects of cell therapy (Jones et al. 2012).

The next step to bring cell therapy for OI closer to the clinic is to find a more suitable cell source. Contrary to fetal organs, extra-embryonic fetal tissues, such as the placenta and amniotic fluid, are readily available either from surplus tissue at routine prenatal diagnostic procedures or at term. Their osteogenic potential has been demonstrated for chorionic stem cells isolated from the first trimester and at term, as well as second trimester amniotic fluid stem cells. Such tissues are available in all clinics, are easy to physically manipulate and their use is unlikely to cause ethical controversy. Future work in rodent models of OI and other disease pathologies should start to assess the use of these cells in regenerative medicine as well as to investigate their potential in cell banking (Abdulrazzak et al. 2010).

Summary

- Osteogenesis imperfecta or brittle bone disease is a heterogenous group of disorders, characterized by bone fragility and fractures caused by mutations in type I collagen or proteins involved in its biosynthesis.
- Current treatments of osteogenesis imperfecta involve a multidisciplinary approach of non-surgical and surgical management and use of pharmacological therapies such as bisphosphonates that cause accumulation of the defective matrix rather than correcting the underlying mutation.
- One alternative treatment of osteogenesis imperfecta is gene therapy to silence mutant genes/supplement normal ones, however the heterogeneity of the disease makes the design of gene therapy both challenging and time consuming.
- A second alternative is cell therapy, which replaces defective cells with healthy ones that regenerate and repair damaged tissues.
- Cell therapy holds promise to treat osteogenesis imperfecta and has already undergone clinical trials with improvements in bone seen in children as well after prenatal trasnplantaion, despite low engraftment levels of donor cells.
- To better understand osteogenesis imperfecta and how best to treat it a number of rodent models of the disorder have been developed, including the *oim*, *Mov13*, *Brtl* and the *OI-transgenic* mouse.
- Initial work in the *OI-transgenic* mouse found donor cells could engraft in bone as well as other tissues, whilst later studies in the *oim* model

showed donor cells migrated to sites of active bone formation and differentiated to functional osteoblasts.

- Human fetal mesenchymal stromal/stem cells have advantages over adult counterparts for cell therapy including smaller size, greater expansion potential and better osteogenic potential.
- When stem cells from fetal blood were transplanted prenatally in *oim* mice a significant reduction in fractures and bone brittleness were recorded, which were attributed to improvements in bone matrix mineralization and lamella organization.
- Current and future work into cell therapy for osteogenesis imperfecta are looking to improve engraftment levels by increasing expression of CXCR4 as well as to identify a more accessible and available cell source, such as extra-embryonic tissues like placenta and amniotic fluid.

Dictionary

- *Allogeneic*: not from the same individual
- *Autologous*: from the same individual
- *col1α1*: Collagen type I alpha 1
- *col1α2*: Collagen type I alpha 2
- *Heterozygote*: has two different alleles for one gene
- *Homozygote*: has two alleles the same for one gene
- *In utero*: environment within the womb
- *In vitro*: environment outside living organisms
- *In vivo*: environment within living organisms
- *Osteogenesis Imperfecta*: brittle bone disease
- *Osteogenic/Osteogenesis*: bone/bone formation
- *Sclera*: white of the eye
- *Scolosis*: curvature of the spine
- *Wild Type*: mouse without genetic mutation.

List of Abbreviations

-/-	:	homozygous for mutation
+/-	:	heterozygous for mutation
AD	:	autosomal dominant inheritance
AR	:	autosomal recessive inheritance
Brtl	:	Brittle mouse
DNA	:	deoxyribonucleic acid
E	:	embryonic day
ES	:	embryonic stem
GFP	:	green fluorescent protein

HLA	:	human leukocyte antigen
HSC	:	hematopoietic stem cells
IFN-γ	:	interferon gamma
i.p.	:	intraperitoneal
IUT	:	*in utero* transplantation
i.v.	:	intravenous
M-MuLV	:	Moloney-murine leukemia virus
MSC	:	mesenchymal stromal/stem cells
OI	:	osteogenesis Imperfecta
Oim	:	osteogenesis imperfecta murine
RNA	:	ribonucleic acid
RNAi	:	interference RNA
siRNA	:	small-interfering RNA

References

Abdulrazzak, H., D. Moschidou, G. Jones and P.V. Guillot. 2010. Biological characteristics of stem cells from foetal, cord blood and extraembryonic tissues. J R Soc Interface 7 Suppl 6: S689–706.

Alanay, Y., H. Avaygan, N. Camacho, G.E. Utine, K. Boduroglu, D. Aktas, M. Alikasifoglu, E. Tuncbilek, D. Orhan, F.T. Bakar, B. Zabel, A. Superti-Furga, L. Bruckner-Tuderman, C.J. Curry, S. Pyott, P.H. Byers, D.R. Eyre, D. Baldridge, B. Lee, A.E. Merrill, E.C. Davis, D.H. Cohn, N. Akarsu and D. Krakow. 2010. Mutations in the gene encoding the RER protein FKBP65 cause autosomal-recessive osteogenesis imperfecta. Am. J. Hum. Genet. 86: 551–559.

Antoniazzi, F., E. Monti, G. Venturi, R. Franceschi, F. Doro, D. Gatti, G. Zamboni and L. Tato. 2010. GH in combination with bisphosphonate treatment in osteogenesis imperfecta. Eur J Endocrinol. 163: 479–487.

Bonadio, J., T.L. Saunders, E. Tsai, S.A. Goldstein, J. Morris-Wiman, L. Brinkley, D.F. Dolan, R.A. Altschuler, J.E. Hawkins, Jr. and J.F. Bateman. 1990. Transgenic mouse model of the mild dominant form of osteogenesis imperfecta. Proc Natl Acad Sci USA. 87: 7145–7149.

Cabral, W.A. and J.C. Marini. 2004. High proportion of mutant osteoblasts is compatible with normal skeletal function in mosaic carriers of osteogenesis imperfecta. Am J Hum Genet. 74: 752–760.

Cabral, W.A., W. Chang, A.M. Barnes, M. Weis, M.A. Scott, S.E. Leikin, E. Makareeva, N.V. Kuznetsova, K.N. Rosenbaum, C.J. Tifft, D.I. Bulas, C. Kozma, P.A. Smith, D.R. Eyre and J.C. Marini. 2007. Prolyl 3-hydroxylase 1 deficiency causes a recessive metabolic bone disorder resembling lethal/severe osteogenesis imperfecta. Nat Genet. 39: 359–365.

Chan, J., S. Kumar and N.M. Fisk. 2008. First trimester embryo-fetoscopic and ultrasound-guided fetal blood sampling for *ex vivo* viral transduction of cultured human fetal mesenchymal stem cells. Hum Reprod. 23: 2427–2437.

Chipman, S.D., H.O. Sweet, D.J. McBride, Jr., M.T. Davisson, S.C. Marks, Jr., A.R. Shuldiner, R.J. Wenstrup, D.W. Rowe and J.R. Shapiro. 1993. Defective pro alpha 2(I) collagen synthesis in a recessive mutation in mice: a model of human osteogenesis imperfecta. Proc Natl Acad Sci USA. 90: 1701–1705.

Christiansen, H.E., U. Schwarze, S.M. Pyott, A. AlSwaid, M. Al Balwi, S. Alrasheed, M.G. Pepin, M.A. Weis, D.R. Eyre and P.H. Byers. 2010. Homozygosity for a missense mutation in SERPINH1, which encodes the collagen chaperone protein HSP47, results in severe recessive osteogenesis imperfecta. Am J Hum Genet. 86: 389–398.

Forlino, A., F.D. Porter, E.J. Lee, H. Westphal and J.C. Marini. 1999. Use of the Cre/lox recombination system to develop a non-lethal knock-in murine model for osteogenesis imperfecta with an alpha1(I) G349C substitution. Variability in phenotype in BrtlIV mice. J Biol Chem. 274: 37923–37931.

Forlino, A., W.A. Cabral, A.M. Barnes and J.C. Marini. 2011. New perspectives on osteogenesis imperfecta. Nat Rev Endocrinol. 7: 540–557.

Glorieux, F.H., F. Rauch, H. Plotkin, L. Ward, R. Travers, P. Roughley, L. Lalic, D.F. Glorieux, F. Fassier and N.J. Bishop. 2000. Type V osteogenesis imperfecta: a new form of brittle bone disease. J Bone Miner Res. 15: 1650–1658.

Glorieux, F.H., L.M. Ward, F. Rauch, L. Lalic, P.J. Roughley and R. Travers. 2002. Osteogenesis imperfecta type VI: a form of brittle bone disease with a mineralization defect. J Bone Miner Res. 17: 30–38.

Gotherstrom, C., O. Ringden, M. Westgren, C. Tammik and K. Le Blanc. 2003. Immunomodulatory effects of human foetal liver-derived mesenchymal stem cells. Bone Marrow Transplant. 32: 265–272.

Granero-Molto, F., J.A. Weis, M.I. Miga, B. Landis, T.J Myers, L. O'Rear, L. Longobardi, E.D. Jansen, D.P. Mortlock and A. Spagnoli. 2009. Regenerative effects of transplanted mesenchymal stem cells in fracture healing. Stem Cells. 27: 1887–1898.

Guillot, P.V., C. Gotherstrom, J. Chan, H. Kurata and N.M. Fisk. 2007. Human first-trimester fetal MSC express pluripotency markers and grow faster and have longer telomeres than adult MSC. Stem Cells. 25: 646–654.

Guillot, P.V., O. Abass, J.H. Bassett, S.J. Shefelbine, G. Bou-Gharios, J. Chan, H. Kurata, G.R. Williams, J. Polak and N.M. Fisk. 2008a. Intrauterine transplantation of human fetal mesenchymal stem cells from first-trimester blood repairs bone and reduces fractures in osteogenesis imperfecta mice. Blood. 111: 1717–1725.

Guillot, P.V., C. De Bari, F. Dell'Accio, H. Kurata, J. Polak and N.M. Fisk. 2008b. Comparative osteogenic transcription profiling of various fetal and adult mesenchymal stem cell sources. Differentiation. 76: 946–957.

Horwitz, E.M., P.L. Gordon, W.K. Koo, J.C. Marx, M.D. Neel, R.Y. McNall, L. Muul and T. Hofmann. 2002. Isolated allogeneic bone marrow-derived mesenchymal cells engraft and stimulate growth in children with osteogenesis imperfecta: Implications for cell therapy of bone. Proc Natl Acad Sci USA. 99: 8932–8937.

Jones, G.N., D. Moschidou, K. Lay, H. Abdulrazzak, M. Vanleene, S.J. Shefelbine, J. Polak, P. De Coppi, N.M. Fisk and P.V. Guillot. 2012. Upregulating CXCR4 in human fetal mesenchymal stem cells enhances engraftment and bone mechanics in a mouse model of osteogenesis imperfecta. Stem Cells Trans Med. 1: 70–78.

Khillan, J.S., A.S. Olsen, S. Kontusaari, B. Sokolov and D.J. Prockop. 1991. Transgenic mice that express a mini-gene version of the human gene for type I procollagen (COL1A1) develop a phenotype resembling a lethal form of osteogenesis imperfecta. J Biol Chem. 266: 23373–23379.

Le Blanc, K., C. Tammik, K. Rosendahl, E. Zetterberg and O. Ringden. 2003. HLA expression and immunologic properties of differentiated and undifferentiated mesenchymal stem cells. Exp Hematol. 31: 890–896.

Le Blanc, K., C. Gotherstrom, O. Ringden, M. Hassan, R. McMahon, E. Horwitz, G. Anneren, O. Axelsson, J. Nunn, U. Ewald, S. Nordon-Lindeberg, M. Jansson, A. Dalton, E. Astrom and M. Westgren. 2005. Fetal mesenchymal stem-cell engraftment in bone after in utero transplantation in a patient with severe osteogenesis imperfecta. Transplantation. 79: 1607–1614.

Letocha, A.D., H.L. Cintas, J.F. Troendle, J.C. Reynolds, C.E. Cann, E.J Chernoff, S.C. Hill, L.H. Gerber and J.C. Marini. 2005. Controlled trial of pamidronate in children with types III and IV osteogenesis imperfecta confirms vertebral gains but not short-term functional improvement. J Bone Miner Res. 20: 977–986.

Li, F., X. Wang and C. Niyibizi. 2007. Distribution of single-cell expanded marrow derived progenitors in a developing mouse model of osteogenesis imperfecta following systemic transplantation. Stem Cells. 25: 3183–3193.

Marino, R., C. Martinez, K. Boyd, M. Dominici, T.J. Hofmann and E.M. Horwitz. 2008. Transplantable marrow osteoprogenitors engraft in discrete saturable sites in the marrow microenvironment. Exp Hematol. 36: 360–368.

Monti, E., M. Mottes, P. Fraschini, P. Brunelli, A. Forlino, G. Venturi, F. Doro, S. Perlini, P. Cavarzere and F. Antoniazzi. 2010. Current and emerging treatments for the management of osteogenesis imperfecta. Ther Clin Risk Manag. 6: 367–381.

Morello, R., T.K. Bertin, Y. Chen, J. Hicks, L. Tonachini, M. Monticone, P. Castagnola, F. Rauch, F.H. Glorieux, J. Vranka, H.P. Bachinger, J.M. Pace, U. Schwarze, P.H. Byers, M. Weis, R.J. Fernandes, D.R. Eyre, Z. Yao, B.F. Boyce and B. Lee. 2006. CRTAP is required for prolyl 3-hydroxylation and mutations cause recessive osteogenesis imperfecta. Cell. 127: 291–304.

Munns, C.F., F. Rauch, R. Travers and F.H. Glorieux. 2005. Effects of intravenous pamidronate treatment in infants with osteogenesis imperfecta: clinical and histomorphometric outcome. J Bone Miner Res. 20: 1235–1243.

Niyibizi, C., S. Wang, Z. Mi and P.D. Robbins. 2004. Gene therapy approaches for osteogenesis imperfecta. Gene Ther. 11(4): 408–416.

Panaroni, C., R. Gioia, A. Lupi, R. Besio, S.A. Goldstein, J. Kreider, S. Leikin, J.C. Vera, E.L. Mertz, E. Perilli, F. Baruffaldi, I. Villa, A. Farina, M. Casasco, G. Cetta, A. Rossi, A. Frattini, J.C. Marini, P. Vezzoni and A. Forlino. 2009. *In utero* transplantation of adult bone marrow decreases perinatal lethality and rescues the bone phenotype in the knock-in murine model for classical, dominant osteogenesis imperfecta. Blood. 114: 459–468.

Pereira, R.F., M.D. O'Hara, A.V. Laptev, K.W. Halford, M.D. Pollard, R. Class, D. Simon, K. Livezey and D.J. Prockop. 1998. Marrow stromal cells as a source of progenitor cells for nonhematopoietic tissues in transgenic mice with a phenotype of osteogenesis imperfecta. Proc Natl Acad Sci USA. 95: 1142–1147.

Sillence, D.O., A. Senn and D.M. Danks. 1979. Genetic heterogeneity in osteogenesis imperfecta. J Med Genet 16: 101–116.

Wang, X., F. Li and C. Niyibizi. 2006. Progenitors systemically transplanted into neonatal mice localize to areas of active bone formation *in vivo*: implications of cell therapy for skeletal diseases. Stem Cells. 24: 1869–1878.

Van Dijk, F.S., I.M. Nesbitt, E.H. Zwikstra, P.G. Nikkels, S.R. Piersma, S.A. Fratantoni, C.R. Jimenez, M. Huizer, A.C. Morsman, J.M. Cobben, M.H. van Roij, M.W. Elting, J.I. Verbeke, L.C. Wijnaendts, N.J. Shaw, W. Hogler, C. McKeown, E.A. Sistermans, A. Dalton, H. Meijers-Heijboer and G. Pals. 2009. PPIB mutations cause severe osteogenesis imperfecta. Am J Hum Genet. 85: 521–527.

Van Dijk, F.S., G. Pals, R.R. Van Rijn, P.G. Nikkels and J.M. Cobben. 2010. Classification of Osteogenesis Imperfecta revisited. Eur J Med Genet. 53: 1–5.

Vanleene, M., Z. Saldanha, K.L Cloyd, G. Jell, G. Bou-Gharios, J.H. Bassett, G.R. Williams, N.M. Fisk, M.L. Oyen, M.M. Stevens, P.V. Guillot and S.J. Shefelbine. 2011. Transplantation of human fetal blood stem cells in the osteogenesis imperfecta mouse leads to improvement in multiscale tissue properties. Blood. 117: 1053–1060.

CHAPTER 17

G-CSF-mobilized Haploidentical Peripheral Blood Stem Cell Transplantation in Childhood Osteopetrosis

Fikret Arpaci[a],* and Asli Selmin Ataergin[b]

ABSTRACT

Autosomal recessive osteopetrosis (classic and with renal tubular acidosis) is a fatal disease and at present, no effective treatment for osteopetrosis exists. Treatment is generally supportive. Because osteoclasts are of hematopoietic origin, hematopoietic stem cell transplantation (HSCT) can be an effective treatment modality. Successful allogeneic HSCT is capable of producing long-term benefit in patients with osteopetrosis. Bone remodeling, restoration of growth and reconstitution of normal hematopoiesis may be supplied by HSCT. Of all the potential sources of allografts, HSCT from HLA-matched sibling has generally produced the best survivals. Seventy-three percent of patients are alive and disease-free at 5 yr after transplantation. Outcomes are better with earlier transplantation, particularly before the age of 3

Department of Medical Oncology and Bone Marrow Transplantation Unit, Gulhane Faculty of Medicine, 06018: Gn. Dr. Tevfik Saglam caddesi, Etlik, Ankara, Turkey.
[a]E-mail: farpaci@gata.edu.tr
[b]E-mail: sataergin@hotmail.com; sataergin@yahoo.com
*Corresponding author

List of abbreviations given at the end of the text.

mon, but only about one-third of candidates for allogeneic HSCT have HLA-matched siblings. For patients who lack HLA-matched siblings, there are three alternative sources of stem cells: 1- unrelated donors, 2- umbilical cord blood and, 3- partially HLA mismatched (haploidentical) related donors. Due to immediate availability, the possibility of using a haploidentical donor independent of HLA-matching would offer some advantages. It has been shown that increasing the number of transplanted mismatched stem cells and completely depleted of T-cells can overcome the HLA barrier and sustained engraftment can be achieved. By using granulocyte-colony stimulating factor (G-CSF), the number of CD 34+ progenitors can be drastically increased compared to that obtained from bone marrow. EBMT and CIBMTR surveys revealed that there are less than 40 patients who have been transplanted from G-CSF mobilized partially mismatched donors. The results of transplantations from haploidentical donors compared to sibling transplantations are inferior due to an increased incidence and severity, graft-versus-host-disease (GVHD), graft rejection and infections. The 100-d mortality rate in the posttransplant period is approximately 25 percent and 70 percent of patients are alive during 11–48-mon of follow-up. The recent advances in effective T-cell depletion and reduced intensity conditioning will decrease the early transplant-related mortality and GVHD, and therefore will enhance the therapeutic benefits of G-SCF-mobilized haploidentical transplantation.

Introduction

Although osteopetrosis (OP) is a rare disease, some forms may be fatal during infancy and early childhood period. Neurologic problems, cranial nerve compression, hydrocephalus, mental retardation, and bone marrow impairment as well as related immunodeficiency are life-threatining complications. The most severe complication is bone marrow suppression. The obliteration of marrow cavities with sclerotic bone can result in pancytopenia, recurrent infections and death. At present, no effective medical treatment for OP exists. Treatment is largely supportive and symptomatic. Because osteoclasts are of hematopoietic origin, autologous and allogeneic hematopoietic stem cell transplantation (HSCT) can be the only effective treatment for bone marrow impairment and might be a solution for other complications. Granulocyte-colony-stimulating-factor (G-CSF)-mobilized HSCTs have been used with great success in patients with benign and malignant bone diseases (Arpaci et al. 2005; Ataergin et al. 2007). HLA-matched sibling, matched unrelated and haploidentical related donors are potential sources of allogeneic HSCT. In this chapter, we have focused on G-CSF mobilized haploidentical HSCT in patients with OP.

General Overlook to OP

Definition

OP has been defined as a class of rare, heritable disease group of the skeleton characterized by excessive bone density. The disorder has three forms: 1—benign (tarda) variety, with autosomal dominant (AD) inheritance representing mild symptoms; 2—the intermediate forms, with autosomal recessive (AR) inheritance and mild symptoms; and 3—the malignant infantile (congenital, MIOP) form with AR inheritance presented with severe early symptoms and poor prognosis.

ARO has 3 subtypes: classic, neuropathic and ARO with renal tubular acidosis (RTA). MIOP is genetically due to defects (absence or malfunction) of osteoclasts.

Epidemiology, aetiology and clinical presentation

The incidence rates are 1:250,000 births and 1:20,000 births for AR and AD varieties, respectively; AR represents as the rarest form. The age of onset for ARO and X-linked OP is perinatal and infancy period, while other forms occur during childhood, late childhood or adolescence (Stark and Savarirayan 2009).

Clinically, major consequences are susceptibility to fractures and nerve compression due to failure of remodeling of the skull base. This causes optic, auditory, and facial nerve compression resulting in blindness, deafness, and facial palsy (Stoker 2002).

The increase in bone density can paradoxically weaken the bone predisposing the bones to fractures and osteomyelitis. The longitudinal growth of bones is impaired leading to short stature. Macrocephaly, frontal bossing developing within the first year and skull changes can result in choanal stenosis and hydrocephalus.

Tooth eruption defects and severe dental caries are also common. These patients are at risk of developing hypocalsemia, with attendant titanic seizures and secondary hyperparatiroidism.

The most severe complication is bone marrow suppression: obliteration of marrow cavities in other words, encroachment of sclerotic bone on the marrow space can result in pancytopenia, recurrent infections, and death in early life of the affected children (Stoker 2002; Stark and Savarirayan 2009). Furthermore, extramedullary hemopoiesis, causes hepatosplenomegaly and, frequently, causes progression to life-threating pancytopenia. There is also increased susceptibility to infection in some patients because of an unexplained defect of neutrophil superoxide function. The combination of problems results in the death of 70 percent of affected children by the age of 6 yr.

Diagnosis and prognosis

Genetic testing can be used to confirm the diagnosis and differentiate between different subtypes of OP. Once the diagnosis of a primary OP is made, it is important to distinguish between different subtypes as they have different responses to treatment, prognosis and recurrence risk (Stark and Savarirayan 2009).

The prognosis of the disease is mostly severe in AR and X-linked OP resulting in fatal outcome and moderate in ARO with RTA with a variable prognosis.

Treatment

At present, no effective medical treatment for OP exists. Treatment is largely supportive and based on symptomatic management of complications together with a multidisciplinary surveillance.

HSCT remains the only curative therapy and forms the cornerstone of treatment for classic (MIOP) and ARO with RTA (Fig.17.1; Table 17.1). Osteoclasts are of hematopoietic origin and hematopoietic stem cell

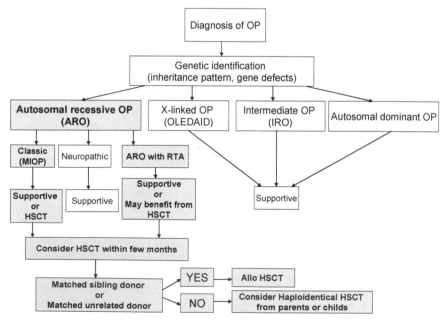

Figure 17.1 The algorithm of treatment in children with osteopetrosis.

Abbreviations; ARO, autosomal recessive osteopetrosis; HSCT, hematopoietic stem cell transplantation; MIOP, malignant infantile osteopetrosis; OP, osteopetrosis; RTA, renal tubular acidosis

Table 17.1 Hematopoietic stem cell transplantation in osteopetrosis.

OP subtype	Benefit of HSCT
Autosomal recessive osteopetrosis	
Classical type (MIOP)	Yes
OP with RTA	May be yes
Other forms of osteopetrosis	No

Abbreviations; HSCT, Hematopoietic stem cell transplantation; MIOP, malignant infantile osteopetrosis; OP, osteopetrosis; RTA, renal tubular acidosis

transplantation (HSCT) represents an effective treatment (Coccia et al. 1980). Successful allogeneic HSCT is currently the only therapy capable of producing long-term benefit in children. This results in bone remodeling, restoration of growth and reconstitution of normal hematopoiesis and neutrophil function.

After successful engraftment, children normalize their bone density and hematologic parameters and have stabilization or prevention of cranial nerve damage and hydrocephalus.

Hematopoietic Stem Cell Transplantation in Children with Osteopetrosis

History

In animal models, Donald Walker showed that microphtalmic osteopetrotic mouse strains could be cured by transferring marrow or spleen cells of the donor to the recipient, and that osteoclasts in transplanted animals were of donor origin (Walker 1975). This was the first basic study for further human transplants. In this way, transplant activities in humans then began in early 1980s and many successful transplants were performed during the 1980s from a range of matched and mismatched donors.

Some successes have been reported with allogeneic bone marrow transplantation where HLA-matched siblings were available (Fisher et al. 1986).

In 1987, Orchard et al. reported the first case without a matched sibling who was treated with a T-cell-depleted bone marrow transplant from her HLA haploidentical mother (Orchard et al. 1987). Finally, Handgretinger et al. reported transplantation of G-CSF-mobilized megadoses of purified haploidentical peripheral HSCT (Handgretinger et al. 1999).

HLA identical hematopoietic stem cell transplantation in osteopetrosis

Allogeneic transplantation of hematopoietic progenitor cells either by myeloablative or reduced intensity conditioning, is the only potentially curative treatment modality for many malignant and nonmalignant diseases and might be able to offer a cure for these patients (Armitage 1994).

Of all the potential sources of allografts, transplantation of stem cells from an HLA-matched sibling has generally produced the best overall and progression-free survivals. Results are excellent after matched sibling allogeneic grafts; 73 percent of patients are alive and disease-free at 5 yr after transplantation (Driessen et al. 2003). Unfortunately, only about one-third of candidates for allo HSCT have HLA-matched siblings.

Alternative donors and haploidentical hematopoietic stem cell transplantation

Due to immediate availability, the possibility of using a haploidentical donor independent of HLA-matching would offer a number of advantages, presents a better logistic and practical alternative to matched unrelated donor transplantations. Almost all patients who may benefit from allogeneic transplantation would have a donor within a certain time-frame. This may be especially important when dealing with a patient suffering from a disease with a rapid tempo where the urgency of transplant does not allow transplant from an unrelated donor to be organized.

For patients who lack HLA-matched siblings, there are three alternative sources of stem cells for all-SCT: (1) volunteer unrelated donors (VUDs); (2) umbilical cord blood; and (3) partially HLA-mismatched, or HLA-haploidentical, related donors. In contrast to the use of cord blood cells as the stem cell source, the haploidentical donor is still available in the case of nonengraftment or rejection and is also readily available for possible future strategies of post-transplant immunotherapy such as donor lymphocyte infusion. Next to HLA-matched related donors, phenotypically matched VUDs are the most widely sought for allo-HSCT. However, the chance of finding an HLA-matched VUD varies significantly depending upon the racial and ethnic background of the recipient ranging from 60–70 percent in Caucasians, to about 10–20 percent for ethnic minorities (Beatty et al. 1995). The search for an HLA-matched VUD is also hindered by the amount of time it takes from search initiation to donor identification. In contrast, a partially HLA-compatible first-degree relative can be identified and mobilized immediately for transplantation in nearly all situations. This is because a patient shares exactly one HLA haplotype with each biological parent or child, and each sibling of the parent has a 50 percent likelihood of sharing one HLA haplotype while being variably mismatched for HLA genes of the other haplotype. Thus, when a patient lacks an HLA-identical sibling, the treating physician must balance the risks that the patient's disease will progress or health will deteriorate while searching for a VUD versus the risk of crossing HLA barriers with the use of an HLA-haploidentical donor (Symons and Fuchs 2008).

However, the results of transplantation with unrelated donors, compared to sibling transplantations, are inferior due to an increased incidence and severity of GVHD, graft rejection and infections. Similarly, haploidentical transplants were associated in the past with an increased incidence of acute and chronic GVHD, nonengraftment or graft rejection and prolonged immunodysregulation increasing the risk of fatal infections and lymphoproliferative disorders. In sum, disease-free HLA-identical and matched unrelated donors are the optimal choice, followed by phenotypically matched family donors (Fig. 17.1). Bone marrow or peripheral blood stem cells from matched unrelated donors are currently preferred over cord donors, although definitive registry publications on cord blood transplantation are awaited.

Table 17.2 shows the haploidentical HSCT of the European Bone Marrow Transplantation (EBMT) registry between 1968 and 2001. Approximately 50 haploidentical HSCT have been done since 1968, and the stem cell source was mostly bone marrow. In contrast to EBMT, the Center for International Blood and Marrow Transplant Registry (CIBMTR) have registered 24 mismatched HSCT for OP between 2000 and 2010; the stem cell source of the was mostly peripheral blood (Table 17.3) (Mary Horowitz (CIBMTR), personal communication). Unfortunately, results are much poorer after non-identical bone marrow transplantation (Gerritsen et al. 1994).

Recent advances with effective T-cell depletion and reduced intensity conditioning have significantly decreased the early transplant-related mortality and GVHD, while enabling robust and prompt engraftment, and hence enhancing the therapeutic benefits of haploidentical transplantation.

It has been shown that increasing the number of transplanted mismatched stem cells completely depleted of T cells, can overcome the

Table 17.2 Haploidentical hematopoietic stem cell transplantation in osteopetrosis: European survey.

Author	Publication year	Period of the survey	Number of cases	Stem cell source
Fischer A	1986	1968–1985	2	BM
Fischer A	1991	1985–1989	11	BM
Driessen GJA	2003	1980–2001	41	Mostly BM

Abbreviations; BM, bone marrow

Table 17.3 Characteristics of patients with osteopetrosis receiving HLA mismatched related donor hematopoietic stem cell transplants (HSCT) registered to the CIBMTR between 2000 and 2010.*

Variable	n (%) Median (range)
N	24
Age (years)	1 (0–5)

Gender
  Male
  Female
	
	12 (50)
	12 (50)

Graft type
  BM
  PB ± BM
	
	4 (17)
	20 (83)

Region
  US
  Canada
  Europe
  Australia/New Zealand
  Mideast/Africa
  Central/South America
	
	11 (46)
	1 (4)
	5 (21)
	1 (4)
	5 (21)
	1 (4)

Year of transplant
  2001–2005
  2006–2010
	
	16 (67)
	8 (33)

Probability of overall survival, % (95% CI)
  100-day
  6 months
	
	60 (38–80)
	44 (23–66)

Abbreviations; BM, bone marrow; PB, peripheral blood

*The data presented here are preliminary and were obtained with permission from the Statistical Center of the Center for International Blood and Marrow Transplant Research. The analysis has not been reviwed or approved by the Advisory or Scientific Committee of the CIBMTR.

HLA barrier and sustained engraftment can be achieved in the absence of acute and chronic GVHD and other severe side-effects (Reisner and Martelli 1995). However, delayed immune recovery was still observed in adult patients transplanted with T-cell depleted PBSC resulting in a high mortality rate due to fatal infections.

G-CSF mobilized peripheral blood stem cells

It has long been recognized that immature myeloid and erythroid precursors circulate in increased numbers in patients with MIOP, producing a leukoerytroblastic blood picture (Marcus et al. 1982). In 1995, Steward et al. began to to investigate spontaneous peripheral blood stem cell (PBSC) numbers in presenting osteopetrotic patients, reasoning that, if they were markedly increased, this could provide an alternative source of cells for rescue in the event of failed engraftment. In a study published in 2005, they presented their data on 10 consecutive patients with characteristic presentations of MIOP, from them two patients had undergone high-dose haploidentical parental HSCT. They investigated whether high PBSC count enable autologous backup before HSCT based on the data that the vast majority of patients with severe MIOP have extremely high numbers of circulating PBSCs (typically in the range of 1–5 percent) and that these have progenitor potential comparable to that of normal donor BM.

Therafter, in allogeneic setting, by using G-CSF-mobilized PBSC, the number of CD34+ progenitors can be drastically increased compared to that obtained from bone marrow (Körbling and Champlin 1996; Arpaci 2011).

The approach using megadose transplantation of G-CSF-mobilized peripheral blood CD34+ progenitor cells from HLA-mismatched parental donors offers a promising therapeutic option for every child without an otherwise suitable donor (Handgretinger et al. 2001).

The transplantation of high numbers of G-CSF-mobilized CD34+ progenitors resulted in a rapid immunological recovery, especially when the target cell dose of transplanted CD34+ cells was in the range of 10–20 X 10^6/kg (Hangretinger 1999, 2001). To mobilize CD34+ progenitor cells, donors were treated with recombinant human G-CSF at 10 µg/kg per day subcutaneously for 4 to 5 d. Donor leukocytes were obtained by apheresis using a COBE Spectra or Fenwall CS3000 cell seperators. There are less than 40 patients who have been transplanted with G-CSF mobilized partially mismatched donors according to CIBMTR and case reports in PubMed (Tables 17.3, 17.4).

Table 17.4 Haploidentical hematopoietic stem cell transplantation in osteopetrosis: Case reports.

Author	Publication year	Period of the survey	Number of cases (n: 23)	Stem cell source
Orchard	1987	NR	1	BM
Martinetti	1991	1988–1991	1	BM
Andolina	2000	1986–2000	4	BM
Eyrich	2001	1997–1999	2	PBSC
Kapelushnik	2001	NR	1	BM
Schulz	2001	1996–1999	7	PBSC
Starỳ	2002	1995–2000	1	PBSC
Handgretinger	2003	1995–2000	2	PBSC
Steward	2005	1995–2003	2	PBSC
Arpaci	2008	2000–2005	2	PBSC
Stepensky	2011	2011	1	PBSC

Abbreviations; BM, bone marrow; OP, osteopetrosis; PBSC; peripheral blood stem cell; NR, not reported

Stem cell purification (CD34+ selection)

Megadose transplantation of purified peripheral blood CD34+ progenitor cells from HLA-mismatched parental donors in children) offered as a realistic therapeutic option (Handgretinger 1999, 2001). Handgretinger et al. especially focused on the methods of CD34+ positive selection for two reasons: first, positive selection of a target cell population is associated with less unspesific cell loss and second, positive selection of CD34+ cells not only offers T-cell depletion for GVHD prevention, but also depletion of B-lymphocytes, which might prevent donor-derived Epstein-Barr virus-associated lymphoproliferative diseases. The authors optimized MACS method for clinical scale using the SuperMACS and later the CliniMACS (Handgretinger et al. 1999; Schumm et al. 1999). The purity of the CD34+ cells with this method was found in the range of 98–99 percent, the depletion of T- and B-lymphocytes was extremely effective, and the average number of transplanted T-lymphocytes was extremely low. No further GVHD prophylaxis was necessary even in the three-loci mismatch situation. An additional advantage of the MACS method is the good recovery (between 70 and 90 percent) of the CD34+ cells after positive selection, which is very important in order to augment the stem cell dose.

This method is based on the positive selection of magnetized CD34+ cells·using a strong magnetic field (Miltenyi et al. 1990). Briefly, magnetic microbeads are conjugated to an anti-CD34 monoclonal antibody. PBSCs (up to 1×10^{11}) are incubated with the antibody/microbead conjugate. The cell suspension is then run through a column placed in a strong magnetic field. The CD34+ cells are coated with magnetic microbeads and are therefore retained in the column, whereas the non-labeled cells are eluted. After removal of the column from the magnetic field, the CD34+ cells can easily be recovered with a purity in the range of 95–99 percent with a median recovery of CD34+ stem cells of 70 percent (Handgretinger et al. 2003). The T cell depletion is > 5 log and the percentage of contaminating B lymphocytes is < 0.5 percent.

Conditioning regimens in haploidentical hematopoietic stem cell transplantation for osteopetrosis

During the 1980s the conditioning regimen was mainly with busulphan and cyclophosphamide as the backbone of the conditioning regimens in order to avoid radiotherapy. The main study was conducted by Schulz et al. included just seven infants with MIOP (Schulz et al. 2002). They were given a conditioning regimen as follows: busulfan 4 to 5 mg/kg orally in divided doses daily for 4 d (total dose 16–20 mg/kg), thiotepa 10 mg/kg IV in two divided doses for 1 d, with IV cyclophosphamide 50 mg/kg once daily for

4 consecutive d (total dose 200 mg/kg; four patients) or cyclophosphamide 60 mg/kg once daily IV for 2 consecutive d (total dose 120 mg/kg; one patient) or fludarabine 40 mg/m2 IV once daily for 5 consecutive d (total dose 200 mg/m2, two patients). For rejection prophylxis, six patients received antihuman thymocyte globulin IV (ATG) for 3 or 4 consecutive d (total dose 20–30 mg/kg), and one patient received OKT-3 0.1 mg/kg once daily IV for 10 consecutive d (total dose 1 mg/kg).

Another study described a patient with OP who received a myeloablative therapy that consited of busulfan based regimen (busulfan on a total dose of 20 mg/kg, thiotepa 10 mg/kg, cyclophosphamide 200 mg/kg, and ATG 60 mg/kg). Conditioning regimens reported in other studies were mostly similar to these ones (Tables 17.5, 17.6).

The current preference in Europe is to try to reduce the risk of VOD by using treosulfan, or intravenous rather than oral busulphan, to substitute fludarabine for cyclophosphamide (because of lower toxicity profile), and to add thiotepa in nongenoidentical transplants (Greystoke et al. 2008; Steward 2010).

Engraftment results in G-CSF mobilized on haploidentical hematopoietic stem cell transplantation in osteopetrosis

G-CSF-mobilized haploidentical PBSCT in a pediatric cohort was first reported in 2001 by Eyrich et al. (Eyrich et al. 2001). This study included two patients with OP and reported a successful engraftment with a sufficient number of CD34+ cells (> 10 X 10^6/kg).

Schulz et al's study is the largest study with seven patients who underwent a HLA-haploidentical G-CSF mobilized HSCT in OP (Schulz et al. 2002). Their study revealed that the mean number of transplanted CD34+ cells was 30.37 X 10^6/kg for the initial graft. Moreover they gave boost CD34+ cells in six patients (mean number 26.71 X 10^6/kg) once in five patients in mean + 22. d post transplant and twice in one patient on +7 and +42 post transplant d. Hematopoietic reconstitution after transplant was achieved in five out of eight patients (mean neutrophil recovery >1,000 on d + 41.8, mean platelet recovery on +130. d post-transplant).

Starỳ et al. reported the engraftment of one patient with OP who underwent a double transplant from her mother (Starỳ et al. 2002). Although this patient received a sufficient number of CD34+ cells had an early rejection, bleeding and infection, therefore underwent a second haplo transplant from the same donor and finally achieved a successful engraftment and a survival over 20 mon.

Handgretinger et al. reported the engraftment results of two children with OP (Handgretinger et al. 1999, 2001, 2003). However, detailed data is present for one of them and reports a successful engraftment.

Table 17.5 Pre-transplant characteristics of G-CSF mobilized haploidentical PBSCT in OP.

Author	Patients (n: 17)	Age at tx (mon)	Donor	Conditioning regimens	Transplanted CD34+ cells (X 10⁶/kg)	T cell depletion (< 1 X 10⁵/kg)
Eyrich	2	6 and 36	Father	Bu (20)/TT(10)/Cy(200)/ATG(30) and FTBI(8)/TT(10)/Cy(200)/ OKT3	Both > 10	Yes
Schulz	7	5.5 6 3.5 2 1.5 5 6	5 fathers 2 mothers 1 aunt	Bu (16)/TT(10)/Cy(200)/ATG or Bu(20)/TT(10)/Cy(120)/ATG or Bu(20)/TT(10)/Cy(120)/ATG or Bu(20)/TT(10)/Cy(200)/ATG or Bu(20)/TT(10)/Flu/ATG or Bu(20)/TT(10)/Flu/OKT-3	All > 10	Yes (all)
Starý	1	10 (1st tx) 12 (2nd tx)	Mother	Bu/TT/Flu and ATG (1st tx) Cy/metylprednisolon/ATG (2nd tx)	Both > 10	Yes
Handgretinger	2	6 mo and NR	Father and NR	Bu/TT/Cy/ATG and NR	> 10 and NR	Yes and NR
Steward	2	62 and 5	NR	Bu/Cy/ATG	Both < 5	NR
Arpaci	2	16 and 7	Father and mother	Flu/Cy/ATG and Bu(16)/TT/Cy/ATG	<10 and > 10	Yes
Stepensky	1	5	Mother	Flu/Bu/TT and campath-1 (1st tx) Treosulfan/Cy/ATG (2nd tx)	Both > 10	Both yes

Abbreviations; ATG (30), antithymocte globulin (3 X 10 mg/kg); Bu, busulphan (4 X 4 mg/kg or 5 X 4 mg/kg); Cy, cyclophosphamide (2 X 60 mg/kg or 2 X 100 mg/kg); Flu, fludarabine; FTBI (8), fractionated total body irradiation (8 Gray); OP, osteopetrosis; PBSC; peripheral blood stem cell transplantation; TT (10), thiotepa (10 mg/kg); OKT3, anti-CD3 antibody

Table 17.6 Post-transplant characteristics of G-CSF mobilized Haploidentical PBSCT in osteopetrosis.

Author	Patients (n: 17)	Immuno-suppression	Engraftment ANC recovery > 0.5 X 109/l (+day)	Plt recovery > 20.000 (+day)	2nd Tx	Complications	Outcome	Survival (month)
Eyrich	2	Yes	10 and 9	NR	-	-	Both alive	>25 and 14
Schulz	7	Yes	27 / - / 22 / 7 / 22 / 17 / 14	49 / - / 44 / 51 / 70 / 53 / 53	1 (father and mother)	Respiratory insufficiency (4), CMV-disease (2), VOD (4), rejection (1)	alive (5), visual impairment (4), dead (2), mental retardation (1)	>43 (mean)
Stary	1	Yes	13 (1st tx) 10 (2nd tx)	NR	Yes	Early rejection, gastrointestinal bleeding, CMV infection, hypercalcemia	alive	>20
Handgretinger	2	Yes	11 and NR	NR	-	NR	alive and NR	>24 and NR
Steward	2	Yes	The 1st not recovered until autologous PBSC and the 2nd +27	1st has still low levels (19–55 X 10⁹/L in the 4th year of HSCT	2 patients underwent autologous PBSC Tx	Respiratory insufficiency, adenovirus and CMV-infection, graft rejection sustained low platelet count(1); fatal pulmonary arterial hypertension (1)	alive and dead	>48
Arpaci	2	Yes	28 (2nd patient)	low platelet (2nd patient)	-	Sepsis, GVHD, and graft failure	dead (1st patient) alive (2nd patient)	>48 (2nd patient)
Stepensky	1	Yes	25 (1st tx) 12 (2nd tx)	37 (1st tx) 14 (2nd tx)	Yes	Sepsis, hypercalcemia, hyperphospatemia, bilateral otitis, mastoiditis	alive	>11

Abbreviations; ANC, absolute neutrophil count; CMV, cytomegalovirus; VOD, veno-occlusive disease; G-CSF, granulocyte-colony stimulating factor; HSCT, hematopoietic stem cell transplantation; NR, not reported; OP, osteopetrosis; Plt, platelet; PBSCT, peripheral blood stem cell transplantation; Tx, transplantation

Steward et al. treated two patients with MIOP with haploidentical HSCT. These two patients needed transfusions of backups. The first one with very low CD34+ cell count (1.26 X 10⁶/kg) experienced severe infections needed to take the backup due to graft rejection on day 15 after haploidentical HSCT. In the other patient recovery was slower who received autologous reinfusion on day 19 after haplo HSCT (Steward et al. 2005).

Arpaci et al reported successful and unsuccessful engraftments of G-CSF-mobilized haploidentical HSCT in two children (Arpaci et al. 2008).

A recent paper from Stepensky et al. reported a successful double haploidentical HSCT in OP (Stepensky et al. 2011). Alhough a sufficient number of CD34+ cells were collected in both transplants, the authors decided to perform the second transplant 7 mon after the first transplant, because they observed that the child had a gradual decrease in donor chimerism after the 2nd month of the transplant.

Complications after haploidentical HSCT in OP

Most studies revealed that rejection, delayed hematopoietic reconstitution, venous occlusive disease, pulmonary hypertension or hypercalcaemic crisis were the common complications after haploidentical HSCT in OP (Steward et al. 2004; Steward 2010).

Schulz et al.'s report included a very detailed description of the complications. In this study, transplant courses were accompanied by a high rate of toxic and infectious complications. Thus, severe respiratory problems requiring mechanical ventilation developed in four cases (due to cytomegalovirus pneumonitis, due to respiratory failure from severe airway obstruction and due to late veno-occlusive disease of the lung. Each of these four patients also developed hepatic VOD, representing an unusually high incidence of this complication. Graft failure has been developed in two patients (the one after 1 mon in the context of CMV disease, which was reversible by a second graft on day +42 from the same donor without further conditioning. In the other patient the graft was rejected on day +15. This child died from CMV pneumonia shortly after a second transplant from the opposite parent). Two patients received late boosts of CD34+ cells at day 42 because of persistent marked tricytopenia in one and because of secondary graft failure in the other. In both patients rapid normalization of blood cell counts was observed, suggesting that delayed boosts may improve hematopoietic functions after HSCT in OP. T-cell immunity by donor cells developed with a delay of 3 to 6 mon, regardless of the total number of CD34+ cells infused. No patient developed acute or chronic GVHD.

Outcome and Conclusion

Historically, the results for haploidentical HSCT in OP were poor, although there has been a trend for better outcomes (Schulz et al. 2002). There are less than 40 patients who were transplanted with G-CSF-mobilized partially mismatched donors according to EBMT and CIBMTR surveys. The outcome of transplantation with haploidentical donors are inferior compared to sibling transplantation due to an increase incidence and severity of GVHD, graft rejection and infections. The 100-d mortality rate in the posttransplant period is approximately 25 percent. In addition, 70 percent of the patients are alive for 11 to 48-mon follow-up period (Tables 17.3 and 17.4). Outcomes are better with earlier transplantation, particularly before the age of 3 mon.

Recent advances with effective T-cell depletion, reduced intensity conditioning and megadoses of CD34+ stem cells will decrease the early transplant-related mortality, GVHD, graft failure and severe infections.

Key Facts of G-CSF-Mobilized Haploidentical Peripheral Blood Stem Cell Transplantation in Childhood Osteopetrosis

- Osteopetrosis is a rare genetic bone disease.
- The incidence rate is 1: 250,000 for autosomal recessive form of OP.
- Osteopetrosis may be fatal during infancy and early childhood periods.
- Neurologic problems, cranial nerve compression, hydrocephalus, mental retardation and bone marrow impairment are life-threating complications.
- The most severe complication is bone marrow suppression.
- The obliteration of marrow cavities with sclerotic bone can result in pancytopenia, recurrent infections and death.
- HSCT can be effective treatment for reversal of the bone marrow impairment and might be a solution for other complications.

Dictionary

- *G-CSF-mobilized PBSC*: Mobilized stem cell from bone marrow into the peripheral blood after administering granulocyte-colony stimulating factor.
- *Haploidentical HSCT*: Allogeneic transplantation from father, mother or > 1 HLA-antigen mismatched related donor.
- *Osteopetrosis*: A genetic metabolic bone disorder chaacterized by excessive bone density.

- *Megadoses of stem cells*: Target stem cell dose of transplanted CD34+ cells more than 10×10^6/kg.
- *Purified stem cells*: Positive selection of CD34+ cells using a cell separator.
- *EBMT*: European Bone Marrow Transplantation Registry.
- *CIBMTR*: Center for International Blod and Marrow Transplant Registry.

Summary Points

- At present, no effective medical treatment for osteopetrosis exists.
- HSCT remains the only curative therapy and is the cornerstone of treatment.
- Osteoclasts are of hematopoietic origin and HSCT provides an effective treatment for their regeneration in osteopetrotic bones.
- Results are excellent after matched sibling allogeneic grafts, but with unrelated donors are inferior due to an increased incidence and severity of GVHD, graft rejection and infections.
- Unfortunately, only about one-third of candidates for allogeneic HSCT have HLA-matched siblings.
- Haploidentical HSCT presents a better logistic and practical alternative to matched unrelated donor transplantations.
- Due to immediate availability, the possibility of using a haploidentical donor independent of HLA-matching would offer a number of advantages.
- The approach using megadose transplantation of G-CSF-mobilized peripheral blood CD34+ progenitor cells from HLA-mismatched parental donors offers a promising therapeutic option for every child without an otherwise suitable donor.
- Outcomes are better with earlier transplantation, particularly before the age of 3 mon.
- HSCT still carries many risks, some unique to this disease, and a high proportion of patients have sequelae of their disease despite transplantation.

List of Abbreviations

AD	:	autosomal dominant
ANC	:	absolute neutrophil count
AR	:	autosomal recessive
ARO	:	autosomal recessive osteopetrosis
ATG	:	antithymocte globulin

BM	:	bone marrow
Bu	:	busulphan
CIBMTR	:	Center for International Blood and Marrow Transplant Registry
CMV	:	cytomegalovirus
Cy	:	cyclophosphamide
EBMT	:	European Bone Marrow Transplantation Registry
Flu	:	fludarabine
FTBI	:	fractionated total body irradiation
G-CSF	:	granulocyte-colony stimulating factor
GVHD	:	graft-versus-host disease
HLA	:	human leucocyte antigen
HSCT	:	hematopoietic stem cell transplantation
MACS	:	magnetic activated cell separator
MIOP	:	malignant infantile osteopetrosis
NR	:	not reported
OKT3	:	anti-CD3 antibody
OP	:	osteopetrosis
PB	:	peripheral blood
PBSC	:	peripheral blood stem cell
Plt	:	platelet
RTA	:	renal tubular acidosis
TT	:	thiotepa
Tx	:	transplantation
VUD	:	volunteer unrelated donor
VOD	:	veno-occlusive disease

References

Armitage, J.O. 1994. Bone marrow transplantation. N Engl J Med. 330: 827–838.

Arpaci, F. 2011. CD 34+ cell selection and purging in hematopoietic cell transplantation. Gulhane Med J. 53: 226–231.

Arpaci, F., S. Ataergin, A. Ozet, K. Erler, M. Basbozkurt, A. Ozcan, S. Komurcu, B. Ozturk, B. Celasun, S. Kilic and O. Kuzhan. 2005. The feasibility of neoadjuvant high-dose chemotherapy and autologous peripheral blood stem cell transplantation in patients with nonmetastatic high grade localized osteosarcoma: results of a phase II study. Cancer. 104: 1058–1065.

Arpaci, F., I. Tezcan, O. Kuzhan, N. Yalman, D. Uckan, A.E. Kurekci, A. Ikinciogullari, A. Ozet and A. Tanyeli. 2008. G- G-CSF-mobilized haploidentical peripheral blood stem cell transplantation in children with poor prognostic nonmalignant disorders. Am J Hematol. 83: 133–136.

Ataergin, S., F. Arpaci, K. Erler, B. Demiralp, I. Cicek, C. Ulutin, L. Solchaga and A. Ozcan. 2007. Successful treatment of osteosarcoma arising in osteogenesis imperfecta with high-dose chemotherapy and autologous peripheral blood stem cell transplantation followed-by limb sparing surgery. Clinical Medicine: Oncology. 1: 99–104

Beatty, P.G., M. Mori and E. Milford. 1995. Impact of racial genetic polymorphism on the probability of finding an HLA-matched donor. Transplantation. 60: 778–783.

Coccia, P.F., W. Krivit, J. Cervenka, C. Clawson, J.H. Kersey, T.H. Kim, M.E. Nesbit, N.K. Ramsay, P.I. Warkentin, S.L. Teitelbaum, A.J. Kahn and D.M. Brown. 1980. Successful bone-marrow transplantation for infantile malignant osteopetrosis. N Engl J Med. 302: 701–708.

Driessen, G.J., E.J. Gerritsen, A. Fischer, A. Fasth, W.C. Hop, P. Veys, F. Porta, A. Cant, C.G. Steward, J.M. Vossen, D. Uckan and W. Friedrich. 2003. Long-term outcome of haematopoietic stem cell transplantation in autosomal recessive osteopetrosis: an EBMT report. Bone Marrow Transplant. 32: 657–663.

Eyrich, M., P. Lang, S. Lal, P. Bader, R. Handgretinger, T. Klingebiel, D. Niethammer and P.G. Schlegel. 2001. A prospective analysis of the pattern of immune reconstitution in a paediatric cohort following transplantation of positively selected human leucocyte antigen-disparate haematopoietic stem cells from parental donors. Br J Haematol. 114: 422–432.

Fischer, A., C. Griscelli, W. Friedrich, B. Kubanek, R. Levinsky, G. Morgan, J. Vossen, G. Wagemaker and P. Landais. 1986. Bone-marrow transplantation for immunodeficiencies and osteopetrosis: European survey, 1968–1985. Lancet. 2: 1080–1084.

Fischer, A., W. Friedrich, A. Fasth, S. Blanche, F. Le Deist, D. Girault, F. Veber, J. Vossen, M. Lopez, C. Griscelli et al. 1991. Reduction of graft failure by a monoclonal antibody (anti-LFA-1 CD11a) after HLA nonidentical bone marrow transplantation in children with immunodeficiencies, osteopetrosis, and Fanconi's anemia: a European Group for Immunodeficiency/European Group for Bone Marrow Transplantation report. Blood. 77: 249–256.

Gerritsen, E.J., J.M. Vossen, A. Fasth, W. Friedrich, G. Morgan, A. Padmos, A. Vellodi, O. Porras, A. O'Meara, F. Porta et al. 1994. Bone marrow transplantation for autosomal recessive osteopetrosis. A report from the Working Party on Inborn Errors of the European Bone Marrow Transplantation Group. J Pediatr. 125: 896–902.

Greystoke, B., S. Bonanomi, T.F. Carr, M. Gharib, T. Khalid, M. Coussons, M. Jagani, P. Naik, K. Rao, N. Goulden, P. Amrolia, R.F. Wynn and P.A.Veys. 2008. Treosulfan-containing regimens achieve high rates of engraftment associated with low transplant morbidity and mortality in children with non-malignant disease and significant co-morbidities. Br J Haematol. 142: 257–262.

Handgretinger, R., M. Schumm, P. Lang, J. Greil, A. Reiter, P. Bader, D. Niethammer and T. Klingebiel. 1999. Transplantation of megadoses of purified haploidentical stem cells. Ann N Y Acad Sci. 872:351–361; discussion 361–362.

Handgretinger, R., T. Klingebiel, P. Lang, M. Schumm, S. Neu, A. Geiselhart, P. Bader, P.G. Schlegel, J. Greil, D. Stachel, R.J. Herzog and D. Niethammer. 2001. Megadose transplantation of purified peripheral blood CD34(+) progenitor cells from HLA-mismatched parental donors in children. Bone Marrow Transplant. 27: 777–783.

Handgretinger, R., T. Klingebiel, P. Lang, P. Gordon and D. Niethammer. 2003. Megadose transplantation of highly purified haploidentical stem cells: current results and future prospects. Pediatr Transplant. Suppl. 3: 51–55.

Körbling, M. and R. Champlin. 1996. Peripheral blood progenitor cell transplantation: a replacement for marrow auto- or allografts. Stem Cells. 14: 185–195.

Marcus, J.R., E. Fibach and M. Aker. 1982. Circulating myeloid and erythroid progenitor cells in malignant osteopetrosis. Acta Haematol. 67: 185–189.

Miltenyi, S., W. Müler, W. Weichel and A. Radbruch. 1990. High gradient magnetic cell separation with MACS. Cytometry. 11: 231–238.

Orchard, P.J., J.D. Dickerman, C.H. Mathews, S. Frierdich, R. Hong, M.E. Trigg, N.T. Shahidi, J.L. Finlay and P.M. Sondel. 1987. Haploidentical bone marrow transplantation for osteopetrosis. Am J Pediatr Hematol Oncol. 9: 335–340.

Reisner, Y. and M.F. Martelli. 1995. Bone marrow transplantation across HLA barriers by increasing the number of transplanted cells. Immunol Today. 16: 437–440.

Schulz, A.S., C.F. Classen, W.A. Mihatsch, M. Sigl-Kraetzig, M. Wiesneth, K.M. Debatin, W. Friedrich and S.M. Müler. 2002. HLA-haploidentical blood progenitor cell transplantation in osteopetrosis. Blood. 99: 3458–3460.

Schumm, M., P. Lang, G. Taylor, S. Kuçi, T. Klingebiel, H.J. Bühring, A. Geiselhart, D. Niethammer and R. Handgretinger. 1999. Isolation of highly purified autologous and allogeneic peripheral CD34+ cells using the CliniMACS device. J Hematother. 8: 209–218.

Stark, Z. and R. Savarirayan. 2009. Osteopetrosis. Orphanet J Rare Dis. 4:5.

Starý, J., P. Sedlácek, S. Vodvárková, A. Poloucková, R. Formánková, Z. Gasová and I. Marinov. 2002. Transplantation of highly purified CD34+ hematologic peripheral stem cells from haploidentical related donors in the treatment of children with non-malignant diseases. Cas Lek Cesk. 141: 176–181.

Stepensky, P., A.S. Schulz, G. Lahr, N. Simanovsky, R. Brooks, S. Samuel, R. Or, M. Weintraub and I. Resnick. 2011. Successful second haploidentical SCT in osteopetrosis. Bone Marrow Transplant. 46: 1021–1022.

Steward, C.G. 2010. Hematopoietic stem cell transplantation for osteopetrosis. Pediatr Clin North Am. 57: 171–180.

Steward, C.G., I. Pellier, A. Mahajan, M.T. Ashworth, A.G. Stuart, A. Fasth, D. Lang, A. Fischer, W. Friedrich and A.S. Schulz. 2004. Working Party on Inborn Errors of the European Blood and Marrow Transplantation Group. Severe pulmonary hypertension: a frequent complication of stem cell transplantation for malignant infantile osteopetrosis. Br J Haematol. 124: 63–71.

Steward, C.G., A. Blair, J. Moppett, E. Clarke, P. Virgo, A. Lankester, S.R. Burger, M.G. Sauer, A.M. Flanagan, D.H. Pamphilon and P.J. Orchard. 2005. High peripheral blood progenitor cell counts enable autologous backup before stem cell transplantation for malignant infantile osteopetrosis. Biol Blood Marrow Transplant. 11: 115–121.

Stoker, D.J. 2002. Osteopetrosis. Semin Musculoskelet Radiol. 6: 299–305.

Symons, H.J. and E.J. Fuchs. 2008. Hematopoietic SCT from partially HLA-mismatched (HLA-haploidentical) related donors. Bone Marrow Transplant. 42: 365–377.

Walker, D.G. 1975. Control of bone resorption by hematopoietic tissue. The induction and reversal of congenital osteopetrosis in mice through use of bone marrow and splenic transplants. J Exp Med. 142: 651–663.

Combining Osteochondral Stem Cells and Biodegradable Hydrogels for Bone Regeneration

Matilde Bongio,[a,#] Wanxun Yang,[b,#] Fang Yang,[c] Sanne K. Both,[d] Sander C.G. Leeuwenburgh,[e] Jeroen J.J.P. van den Beucken[f] and John A. Jansen[g,*]

ABSTRACT

The increase of life-expectancy and dynamism in daily activities has led to a dramatic boost of musculoskeletal malconditions with a staggering social and economical impact. The shortcomings of conventional treatment approaches, such as autografts and allografts, forced the scientific community to explore "man-made" alternatives

Radboud University Nijmegen Medical Center, Department of Biomaterials (309), PO Box 9101, 6500 HB Nijmegen, The Netherlands.
[a]E-mail: M.Bongio@dent.umcn.nl
[b]E-mail: W.Yang@dent.umcn.nl
[c]E-mail: F.Yang@dent.umcn.nl
[d]E-mail: S.Both@dent.umcn.nl
[e]E-mail: S.Leeuwenburgh@dent.umcn.nl
[f]E-mail: J.vandenBeucken@dent.umcn.nl
[g]E-mail: J.Jansen@dent.umcn.nl
*Corresponding author
#These authors contributed equally to this book chapter.

List of abbreviations given at the end of the text.

as effective solutions to bring major benefits for patients, including reduced need and cost of medical attention, and short-term recovery. Specifically, regenerative medicine (RM) research is currently devoted to the development of natural tissue-like materials, the so-called biomaterials, either combined with stem cells, stimulating factors or a combination thereof. To that end, such constructs are capable not only to replace the injured tissue, but also to induce and direct the healing process, and completely restore the native tissue functions. So far, the most common approaches to create bone have focused on intramembranous ossification (i.e., bone formation through direct osteogenic differentiation), without however obtaining promising results. Conversely, alternative approaches aiming at endochondral ossification (i.e., bone formation via formation of hypertrophic cartilage remodeling) seem to be more promising for bone repair. Compared to osteogenic cells, chondrocytes are less oxygen sensitive and thus can easily survive within the scaffolds in long-term culture, and can induce vascularization culminating in new bone formation. Biomaterial scaffolds that have been developed for bone regeneration include ceramics and natural or synthetic polymers. Among the polymer-based materials, hydrogels are currently gaining significant popularity in the medical community as appealing solutions for treatment of bone tissue defects. Owing to their viscoelastic properties and hence similarities with the extracellular matrix (ECM), hydrogels represent suitable scaffolds to support cell functions and can form an ideal healing environment. The present chapter will first provide a description of bone tissue and osteogenesis mechanisms, in order to reach the fundamental understanding of the processes involved in RM-based approaches for bone regeneration utilizing biomaterials and cell culture strategies. Subsequently, current results in basic research on biomaterials for bone repair, with emphasis on hydrogels, will be discussed.

Introduction

Musculoskeletal malconditions (e.g., arthritis, back pain, spinal disorders, osteoporosis, bone fractures, trauma, and congenital problems) affect hundreds of millions of people across the world. According to "The Burden of Musculoskeletal Diseases in the United States" (BMUS), over 130 million patient visits to healthcare providers occur annually, and the economic impact of these conditions is staggering: in 2004, the annual costs for bone and joint health was estimated to be US$ 849 billion, and the burden of musculoskeletal conditions is expected to escalate in the next 10–20 yr as the population ages and lifestyles become more active (http://www. boneandjointburden.org).

The most favorable choice for treatment of bone defects is the graft of the patient's own healthy bone (so-called autograft) harvested from a donor site, usually iliac crest; or, alternatively, from a cadaveric bone typically sourced from a bone bank (allograft). In addition to the respective benefits of these approaches, several drawbacks limit their clinical applications. Autografts reduce the risk of rejection but require an additional surgical site, thus increasing post-operative pain and complications for the patient. Allograft eliminates the risk of donor site morbidity and is hence associated with less pain and discomfort for the recipient; however, rejection or disease transmission cannot be ruled out.

In view of these drawbacks associated with autografts and allografts, the multidisciplinary field of regenerative medicine (RM) holds the promise of creating appealing new possibilities regarding the quality and duration of life. Currently, RM-based treatment strategies for critical-size bone defects (i.e., defects that will not heal spontaneously) propose the design of biomaterials either alone or in combination with stem cells and signaling molecules, which resemble the three dimensional *in vivo* tissue arrangements and promote restoration of normal tissue functions following grafting in the body (Dominici et al. 2001). To that end, a deep knowledge of bone biology will lead to better understanding of the mechanisms of interaction between cells and scaffolds, including cell adhesion, proliferation, and differentiation, which are all important steps for the success of a man-made bone graft. Up to now, thousands of scaffolds with variable properties and chemical composition, including ceramics, natural and synthetic polymers, as well as cell culture strategies have been proposed. Nevertheless, as for most of these approaches, the developmental stage is in its infancy and the challenge is to obtain proof of clinical efficacy in order to proceed towards routine clinical implementation.

The present chapter will first describe bone tissue and its developmental mechanisms, in order to reach a fundamental understanding of the processes involved in RM-based approaches for bone regeneration utilizing biomaterials and cell culture strategies. Subsequently, current results in basic research on biomaterials for bone repair, with emphasis on hydrogels, will be discussed.

Bone Tissue

Bone is a dynamic and vascularized tissue with a remarkable ability to heal and remodel, and is responsible for providing structural support to the body and protection of the internal organs. In addition, bone plays a key role in supporting muscle contraction, blood production and the storage of mineral components (Sommerfeldt and Rubin 2001).

Bone is not a uniformly solid structure, but is rather composed of two distinct parts, namely compact and trabecular bone. Compact bone (also called cortical bone) forms the hard outer component of the bones, whereas trabecular bone (also called cancellous bone) fills the inner side. Compact bone contributes approximately 80 percent of the weight of a human skeleton with a very dense and stiff structure and is usually located in the shaft of the long bones and in the peripheral layer of flat bones. In contrast, trabecular bone has a lower density, larger surface area, and is composed of a highly porous network that creates space for bone marrow and blood vessels.

From a material perspective, bone is composed of an inorganic and organic component. The inorganic matrix represents ~70 percent of bone mass and is mainly composed of hydroxyapatite, which provides stiffness and strength. The organic matrix, which represents ~30 percent of bone mass, provides flexibility and tensile strength, and is composed of proteins and cells. Collagen type I is the main component of the organic matrix; whereas other proteins, such as osteocalcin, bone sialoprotein and osteonectin, play important signaling functions and are involved in the mineralization process.

The three types of cells that populate bone tissue are osteoblasts, osteocytes and osteoclasts. Osteoblasts are bone-forming cells that originate from osteoprogenitor cells located in the deeper layer of periosteum and bone marrow, which produce an osteoid matrix that successively mineralizes into bone. Although the term osteoblast implies an immature cell type, osteoblasts are the mature cells of bone that generate bone tissue. At a later stage, osteoblasts can either become bone lining cells or differentiate into osteocytes. Bone lining cells are inactive osteoblasts that reside along the quiescent bone surface forming a protective layer. Although their function is not well understood, they may play a role in the activation of bone remodeling (Bord et al. 1996). Osteocytes are embedded within the bone matrix in the so-called bone lacunae. These mature bone cells are involved in different functions, including matrix maintenance, calcium homeostasis and regulation of bone response to mechanical stimuli. Osteoclasts are responsible for bone resorption by (locally) decreasing the pH and activating lytic enzymes, which in turn break down the mineralized matrix.

Osteogenesis

During osteogenesis, two different processes can result in the formation of normal, healthy bone tissue, i.e., intramembranous ossification and endochondral ossification (Fig. 18.1).

Intramembranous ossification

Intramembranous ossification occurs in a vascularized environment and involves the direct differentiation of mesenchymal stem cells (MSCs) into

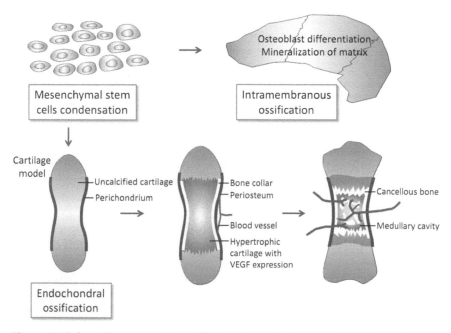

Figure 18.1 Schematic representation of the two types of osteogenesis. 1) Intramenbranous ossification involves bone formation through the direct osteoblastic differentiation; 2) endochondral ossification generates bone via transition from an avascular hypertrophic cartilage to a vascularized and mineralized matrix.

Color image of this figure appears in the color plate section at the end of the book.

osteoblasts. Initially, MSCs are widely dispersed within an ECM that is devoid of collagen. Groups of adjacent MSCs proliferate and form a dense nodule of cell aggregation, which undergoes morphological changes toward the osteoblastic phenotype. Subsequently, cells start to produce new organic matrix (i.e., osteoid), mainly consisting of type I collagen fibrils, and eventually develop into mature osteoblasts. Some osteoblasts align along the periphery of the nodule and keep forming osteoids, whereas others are incorporated into the nodule and become osteocytes. At this point, the osteoid becomes mineralized and forms the rudimentary bone tissue.

Endochondral ossification

Endochondral ossification is the type of osteogenesis that accounts for the development of most bones of the vertebrate skeleton (e.g., long bones) and is the most commonly occurring process in fracture healing. As such, endochondral ossification takes place in a poorly or non-vascularized environment, in which loosely connected MSCs condense at the sites pre-determined for bone formation. The increased cell–cell contacts and low oxygen condition (i.e., hypoxia) trigger the differentiation of MSCs into the chondrogenic lineage. The cartilage template takes shape through the deposition of abundant cartilaginous matrix, and the chondrocytes gradually differentiate until reaching a hypertrophic stage, during which the chondrocytes mineralize and eventually become apoptotic. Simultaneously, these chondrogenic cells release vascular endothelial growth factor (VEGF), which induces blood vessel formation (Farrell et al. 2009). At this point, osteoclasts and osteoblasts, which receive nutrition and oxygen supply from the blood vessels, reach the mineralized matrix and promote the transformation of the cartilage into bone, respectively.

Basic Tools for RM Purposes—Stem Cells and Biomaterial Scaffolds

Stem cells

Stem cells are undifferentiated cells with a high proliferation capability, which have the capacity of self-renewal or differentiate into specialized cell types of our body (Blau et al. 2001). The central focus of RM is human stem cells, for which two main types of stem cells exist based on their origin: embryonic stem cells (ESCs) and adult stem cells. ESCs are derived from the inner cell mass of the blastocyst (i.e., an early-stage embryo) and are able to differentiate into all cell types of the adult organism. The multi-differentiation potential of ESCs has been reported by several researchers (Nakayama et al. 2003; Wiles and Keller 1991), and has attracted wide interest for many medical applications, including regeneration of bone tissue. Indeed, ESCs have demonstrated the capacity to differentiate into osteoblasts when cultured in the presence of osteogenic differentiation-inducing factors, such as dexamethasone (zur Nieden et al. 2003). However, many issues regarding the clinical use of ESCs need to be addressed, including the risk of tumorogenesis, the labor-intensive procedures, and important ethical considerations (Wobus 2001). Adult stem cells, also called somatic stem cells, are present in the body at many sites and are responsible for growth and maintenance of normal tissues, as well as restoration of injured tissues. Initially, adult stem cells were discovered in the bone marrow and, successively, in many other different tissues, including heart, liver, pancreas, brain, epidermis, dental pulp and spinal cord. In the last 30 yr, substantial efforts have been devoted to studies of adult stem cells for a broad range of applications for tissue regeneration. Among the different sources for adult stem cells, mesenchymal stem cells (MSCs) are of great interest in RM applications. MSCs were initially identified in bone marrow as non-hematopoietic stem cells that can differentiate into tissues of mesodermal origin, such as adipocytes, osteoblasts, chondrocytes, tenocytes, skeletal myocytes and visceral stromal cells (Krampera et al. 2006). Friedenstein and co-workers were the first to reveal the bone-forming capacity of MSCs (Friedenstein et al. 1966).

Biomaterial scaffolds

As mentioned earlier in the introduction, a promising strategy to optimize and simplify the treatment of muscoloskeletal disorders is represented by bone grafts from man-made source. Several decades of progress in techniques and understanding of the mechanisms involved in the interaction between synthetic materials and living surrounding tissues resulted in the establishment of "biomaterials" as a scientific discipline, and a rapid progression in the design of increasingly sophisticated materials that provide a template for tissue regeneration. As a result of this evolution, the definition of *"biomaterial"* has constantly changed over the years. Currently, the most appropriate and used definition depicts biomaterial as *"a substance that has been engineered to take a form which, alone or as part of a complex system, is used to direct, by control of interactions with components of living systems, the course of any therapeutic or diagnostic procedure, in human or veterinary medicine"* (Williams 2009).

The history of biomaterials for bone repair is marked by three generations, namely, from tolerant and bioinert (the first generation), to bioactive and osteoconductive (the second generation), and to osteoinductive biomaterials (the third generation) that induce neighboring tissue to become bone (Bongio et al. 2010). Primordial biomaterials were meant to be tolerated by the body and to passively replace missing or damaged tissue. The difficult adaptation of the human body to these foreign materials raised the need to create improved artificial substitutes. These scaffolds reproduce the complex three dimensional environment to which cells are exposed to and actively participate in the dialog with the natural environment (i.e., bioactive). To this end, bone healing requires scaffolds that provide a porous and biologically compatible framework onto which (precursors of) bone-forming cells migrate (i.e., osteoconduction) and form new bone tissue, or even direct differentiation of stem cells into mature osteogenic cells (i.e., osteoinduction) (Hutmacher et al. 2007). Last, but not least, biomaterials should degrade over time in concert with the growth of new bone, and the degradation products should be non-toxic and completely eliminated from the body. Table 18.1 lists the most important properties and respective definitions for appropriate bone graft materials.

Table 18.1 Properties and definitions for a bone graft material.

Properties	Definitions
Biocompatibility	Behavior of a biomaterial and its byproducts to elicit little or no immune response into the body
Bioactivity	Ability of a biomaterial to interact with the natural environment of the body and determine a biological effect
Osteointegration	The full process of obtaining firm anchoring and incorporation of a biomaterial into living bone without fibrous tissue formation at the interface
Osteoconductivity	The ability of a biomaterial to stimulate the migration of adjacent bone-forming cells in order to induce the growth of bone tissue
Osteoinductivity	The ability of a biomaterial to stimulate the migration of undifferentiated cells and induce their differentiation into active osteoblasts, in order to promote *de novo* bone formation
Biodegradability	The ability of a biomaterial to degrade in concert with the natural bone remodeling process

In order to be successfully promote and support bone regeneration, a biomaterial should encounter specific properties, including biocompatibility, bioactivity, osteointegration, osteoconductivity, osteoinductivity and biodegradability.

Ceramic-based materials

In the past three decades, ceramics, such as hydroxyapatite (HA), tricalcium phosphate (TCP) and bioglass, were the main scaffolds studied for bone regeneration, being similar in their chemical composition and crystallographic structure to the mineral component of bone tissue. These characteristics govern the bioactive and osteoconductive behavior, which in turn favor the formation of new bone. However the brittleness and weak mechanical properties represent the major drawbacks that still require optimization (Barrere et al. 2006).

Polymer-based materials

An alternative approach in RM is to engineer new bone tissues by using selective cell transplantation on polymer scaffolds (Langer and Vacanti, 1993). Some of the most commonly explored polymers as scaffold materials include poly(lactic acid) (PLA), poly(lactic-co-glycolic acid) (PLGA), poly(2-hydroxyethyl methacrylate) (pHEMA), poly(ε-caprolactone) (PCL), poly(ethylene glycol) (PEG), polyglycolic acid (PGA), polyurethanes and composites (Nair and Laurencin 2006). Polymers exhibit biocompatibility, degradability and have good processability. These properties confer to polymers a great potential to serve as artificial ECM, thus mimicking and recreating to some extent the structural organization of the bone. A novel and promising class of polymer materials explored as cell carriers and delivery systems for regenerative medicine of bone are hydrogels. As shown in Fig. 18.2, the number of reports related to the research of hydrogels as bone substitutes increased over the last decade, demonstrating the growing interest for the class of biomaterials in bone regeneration.

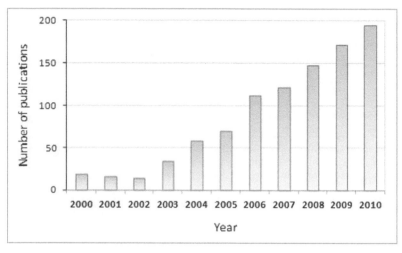

Figure 18.2 Publications in pubmed related to hydrogels for bone tissue from 2000 until 2010. The total number of publications cited in pubmed related to applications of hydrogels for bone tissue increased substantially in the last decade, thus demonstrating a growing interest for these types of polymer-based materials.

Hydrogels

Hydrogels are networks of hydrophilic, crosslinked polymer chains with a distinctive quality to absorb water in aqueous fluids, and hence to exhibit mechanical and visco-elastic properties that closely resemble the structural three dimensional environment of natural ECM. These attributes make hydrogels appealing substrates for cells to adhere, proliferate, secrete new ECM and eventually restore the damage tissue (Nicodemus and Bryant 2008). Most importantly, from a clinical perspective, hydrogels are available ready-to-use grafts, can be injected into the wound site using minimally invasive techniques and allow complete filling of irregularly shaped defects, thus avoiding an open surgery procedure (Fig. 18.3). In combination with stem cells, these scaffolds have been evaluated for their suitability as potential replacements for bone and cartilage.

Hydrogels can be broadly classified into natural and synthetic, based on their origin (Drury and Mooney 2003). Natural hydrogels (e.g., agarose,

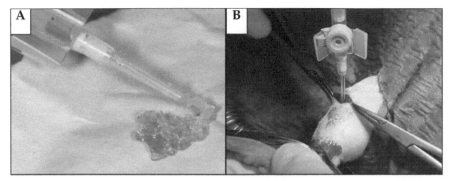

Figure 18.3 Application of hydrogels as injectable bone substitute materials. (A) Hydrogels ejection from syringe through a needle cannula. (B) Injection of a hydrogel into a 1.8mm-bone defect of a guinea pig tibia model.

Color image of this figure appears in the color plate section at the end of the book.

alginate, chitosan, collagen/gelatin, fibrin, hyaluronan, and silk) exhibit excellent bioactivity, as they mimic the native extracellular environment structurally, have unique mechanical properties and are biodegradable by enzymatic or hydrolytic mechanisms. However, natural hydrogels derived from natural polymers have inherent disadvantages associated with unstable material supply, batch-to-batch variations and rapid degradation upon contact with body fluids or culture medium. All these limitations have motivated researchers to modify these natural materials as well as to develop synthetic polymers, which show tremendous advantages over the natural polymers. Synthetic hydrogels can be formed by crosslinking of polymer chains through physical (temperature, UV light or pH) or chemical reactions under controlled conditions. Therefore, variations in the chemical structure and crosslinking density allow to predict and reproduce the mechanical and physical properties, degradation rate (Lutolf and Hubbell 2005), and even to influence stem cell behavior and differentiation (Engler et al. 2006). The major shortcoming of synthetic hydrogels is the lack of informational structures (e.g., cell-binding peptides), which may be pivotal to support cellular responses. However, bioactive cues can be easily incorporated through chemical modifications of the polymer backbone. Other issues include the biocompatibility of the chemical initiator agents, which are required for the crosslinking, and the potential toxicity of byproducts resulting from the degradation. Representative synthetically derived polymers explored in combination with stem cells for bone regeneration purposes include poly(ethylene glycol) (PEG), oligo(poly(ethylene glycol) fumarate) (OPF) hydrogels, and poly(ethylene glycol)-diacrylate (PEG-DA) hydrogels (Betz et al. 2008; Bongio et al. 2011; Burdick and Anseth 2002; Temenoff et al. 2004).

Cell Culture Strategies for Bone Tissue Regeneration

Initial *in vitro* culture set-up, including selection of cell types (mono- or co-culture), growth factors, medium composition and timing are pivotal to determine the possible benefits of well-timed implantation (Farrell et al. 2009). Figure 18.4 depicts all the steps required for the synthesis of a cell-scaffold construct for bone tissue regeneration, which include (1) cell isolation and expansion in complete medium, (2) cell seeding and cell maturation (via intramembranous or endochondral ossification), either prior to or after loading into 3D scaffolds, and (3) *in vivo* grafting.

Isolation and expansion

As described above, natural bone regeneration is a complex process, during which multiple cell types, dispersed in a three dimensional environment, are spatially and temporally regulated by several environmental stimuli. In order to re-create the natural situation, millions of cells are necessary for incorporation into an artificial matrix. From a clinical point of view, it is possible to isolate osteoblasts from biopsies taken from patients, however, this procedure is time consuming and allows to obtain only a limited number of cells. Therefore, stem cells represent a favorable alternative. Stem cells, derived from bone marrow or other tissues, are usually expanded with complete medium in culture plates or flasks, where they adhere, start proliferating and form colony-forming cell clusters. Before these cells become completely confluent, they are passaged and expanded in multiple flasks, thus becoming a more and more homogenous adherent cell population that may proliferate without differentiating up to several generations (Lennon and Caplan 2006).

Seeding and differentiation

Intramembranous ossification

Except for a few reports, regenerative medicine efforts in the field of bone tissue repair have focused on generating bone directly from osteoprogenitor cells in a manner akin to intramembranous ossification *in vivo*. The standard procedure of this strategy involves stem cell differentiation into osteoblasts. To that end, MSCs are exposed to specific culture supplements, such as dexamethasone, ascorbic acid, and beta glycerolphosphate (Petite et al. 2000). Cells can either be differentiated in flasks (2D-environment) and loaded into scaffolds just prior to grafting *in vivo*, or be loaded into scaffolds when they are still uncommitted and undergo maturation in a 3D-environment. Although the second approach is more appealing, as it

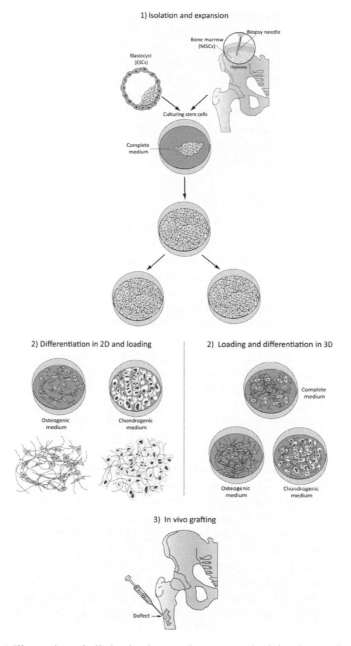

Figure 18.4 Illustration of all the fundamental steps required for the synthesis and application of a cell-material system for bone tissue regeneration. (1) cell isolation and expansion, (2) cell seeding and differentiation, and (3) *in vivo* grafting.

Color image of this figure appears in the color plate section at the end of the book.

resembles the *in vivo* situation, its success has not been satisfactory to date. A major issue with this approach is the limited long-term cell survival and necrosis, arising from lack of nutrient delivery to and waste removal from the center of tissue-engineered constructs, and lack of blood supply (Farrell et al. 2011). First, blood vessels formation is not an instantaneous process and usually occurs too slowly for cells to survive without nutrients. Second, culturing a cell-laden scaffold *in vitro* for several weeks can lead to the formation of an extensive extracellular matrix, which can prevent or seriously hamper blood vessel infiltration *in vivo* by sealing the pores of a scaffold (Meijer et al. 2007).

Endochondral ossification

Because chondrocytes can survive in a hypoxic environment and because they produce anabolic and catabolic factors important for the conversion of avascular tissue into vascularized tissue, it seems straightforward to exploit the chondrocyte-mediated route (i.e., endochondral) for ossification in RM-based approaches, rather than intramembranous ossification (Farrell et al. 2009). Either human ESCs and MSCs are able to form bone via the endochondral pathway, i.e., differentiate to chondrocytes *in vitro* and turn into bone when implanted *in vivo* (Farrell et al. 2011; Jukes et al. 2008). Chondrogenic priming of mesenchymal stem cells can be obtained by culturing cells in differentiation medium containing a mixture of specific supplements, such as dexamethasone, ascorbic acid, insulin, transferrin, selenious acid, sodium pyruvate, transforming growth factor-beta (TGF-β), and a phosphate donor (Mackay et al. 1998). The main idea would be to induce pre-vascularization *in vitro* in order to facilitate blood vessel formation when implanted *in vivo*. However, no *in vitro* system is known to support vascularization necessary for the formation of the bone following the endochondral ossification pathway. Different culture systems can offer a solution for the development of *in vitro* tissue engineering approaches for the creation of vascularized bone *in vivo*. Among these strategies, the use of bioreactors (Montufar-Solis et al. 2004) and an environment with low oxygen tension have demonstrated to support growth and differentiation of cartilage that forms bone (Robins et al. 2005).

   Co-culture systems, i.e., combined culturing of two or more types of cells, represent a new center of interest in the fields of bone tissue engineering and regenerative medicine because they are one step closer to natural conditions and allow elucidation of some aspects of the complex interactions between bone-forming and bone-resorbing cells. As discussed, long-term cell viability is hampered by insufficient vascularization within engineered constructs. Considering that bone development and remodeling are tightly regulated by the vascular system, and endothelial cells modulate activity of osteoblastic and osteoclastic cells (Hauge et al. 2001), co-cultures of osteoprogenitor cells and endothelial cells were initiated in order to simultaneously establish a functional microvascular network and bone formation.

Other approaches include addition of key regulators of the *in vivo* endochondral ossification, such as vascular endothelial growth factor (VEGF) and matrix metalloproteinases (MMPs). Specifically, VEGF is known to play a critical role in the establishment of vessel in-growth (Street et al. 2002), whereas MMPs, expressed by chondrocytes and hypertrophic chondrocytes, are crucial for the degradation of the cartilage. Together, these regulatory molecules coordinate the recruitment and differentiation of endothelial cells, osteoclasts, chondrocytes, and osteoprogenitors (Oliveira et al. 2009).

In vivo grafting

Osteo- or chondroprogenitor cells, previously differentiated from stem cells under prolonged conventional culture conditions, are implanted *in vivo* in conjunction with natural or synthetic scaffolds. While osteoprogenitor cells differentiate directly into osteoblasts that will deposit new bone matrix, chondrogenic-committed cells become hypertrophic and deposit cartilaginous extracellular matrix that becomes calcified, and eventually undergo apoptosis. Subsequently, blood vessels and perichondrial cells surrounding the cartilage differentiate into osteoblasts, invade the hypertrophic cartilage and give rise to bone (Solursh 1989).

Preliminary Studies on Cell-biomaterial Constructs for Endochondral Bone Formation

Ceramics

Ceramics have been investigated extensively in combination with stem cells for tissue engineering of bone. However, the development of three-dimensional *in vitro* models of bone using bioceramics remains an important challenge that is still to be overcome. Calcium phosphates are reactive and their reactivity depends on their characteristics (such as composition, dissolution, sintering temperature, microstructure). When ceramics are cultured *in vitro*, calcium and phosphates are released in the medium and, as a result, can hamper cellular function (Habibovic et al. 2006). Since testing the *in vitro* cellular activity is a pertinent step, and scaffolds and cells are in contact for several days *in vitro* before being implanted, the majority of studies have been focusing on solving the difficulties of removal of excessive ions, as well as oxygen and nutrients supply to cells within ceramic scaffolds (Tortelli and Cancedda 2009). Bioreactors are applied to convey the medium directly throughout the interconnected pores to continuously introduce nutrients and remove wastes. Therefore, the use of such devices is an absolute requirement to favor cell viability in long-term culture and enhance cell activities within 3D ceramic scaffolds (Du et al. 2008). Co-culture of osteoprogenitor and endothelial cells within 3D porous ceramic scaffolds using a perfusion-based bioreactor device showed the pattern of matrix deposition and resorption *in vitro*, followed by formation of bone-like tissue and blood vessels *in vivo* (Papadimitropoulos et al. 2011). A recent study compared the biological responses between murine MSCs and mature osteoblasts, when seeded in ceramic scaffolds and subsequently implanted in immunocompromised mice. The results showed that new vascularized bone was formed through the activation of an endochondral ossification process in the MSCs-scaffolds complex. Conversely, osteoblasts directly formed bone via an intramembranous ossification, without however any signs of vascularization (Tortelli et al. 2010). In another study, mouse ESCs seeded onto ceramic scaffolds and cultured in osteogenic medium, failed to form bone tissue upon implantation in subcutaneous pockets of nude mice. However, when ESCs-ceramic constructs were cultured in chondrogenic differentiation medium, they showed cartilage maturation toward the hypertrophic stage, calcification and ultimately bone formation *in vivo* (Jukes et al. 2008). All together, these studies indicated that the origin of bone cells and the ossification type are dependent on the nature and commitment of the seeded cells.

Hydrogels

In regenerative medicine and tissue engineering of bone, a vast majority of studies have focused on the formation of new bone directly from stem cells or mature osteogenic cells encapsulated within hydrogel matrices. However, more recently, stimulation of cells toward the endochondral ossification using cell-laden hydrogel systems has shown to be not only appealing for the understanding of the regulatory mechanisms of cartilage and bone formation, but is also a superior approach for bone repair.

The following section describes the cell-hydrogel systems that have been explored so far for the fabrication of bone tissue via endochondral ossification.

Alginate

Alginate is a linear heteropolysaccharides composed of D-mannuronic acid and L-guluronic acid, and is derived primarily from brown algae. Due to the presence of the carboxyl groups along the polymer chain, alginates can form gels in the presence of divalent ions such as calcium ions (Smidsrod and Skjak-Braek 1990). Due to easy preparation under gentle conditions, low toxicity and easy availability, alginate hydrogels have been extensively investigated as matrices for encapsulating cells and delivering drugs. Alginate hydrogels have been shown to be suitable matrices for generation of bone tissue via endochondral ossification when associated with committed stem cells and appropriate combinations of regulatory signals (e.g., BMP-2, VEGF, TGF-β3) (Alsberg et al. 2002). Chang et al. followed the progression of osteogenically induced MSCs along the endochondral pathway. At 2 wk after subcutaneous implantation, cell-alginate constructs presented islands of cartilage formation and cells had the typical round morphology of the chondrocyte phenotype. By 6 wk, endochondrosis with trabecular bone deposition started to appear, whereas osteoblasts and osteocytes were seen in the new bone at wk 8 and at wk 12, respectively. Simultaneously, increasing calcification was detected over time. Similarly, in another study, hMSC were suspended in alginate beads with a chondrogenesis-induction medium containing transforming growth factor (TGF)-β3. During the first stage of culture, specific chondrogenic markers, such as collagen type II, type X and proteoglycan, were detactable, and their expression decreased over time in concert with the increase of osteogenic markers, such as osteocalcin (Ichinose et al. 2005).

Chitosan

Chitosan is a linear polymer of (1-4) β-linked d-glucosamine residues with N-acetyl glucosamine groups derived from chitin, a naturally occurring polysaccharide which forms the outer shell of crustaceans, insect exoskeletons, and fungal cell walls. Chitosan is completely soluble in aqueous solutions with pH lower than 5.0 and it undergoes biodegradation *in vivo* enzymatically by lysozyme to nontoxic products (Khor and Lim 2003). Biocompatibility, biodegradability and structural similarity to natural GAGs make crosslinked chitosans hydrogels attractive materials for tissue engineering applications. Moreover, the feasibility of forming porous scaffolds was able to promote osteoblastic differentiation *in vitro* and might support angiogenesis and osteogenesis *in vivo* (Guzman-Morales et al. 2009). Engineered chondrocyte/chitosan templates were shown to carry all the signals necessary to induce endochondral bone formation *in vivo* (Oliveira et al. 2009).

Collagen

Collagen represents the most abundant protein in the human body, being the major component of bone, skin, ligament, cartilage and tendon. It also forms the structural framework for other tissues such as blood vessels. There are at least 19 different types of collagen, but the basic unit of all is a polypeptide consisting of a three-amino-acid sequence (glycine, proline, and hydroxyproline) forming polypeptide chains, which wrap around one another to form the left-handed triple helix structure (Lee et al. 2001). Due to its excellent biocompatibility, enzymatic degradability and resemblance to the organic composition of natural bone, collagen-based matrices find wide application in tissue engineering and have been extensively investigated in combination with cells and growth factors (Heinemann et al. 2011). MSCs-collagen scaffolds pre-cultured on chondrogenic culture medium and subsequently implanted subcutaneously in the dorsum of nude mices showed good cell survival and progression along the endochondral ossification route, and more interestingly, blood vessels formation. Conversely, the osteogenically primed scaffolds showed a mineralized matrix of poor quality, no vascularization, probably due to too much matrix deposition or a lack of release of inductive factors, and few surviving cells, as a result of absence of oxygen and nutrients (Farrell et al. 2011; Farrell et al. 2009). In another study, chondrocytes isolated from chick embryo were cultured in collagen sponges treated with retinoic acid to induce chondrocyte maturation and extracellular matrix deposition. Biological properties and stiffness of collagen matrices supported chondrocytes attachment, proliferation and endochondral bone maturation after implantation (Oliveira et al. 2010).

Conclusion and Future Perspective

Regenerative medicine is a multidisciplinary field aiming at restoration of normal function to injured tissues, by the development of increasingly sophisticated biomaterials, either alone or in combination with cells and/ or stimulating factors. Among these materials, hydrogels are now at the forefront of this research as their biological and structural properties make these materials suitable for cell encapsulation and drug delivery. More importantly, hydrogel formulations are injectable, ready-to-use, and can be molded in irregularly shaped defects, thus allowing surgeons to perform faster and using minimally invasive procedures. The enormous progress achieved so far at the laboratory stage and the continuous worldwide effort towards new insights and technologies, direct researchers to face new challenges. In order to translate basic scientific advantages into approved medical products, all the materials requirements (e.g., safety and efficacy) must comply with the guidelines established by regulatory agencies, including the Food and Drug Administration Agency (FDA) and the European Medicinal Agency (EMEA), and clinical trials must be carried out within acceptable clinical practice and due ethical considerations.

In conclusion, although still in its infancy, research on combining stem cells with hydrogel materials is at the precipice of major breakthroughs that likely will change and simplify the treatment of bone disorders, thus bring major benefits to patients, including reduced need and cost of medical attention, and short-term recovery.

Key Facts

Key Features of Hydrogels

- Bone is a dynamic and vascularized tissue made of an inorganic and an organic matrix. The inorganic matrix represents ~70 percent of bone mass and is mainly composed of hydroxyapatite. The organic matrix represents ~30 percent of bone mass and is composed of proteins (mainly collagen type I) and cells.
- Hydrogels are highly hydrated polymers, which exhibit mechanical and visco-elastic properties that resemble the structural three dimensional environment of organic matrix of bone.
- RM is a multidisciplinary field devoted to the development of artificial materials as temporary supports for cells in order to induce, guide and control the formation of functional tissues.
- Hydrogels form suitable scaffolds to host cells and support cell function.
- When cultured in proper conditions, ESCs and MSCs can differentiate into bone forming-cells either prior to or after incorporation into hydrogels matrix.
- Hydrogels are off-the-shelf materials and can be injected into irregular defect through a small incision, thus to be beneficial materials for minimally invasive approaches.
- Cell-hydrogel constructs can be injected *in vivo* and promote bone formation.

The key facts on hydrogels include a general description of bone, the similarities of hydrogels with the organic matrix of bone tissue, the basic concept of the regenerative medicine, the application of hydrogels materials in combination with stem cells, and the injectable properties of hydrogels. RM: regenerative medicine; ESCs: embryonic stem cells; MSCs: mesenchymal stem cells.

Dictionary of Key Terms

- *Musculoskeletal malconditions*: disorders and diseases that cause the physical disabilities of bones, muscles, joints, tendons, ligaments and nerves, such as arthritis, back pain, spinal disorders, osteoporosis, bone fractures, trauma, and congenital problems.

- *Regenerative medicine (RM)*: a multidisciplinary field that seeks strategies to replace and repair impaired tissues or organs, and to establish the normal functions of the body. RM approaches include, for instance, organ or stem cell transplantation, gene therapy, tissue engineering.
- *Autograft*: graft of tissue harvested from a donor site of patient's own healthy part.
- *Allograft*: graft of tissue from other individuals of the same species, e.g., cadaveric bone.
- *Intramembranous ossification*: osteogenesis process typical of skull and jaw, which occurs in a vascularized environment and involves the direct differentiation of mesenchymal stem cells into osteoblasts.
- *Endochondral ossification*: the most commonly occurring osteogenesis process in the vertebrate skeleton (e.g., long bones), which culminates in bone tissue via the transition from avascular hypertrophic cartilage to vascularized and mineralized matrix.
- *Biomaterials*: artificial materials that are able to interact with components of living system. Biomaterials can be produced in nature or engineered in the laboratory using different approaches.
- *Hydrogels*: networks of hydrophilic, crosslinked polymer chains, which are able to absorb water in aqueous fluids. Hydrogels exhibit mechanical and visco-elastic properties that closely resemble the structural three dimensional environment of natural extracellular matrix.

Summary Points

- Musculoskeletal disorders represent a social and economical burden.
- Conventional treatments for bone defects have shortcomings that limit their applications.
- Regenerative medicine (RM) is a booming multidisciplinary field, which seeks man-made strategies to replace and restore the anatomic structure and function of damaged or missing tissues.
- In the past decades, major efforts have been focused on synthesis of biomaterials of different origin, including ceramics, synthetic and natural polymers.
- Biomaterials have the primary function to offer an environment to host cells and support cellular functions.
- *In vitro* strategies, using mono- or co-culture, as well as differentiation scheme (e.g., culture medium supplements and signaling molecules) are the core issues to obtain successful outcomes *in vivo*.

- Ceramic scaffolds have been extensively investigated in combination with stem cells for bone tissue engineering. However, the reactivity of these materials and lack of interconnectivity represent the major causes of failure.
- Polymer scaffolds exhibit biocompatibility, degradability and good processability. Therefore, they are suitable candidates for cell-based RM strategies.
- Among polymer materials, hydrogels are gaining significant importance as cell carrier systems for RM application in bone tissue regeneration.

List of Abbreviations

BMUS   :   burden of musculoskeletal diseases in the United States
ECM   :   extracellular matrix
ESCs   :   embryonic stem cells
FDA   :   food and drug administration agency
GAGs   :   glycosaminoglycans
HA   :   hydroxyapatite
MMPs   :   matrix metalloproteinases
MSCs   :   mesenchymal stem cells
OPF   :   oligo(poly(ethylene glycol) fumarate)
pHEMA   :   poly(2-hydroxyethyl methacrylate)
PCL   :   poly(ε-caprolactone)
PEG   :   poly(ethylene glycol)
PEG-DA   :   poly(ethylene glycol)-diacrylate
PGA   :   polyglycolic acid
PLA   :   poly(lactic acid)
PLGA   :   poly(lactic-co-glycolic acid)
RM   :   regenerative medicine
TCP   :   tricalcium phosphate
TGF-β :   transforming growth factor-beta
VEGF   :   vascular endothelial growth factor

References

Alsberg, E., K.W. Anderson, A. Albeiruti, J.A. Rowley and D.J. Mooney. 2002. Engineering growing tissues. Proc Natl Acad Sci USA. 99: 12025–30.

Barrere, F., C.A. Van Blitterswijk and K. De Groot. 2006. Bone regeneration: molecular and cellular interactions with calcium phosphate ceramics. Int J Nanomedicine. 1: 317–32.

Betz, M.W., P.C. Modi, J.F. Caccamese, D.P. Coletti, J.J. Sauk and J.P. Fisher. 2008. Cyclic acetal hydrogel system for bone marrow stromal cell encapsulation and osteodifferentiation. J Biomed Mater Res A. 86: 662–70.

Blau, H.M., T.R. Brazelton and J.M. Weimann. 2001. The evolving concept of a stem cell: entity or function? Cell. 105: 829–41.

Bongio, M., J.J.J.P. Van den Beucken, S.C.G. Leeuwenburgh and J.A. Jansen. 2010. Development of bone substitute materials: From 'biocompatible' to 'instructive'. J Mater Chem. 20: 8747–8759.

Bongio, M., J.J.J.P. Van den Beucken, M.R. Nejadnik, S.C.G. Leeuwenburgh, L.A. Kinard, F.K. Kasper, A.G. Mikos and J.A. Jansen. 2011. Biomimetic modification of synthetic hydrogels by incorporation of adhesive peptides and calcium phosphate nanoparticles: in vitro evaluation of cell behavior. Eur Cell Mater. 22: 359–76.

Bord, S., A. Horner, R.M. Hembry, J.J. Reynolds and J.E. Compston. 1996. Production of collagenase by human osteoblasts and osteoclasts *in vivo*. Bone. 19: 35–40.

Burdick, J.A. and K.S. Anseth. 2002. Photoencapsulation of osteoblasts in injectable RGD-modified PEG hydrogels for bone tissue engineering. Biomaterials. 23: 4315–23.

Dominici, M., T.J. Hofmann and E.M. Horwitz. 2001. Bone marrow mesenchymal cells: biological properties and clinical applications. J Biol Regul Homeost Agents. 15: 28–37.

Drury, J.L. and D.J. Mooney. 2003. Hydrogels for tissue engineering: scaffold design variables and applications. Biomaterials. 24: 4337–51.

Du, D., K. Furukawa and T. Ushida. 2008. Oscillatory perfusion seeding and culturing of osteoblast-like cells on porous beta-tricalcium phosphate scaffolds. J Biomed Mater Res A. 86: 796–803.

Engler, A.J., S. Sen, H.L. Sweeney and D.E. Discher. 2006. Matrix elasticity directs stem cell lineage specification. Cell. 126: 677–89.

Farrell, E., O.P. Van der Jagt, W. Koevoet, N. Kops, C.J. Van Manen, C.A. Hellingman, H. Jahr, F.J. O'brien, J.A. Verhaar, H. Weinans and G.J. Van Osch. 2009. Chondrogenic priming of human bone marrow stromal cells: a better route to bone repair? Tissue Eng Part C Methods. 15: 285–95.

Farrell, E., S.K. Both, K.I. Odorfer, W. Koevoet, N. Kops, F.J. O'Brien, R.J. De Jong, J.A. Verhaar, V. Cuijpers, J.A. Jansen, R.G. Erbed and G.J. Van Osch. 2011. *In vivo* generation of bone via endochondral ossification by *in vitro* chondrogenic priming of adult human and rat mesenchymal stem cells. BMC Musculoskelet Disord. 12: 31.

Friedenstein, A.J., S. Piatetzky II and K.V. Petrakova. 1966. Osteogenesis in transplants of bone marrow cells. J Embryol Exp Morphol. 16: 381–90.

Guzman-Morales, J., H. El-Gabalawy, M.H. Pham, N. Tran-Khanh, M.D. Mckee, W. Wu, M. Centola and C.D. Hoemann. 2009. Effect of chitosan particles and dexamethasone on human bone marrow stromal cell osteogenesis and angiogenic factor secretion. Bone. 45: 617–26.

Habibovic, P., T. Woodfield, K. De Groot and C.A. van Blitterswijk. 2006. Predictive value of *in vitro* and *in vivo* assays in bone and cartilage repair—what do they really tell us about the clinical performance? Adv Exp Med Biol. 585: 327–60.

Hauge, E.M., D. Qvesel, E.F. Eriksen, L. Mosekilde and F. Melsen. 2001. Cancellous bone remodeling occurs in specialized compartments lined by cells expressing osteoblastic markers. J Bone Miner Res. 16: 1575–82.

Heinemann, C., S. Heinemann, H. Worch and T. Hanke. 2011. Development of an osteoblast/osteoclast co-culture derived by human bone marrow stromal cells and human monocytes for biomaterials testing. Eur Cell Mater. 21: 80–93.

Hutmacher, D.W., J.T. Schantz, C.X. Lam, K.C. Tan and Lim. 2007. State of the art and future directions of scaffold-based bone engineering from a biomaterials perspective. J Tissue Eng Regen Med. 1: 245–60.

Ichinose, S., K. Yamagata, I. Sekiya, T. Muneta and M. Tagami. 2005. Detailed examination of cartilage formation and endochondral ossification using human mesenchymal stem cells. Clin Exp Pharmacol Physiol. 32: 561–70.

Jukes, J.M., S.K. Both, A. Leusink, L.M. Sterk, C.A. Van Blitterswijk and J. De Boer. 2008. Endochondral bone tissue engineering using embryonic stem cells. Proc Natl Acad Sci USA. 105: 6840–5.

Khor, E. and L.Y. Lim. 2003. Implantable applications of chitin and chitosan. Biomaterials. 24: 2339–49.

Krampera, M., G. Pizzolo, G. Aprili and M. Franchini. 2006. Mesenchymal stem cells for bone, cartilage, tendon and skeletal muscle repair. Bone. 39: 678–83.

Langer, R. and J.P. Vacanti. 1993. Tissue engineering. Science. 260: 920–6.

Lee, C.H., A. Singla and Y. Lee. 2001. Biomedical applications of collagen. Int J Pharm. 221: 1–22.

Lennon, D.P. and A.I. Caplan. 2006. Isolation of human marrow-derived mesenchymal stem cells. Exp Hematol. 34: 1604–5.

Lutolf, M.P. and J.A. Hubbell. 2005. Synthetic biomaterials as instructive extracellular microenvironments for morphogenesis in tissue engineering. Nat Biotechnol. 23: 47–55.

Mackay, A.M., S.C. Beck, J.M. Murphy, F.P. Barry, C.O. Chichester and M.F. Pittenger. 1998. Chondrogenic differentiation of cultured human mesenchymal stem cells from marrow. Tissue Eng. 4: 415–28.

Meijer, G.J., J.D. De Bruijn, R. Koole and C.A. Van Blitterswijk. 2007. Cell-based bone tissue engineering. PLoS Med. 4: e9.

Montufar-Solis, D., H.C. Nguyen, H.D. Nguyen, W.N. Horn, D.D. Cody and P.J. Duke. 2004. Using cartilage to repair bone: an alternative approach in tissue engineering. Ann Biomed Eng. 32: 504–9.

Nair, L.S. and C.T. Laurencin. 2006. Polymers as biomaterials for tissue engineering and controlled drug delivery. Adv Biochem Eng Biotechnol. 102: 47–90.

Nakayama, N., D. Duryea, R. Manoukian, G. Chow and C.Y. Han. 2003. Macroscopic cartilage formation with embryonic stem-cell-derived mesodermal progenitor cells. J Cell Sci. 116: 2015–28.

Nicodemus, G.D. and S.J. Bryant. 2008. Cell encapsulation in biodegradable hydrogels for tissue engineering applications. Tissue Eng Part B Rev. 14: 149–65.

Oliveira, S.M., D.Q. Mijares, G. Turner, I.F. Amaral, M.A. Barbosa and C.C. Teixeira. 2009. Engineering endochondral bone: *in vivo* studies. Tissue Eng Part A. 15: 635–43.

Oliveira, S.M., R.A. Ringshia, R.Z. Legeros, E. Clark, M.J. Yost, L. Terracio and C.C. Teixeira. 2010. An improved collagen scaffold for skeletal regeneration. J Biomed Mater Res A. 94: 371–9.

Papadimitropoulos, A., A. Scherberich, S. Guven, N. Theilgaard, H.J. Crooijmans, F. Santini, K. Scheffler, A. Zallone and I. Martin. 2011. A 3D *in vitro* bone organ model using human progenitor cells. Eur Cell Mater. 21: 445–58; discussion, 458.

Petite, H., V. Viateau, W. Bensaid, A. Meunier, C. De Pollak, M. Bourguignon, K. Oudina, L. Sedel and G. Guillemin. 2000. Tissue-engineered bone regeneration. Nat Biotechnol. 18: 959–63.

Robins, J.C., N. Akeno, A. Mukherjee, R.R. Dalal, B.J. Aronow, P. Koopman and T.L. Clemens. 2005. Hypoxia induces chondrocyte-specific gene expression in mesenchymal cells in association with transcriptional activation of Sox9. Bone. 37: 313–22.

Smidsrod, O. and G. Skjak-Braek. 1990. Alginate as immobilization matrix for cells. Trends Biotechnol. 8: 71–8.

Solursh, M. 1989. Cartilage stem cells: regulation of differentiation. Connect Tissue Res. 20: 81–9.

Sommerfeldt, D.W. and C.T. Rubin. 2001. Biology of bone and how it orchestrates the form and function of the skeleton. Eur Spine J. 10 Suppl. 2: S86–95.

Street, J., M. Bao, L. Deguzman, S. Bunting, F.V. Peale, Jr., N. Ferrara, H. Steinmetz, J. Hoeffel, J.L. Cleland, A. Daugherty, N. Van Bruggen, H.P. Redmond, R.A. Carano and E.H. Filvaroff. 2002. Vascular endothelial growth factor stimulates bone repair by promoting angiogenesis and bone turnover. Proc Natl Acad Sci USA. 99: 9656–61.

Temenoff, J.S., H. Park, E. Jabbari, T.L. Sheffield, R.G. Lebaron, C.G. Ambrose and A.G. Mikos. 2004. *In vitro* osteogenic differentiation of marrow stromal cells encapsulated in biodegradable hydrogels. J Biomed Mater Res A. 70: 235–44.

Tortelli, F. and R. Cancedda. 2009. Three-dimensional cultures of osteogenic and chondrogenic cells: a tissue engineering approach to mimic bone and cartilage *in vitro*. Eur Cell Mater. 17: 1–14.

Tortelli, F., R. Tasso, F. Loiacono and R. Cancedda. 2010. The development of tissue-engineered bone of different origin through endochondral and intramembranous ossification following the implantation of mesenchymal stem cells and osteoblasts in a murine model. Biomaterials. 31: 242–9.

Wiles, M.V. and G. Keller. 1991. Multiple hematopoietic lineages develop from embryonic stem (ES) cells in culture. Development. 111: 259–67.

Williams, D.F. 2009. On the nature of biomaterials. Biomaterials. 30: 5897–909.

Wobus, A.M. 2001. Potential of embryonic stem cells. Mol Aspects Med. 22: 149–64.

Zur Nieden, N.I., G. Kempka and H.J. Ahr. 2003. *In vitro* differentiation of embryonic stem cells into mineralized osteoblasts. Differentiation. 71: 18–27.

CHAPTER 19

Rabbit Adipose-derived Stem Cells and Tibia Repair

Elena Arrigoni,[a] Stefania Niada[b] and Anna T. Brini[c,*]

ABSTRACT

Adipose-derived Stem Cells (ASCs) may represent, alone or in combination with different scaffolds, a novel and efficient approach for bone regeneration. Here, we describe how autologous rabbit ASCs (rbASCs) isolated from interscapular adipose tissue, expanded and characterized *in vitro*, are used to regenerate a full-thickness bone defect in the tibial crest of New Zealand rabbits.

The animals have been divided in four groups: one group where the lesions have been treated with rbASCs seeded on hydroxyapatite-disk (rbASCs-HA), one group with only rbASCs, one with just HA, and one untreated group (just defect). The follow-up was of eight weeks.

Meanwhile, rbASCs have been characterized *in vitro*: these progenitor cells show a homogenous high proliferation rate and a marked clonogenic ability. Moreover, rbASCs demonstrate an osteogenic potential that has been evaluated by the expression of specific bone markers such as alkaline phosphatase, collagen, osteonectin and extracellular calcified matrix deposition, both in the absence and in the presence of hydroxyapatite.

Department of Biomedical, Surgical and Dental Sciences, University of Milan, Via Vanvitelli, 32, 20129 Milan-Italy. IRCCS Galeazzi Orthopaedic Institute, Via R. Galeazzi, 4, 20161 Milan-Italy.
[a]E-mail: elena.arrigoni@unimi.it
[b]E-mail: stefania.niada@unimi.it
[c]E-mail: anna.brini@unimi.it
*Corresponding author

List of abbreviations given at the end of the text.

Eight weeks after surgical interventions, gross appearance, X-rays, histological analyses and biomechanical tests were performed on all the animals.

The macroscopic analyses of all the tibias show a satisfactory filling of the lesions without any significant difference in terms of stiffness. By X-rays, a good osteo-integration appears in both scaffold-treated groups; despite the fact that HA was not completely resorbed, cells-HA treated bones show a more efficient scaffold resorption than the other group. In addition, the scaffold-treated defects show a better bone formation compared to the control samples. In particular, the new bone, formed in the presence of rbASCs-HA, is more mature and similar to the native one showing an improvement in bone mechanical properties.

These results indicate that autologous ASCs-hydroxyapatite bioconstruct may be a potential treatment for the regeneration of bone defects.

Introduction

Until recently, fat has been considered an inert tissue and usually lipoaspirates from aesthetic surgery were discarded as surgical waste. Instead, adipose tissue is a highly complex system which consists of mature adipocytes, preadipocytes, fibroblasts, vascular smooth muscle cells, endothelial cells, resident macrophages and lymphocytes and Adipose-derived Stem Cells (ASCs) (Zuk et al. 2002; de Girolamo et al. 2008). These ASCs are progenitor cells that possess the potential to differentiate into cells of mesodermal origin. Furthermore, there are evidences of the ability of ASCs to transdifferentiate into cells of non-mesodermal origin, such as neurons, endocrine pancreatic cells, hepatocytes, endothelial cells, and cardiomyocite (Schaffler and Buchler 2007).

ASCs are quite similar to mesenchymal stem cells isolated from bone marrow (BMSCs) regarding morphology, immunophenotype, colony frequency and differentiation capacity (Kern et al. 2006).

ASCs sampling is a simple surgical procedure due to the easy and repeatable access to subcutaneous adipose tissue as well as the uncomplicated enzyme-based isolation procedures. Therefore, this tissue offers a great alternative source of autologous adult stem cells that can be obtained repeatedly in large quantities under local anaesthesia, with minimum discomfort for the patient (Schaffler and Buchler 2007).

Other relevant ASCs features are their immunoregolatory properties both *in vivo* and *in vitro* and their low immunogenicity; indeed, they do not express MHC class II and express low levels of MHC class I. Both ASCs (Najar et al. 2010) and BMSCs, are able to inhibit T-cell proliferation in a dose-dependent manner (Di Nicola et al. 2002); ASCs effect appears to be mediated by high levels of secreted leukaemia inhibitory factor (LIF) (Najar et al. 2010).

Following these observations, in recent years the broad immunoregulatory activities of ASCs have been studied and their use in clinical therapies, such as the prevention and/or reduction of graft-versus-host disease (GVHD) and the treatment of autoimmune diseases, seems to be particularly promising (Yanez et al. 2006; Choi et al. 2012; Gonzalez et al. 2009; Constantin et al. 2009).

In parallel to these innovative and very interesting applications, ASCs have been largely studied for their regenerative ability.

Above all, to optimize tissue reconstruction, stem or stromal cells such as ASCs have been used in association with biocompatible scaffolds in the emerging field of tissue engineering. Stem cells adhere to the scaffold, replicate, differentiate and then organize into the new tissue. Generally, an ideal scaffold should be biocompatible, absorbable and highly porous. In particular, for bone regeneration, the scaffold associated with ASCs should also possess osteoconductive and osteoinductive properties and should promote cellular adherence and recruitment (Spencer et al. 2011). Currently, investigations are underway to identify scaffold properties that will optimize stem cell activity.

Although bone is a very dynamic tissue, the spontaneous regeneration is limited to relatively small defects, while large bone defects resulting from trauma, tumours, osteitis, implant loosening or corrective osteotomies require surgical treatment. Actual therapies for skeletal reconstruction include autografts or allografts, mineral bone substitutes, and callus distraction (Sarkar et al. 2005). These techniques present the risk of complications and nonunions, although with a high percentage of success. Moreover, the harvesting of autologous bone often results in donor site morbidity, whereas allografts, either from human cadaver or animals, present potential risks of infection, immune response, inadequate supply, difficulties in obtaining and processing tissue, and rapid resorption (Torroni 2009).

With this background, the possible use of ASC-scaffold constructs for treatment of bone defects constitutes a promising technique, considering also the harvesting site which allows low donor-site morbidity. In this chapter, we present an approach of tissue engineering that use autologous ASCs in a rabbit model of critical bone defect.

Rabbit is a good animal model used in approximately 35 percent of musculoskeletal research studies (Neyt et al. 1998). This is partially due to its easy handling, the use of compact caging and the relative low cost for animal purchase and care. Similar to other small animal models, the rabbit also affords the opportunity to use an adequate number of genotypically similar subjects (Chu et al. 2010). Interestingly for our research field, rabbit reaches skeletal maturity at around 6 mon of age, shortly after it is sexually mature (Gilsanz et al. 1988).

Concerning bone composition, some similarities between humans and rabbits are reported both in bone mineral density (BMD) and in the fracture toughness of mid-diaphyseal bone (Wang et al. 1998). One drawback in using this animal model is the difference in the bone anatomy between rabbits and humans, regarding size, shape and loading, due to the differences in weight and stance. Other differences with humans are in the rabbit bone microstructure (Wang et al. 1998) as well as the skeletal change and bone turnover, which are faster in rabbits (Castaneda et al. 2006; Gilsanz et al. 1988).

Even though these issues make extrapolation of studies performed in rabbits onto the human clinical response quite difficult, this preclinical model remains an accepted and a good tool for testing new biomedical constructs before moving towards a larger animal model.

The main sources of mesenchymal stem cells for bone regeneration studies in rabbit are bone marrow and adipose tissue. However, the type of support used is quite variable; modified or unmodified hydroxyapatite (HA) is one of the principal scaffolds employed (de Girolamo et al. 2011; Oshima et al. 2010; Li et al. 2009; Hao et al. 2010b). Other supports such as β-tricalcium phosphate (Wang et al. 2010), collagen I gel with PLGA-β-TCP (Hao et al. 2010b), titanium dome (Pieri et al. 2010) and modified polycaprolactone (Im and Lee 2010) scaffolds have also been used. Even allogenic (Dudas et al. 2006) or xenogenic (Zhao et al. 2011) bone graft represent valid supports for MSCs.

Since vascularization is required to maintain cell survival, scaffold design, use of bioreactor combined with angiogenic factors and pre-vascularization procedure are currently under investigation for enhancing blood vessels formation in tissue engineered bone grafts (Lovett et al. 2009). Two recent studies by Zhao et al. and Wang et al. show indeed a faster capillary infiltration of the graft when a vascular bundle was previously inserted (Zhao et al. 2011; Wang et al. 2010).

Another aspect under debate is the need or not to differentiate MSCs before the construct is implanted; there are contrasting data about this issue that still needs to be elucidated (de Girolamo et al. 2011; Dudas et al. 2006; Dashtdar et al. 2011).

An interesting alternative experimental approach is to osteo-differentiate MSCs when they are already implanted into the defect. Bone Morphogenic Proteins (BMPs) are among the most potent osteoinductive factors ever discovered (Albrektsson and Johansson 2001). MSCs infected with recombinant adenovirus vector coding for human BMPs produce high levels of BMPs at target sites, obtaining a faster bone defect repair (Hao et al. 2010a; Li et al. 2010).

Even though these strategies, to enhance osteo-differentiation of MSCs, can be useful to accelerate bone repair, however we need to keep in mind that manipulation of bioengineered constructs makes their use more difficult for future clinical application.

Isolation of rabbit Adipose-derived Stem Cells (rbASCs)

Rabbit ASCs (rbASCs) can be derived from surgically resected adipose tissue of New Zealand White Rabbits (Oryctolagus cuniculus). The source is a deposit of adipose tissue located along the dorsomedial line, approximately 5 cm from the skull in the craniocaudal direction (Fig. 19.1A).

The adipose tissue is easily accessible through a 2- to 3-cm sagittal incision in the dorsomedial line over the tissue area (Fig. 19.1B). The amount of adipose tissue varies according to the animal mass, in a proportion of 5 to 7 g per kilogram of body mass: this corresponds to a mass of 20 to 30 g in an adult animal weighing nearly 4.5 kg.

Once harvested, adipose specimens are finely minced and ASCs are purified with 0.1 percent collagenase type I (Dudas et al. 2006; Torres et al. 2007; Arrigoni et al. 2009). The isolated Stromal Vascular Fraction (SVF) is plated and Adipose-derived Stem Cells (ASCs) adhere to the Petri dish

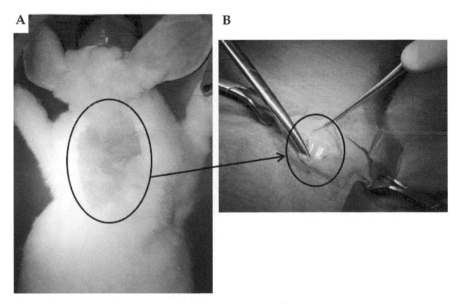

Figure 19.1 Adipose tissue localization. Rabbit dorsal region where the adipose tissue pouch is located (A). Surgical procedure for withdrawing the adipose tissue pouch (B) (unpublished data).

(Fig. 19.2). From an average of 5.2 ± 1.6 ml of raw adipose tissue 2.8x10⁵ ± 1.9x10⁵ cells/ml are purified (Fig. 19.2E).

Rabbit ASCs are cultured at a density of 10^5 cells/cm² in DMEM supplemented with 10 percent fetal bovine serum (FBS), 2 mM L-glutamine, 50 U/mL penicillin, and 50 µg/mL streptomycin (CTRL medium), and maintained at 37°C in a humidified atmosphere with 5 percent CO_2. After 48–72 hr, non-adherent cells and erythrocytes are discarded, and, when rbASCs reach 80–90 percent confluence, cells are detached by 0.5 percent trypsin/0.2 percent EDTA and plated at a density of $5x10^3$ cells/cm² (Arrigoni et al. 2009).

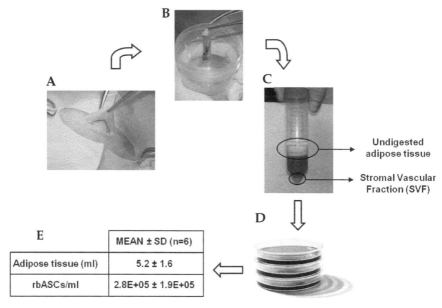

Figure 19.2 Isolation of rabbit Adipose-derived Stem Cells. Adipose tissue is harvested from interscapular region (A) and maintained in sterile PBS (phosphate-buffered saline) until processed (B). Stromal Vascular Fraction is separated by centrifugation (C), then seeded in Petri dishes and cultured in humidified atmosphere at 37°C with 5 percent CO_2 (D) (unpublished data). Cellular yield of rbASCs per ml of adipose tissue (E) (with permission from de Girolamo et al. 2011).

In vitro Culture of rbASCs

One week later, rbASCs start to rapidly proliferate and, at passage 1, the variability of cell number among the rbASCs derived from different animals is reduced, reaching, in 30 d, 2.6x10⁸ ± 9.9x10⁷ cells starting from 1.5x10⁵ rbASCs. Usually every 7 d these cells reach the confluence: the doubling time of 56.9 ± 14.8 hr is constant from passage 2 to 6 (Fig. 19.3A). These cells

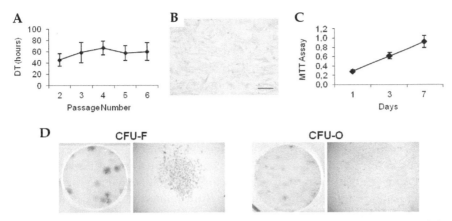

Figure 19.3 Characterization of rbASCs. Proliferation of rbASCs at early passages (A). Microphotographs of fixed rbASCs stained with haematoxylin-eosin (scale bar 100 µm) (B). Viability of rbASCs maintained for one week in undifferentiated condition (mean ± SD, n=6) (C). Microphotographs of CFU-F and CFU-O assays of rbASCs stained with Crystal Violet and Alizarin Red-S, respectively (D) (modified by de Girolamo et al. 2011).

show the characteristic fibroblast-like morphology (Fig. 19.3B), without any sign of cellular senescence. During one week of culture, we observed an exponential growth of rbASCs, one of the main features that characterize mesenchymal stem cells. These data are also confirmed by MTT cell viability assay (Fig. 19.3C). Cell viability is monitored by the reduction of tetrazolium salts by metabolically active cells, through the action of dehydrogenase enzymes in the mitochondria, to generate reducing agents such as NADH and NADPH.

As originally indicated by Friedenstein et al. (Friedenstein et al. 1974), another prominent feature of MSCs is their ability to generate colonies, evaluated by Colony Forming Unit (CFU) assay. rbASCs show a great clonogenic ability maintained quite constant from passage 1 to 4, with a mean value of 3.0 ± 1.6 percent (mean ± SD). Moreover, these rbASCs possess the ability to produce osteoblast (CFU-O) colonies with a frequency of 4.3 ± 1.9 percent (mean ± SD), highlighting the presence of osteogenic progenitors able to differentiate towards cells of the osteogenic lineage (Fig. 19.3D).

Furthermore, cryopreservation does not affect rabbit stem cells features, as already reported for human ASCs. At low temperatures any biological activity is efficiently blocked and storage at very low temperatures is presumed to provide an indefinite cellular longevity although the actual "shelf life" is rather difficult to presume.

Osteogenic Differentiation and Evaluation of Tissue Specific Markers

Osteogenesis starts with a commitment by mesenchymal stem cells to differentiate, showing an up-regulation of osteoblast-specific genes. ASCs can be induced to differentiate *in vitro*, by maintaining them in CTRL medium supplemented with 0,15 mM ascorbic acid, 0.01 μM dexamethasone, 10 mM β-glycerolphosphate, 10 nM cholecalciferol (OSTEO).

Multiple signalling pathways have been demonstrated to participate in the differentiation of an osteoblast progenitor to a committed osteoblast, including Transforming Growth Factor β (TGF-β)/BMP, Wnt/β-catenin, Notch and Hedgehog. BMPs activate Runt-related protein 2 (Runx2) and Osterix (Osx), considered the master regulators for bone formation and the key transcriptional activators of osteogenic differentiation (Yoshimura et al. 2006). Binding sites for Runx2 are present in genes whose products play a role in extracellular matrix mineralization, such as collagen type I and osteopontin (OPN), as well as in genes involved in cell growth and angiogenesis (Lian et al. 2006).

To assess the osteogenic differentiation of rbASCs, specific osteogenic proteins have been analyzed over time: alkaline phosphatase (ALP) and OPN for early osteogenesis, collagen type I (Coll I) and osteonectin (ONC) as intermediate phase markers, and osteocalcin (OC) and bone mineralization for late osteogenesis.

rbASCs efficiently differentiate into osteoblast-like cells and a morphological change is observed already after 7 d of differentiation (Fig. 19.4A). ALP activity increases of about 168 percent compared to undifferentiated cells (Fig. 19.4B), and osteo-rbASCs show an induction of ONC at 7 d (+157.6 percent), which is further up-regulated after 14 d (+1193.7 percent) (Fig. 19.4C). A similar pattern of expression is depicted for collagen type I (Fig. 19.4C). Indeed, densitometric analysis indicate a significant increase of 432.2 and 1233.9 percent after 7 and 14 d, respectively (Fig. 19.4C).

These data are also confirmed by Sirius Red assay, which detects secreted collagen: this method is based on the ability of Sirius Red compound to selectively bind to fibrillar collagens (types I to V), specifically to the $[Gly-X-Y]_n$ helical structure. Osteogenic differentiated cells produce significant amount of collagen, with an average increase of 105.9 percent respective to the same cells maintained in CTRL medium (Fig. 19.4D). Moreover, cells differentiated for 2 wk produce calcified extracellular matrix (+161.6 percent versus undifferentiated rbASCs) (Fig. 19.4E).

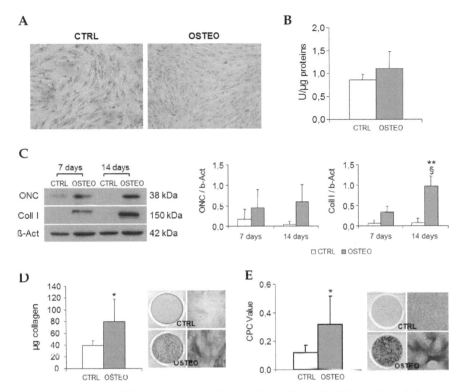

Figure 19.4 **rbASCs differentiation towards osteoblasts-like cells.** Morphological changes of rbASCs undifferentiated and differentiated for 7 d (A). ALP activity of rbASCs after 7 d of differentiation (mean ± SD, n=6) (B). Osteonectin (ONC) and Coll I expression at 7 and 14 d in undifferentiated (CTRL) and osteogenic differentiated (OSTEO) rbASCs; western blot (left panel) and densitometric analysis (mean ± SD, n=2) (right panel) (C). Collagen production by rbASCs cultured for 14 d in undifferentiated and osteogenic conditions (means ± SD, n=4) and microphotographs of rbASCs maintained in CTRL or OSTEO medium stained with Sirius Red (D). Calcified extracellular matrix quantification (means ± SD, n=6) produced by rbASCs cultured on polystyrene for 14 d in CTRL and OSTEO conditions (CPC, cetylpyridinium chloride extraction) and microphotographs of rbASCs stained with Alizarin Red-S (E). OSTEO vs CTRL: *p<.05; **p<.01; 14 d vs 7 d: §p<.05 (modified by de Girolamo et al. 2011).

In vitro rbASCs—hydroxyapatite Constructs

Stem cells exist in tightly controlled niches and alteration in this microenvironment can modify their behaviour. Furthermore, stem cells are often involved in a disease or injury setting where inflammatory signals may be prevalent in recruiting them and in modulating their function. Biomaterials as scaffolds may potentially provide a controlled environment, and several studies have demonstrated the benefits of using mesenchymal stem cell-scaffold constructs in regenerative medicine (Levi and Longaker 2011).

In the orthopaedic field, various studies have established the efficacy of hydroxyapatite as a bone substitute. Hydroxyapatite (HA) is a biocompatible and bioactive material that can be used to restore or repair damaged bone tissue. This scaffold exhibits a strong bond to the bone, and possesses suitable mechanical properties and interconnecting pores which allow cellular infiltration, graft integration and vascularization. Other important characteristics of this scaffold are its osteoinductive and osteoconductive properties.

To confirm all these effects on rbASCs, undifferentiated cells are maintained and grown in the presence of HA ($Ca_{10}(PO_4)_6(OH)_2$, 70–80% of porosity, pore size <10μm, ≈3% vol; 10–150 μm, ≈11% vol; >150μm, ≈86% vol, kindly provided by Finceramica S.p.A., Italy) (Fig. 19.5A). 97.1 ± 2.4 percent of rbASCs finely adhere to the scaffold (Fig. 19.5B and D) and they produce greater levels of collagen in comparison with cells maintained on monolayer (PA, plastic adherence), with an average increase of 48.2 percent (Fig. 19.5C).

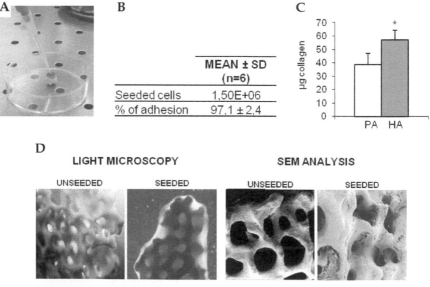

Figure 19.5 rbASCs-hydroxyapatite interaction. Loading of 1.5x10⁶ undifferentiated rbASCs on hydroxyapatite disks (A). Percentages of adhesion of rbASCs in contact with the HA disk *in vitro* for 16 hr (mean ± SD, n=6) (modified by de Girolamo et al. 2011) (B). Quantification of collagen produced by undifferentiated rbASCs cultured on polystyrene (PA) or on hydroxyapatite (HA) for 14 d (means ± SD, n=4) (C). Light microscopy and SEM pictures of HA granules unseeded or seeded with rbASCs (with permission from Arrigoni et al. 2009) (D). HA vs PA: *p<.05.

Recently, Hao et al. (Hao et al. 2010a), fabricated a novel bioconstruct combining a nano-hydroxyapatite/recombinant human-like collagen/poly(lactic acid) scaffold and rbASCs transfected with recombinant adenovirus vector producing human BMP-2. Scanning electron microscope (SEM) demonstrated integration of rbASCs within the scaffold and analysis of gene expression showed an up-regulation of collagen type I, osteonectin and osteopontin, indicating the osteogenic potential of rabbit stem cells.

In vivo Models

To adequately translate *in vivo* findings to the clinic, important pre-clinical data must be obtained in order to demonstrate the osteogenic ability of rbASCs to regenerate bone tissue. Many groups use either femoral or tibia defects as a bone load-bearing basis, seeding ASCs on osteoconductive scaffolds. Some reports demonstrated failed repair of bone defects in rat models by using ASCs incorporated in biomaterials (Li et al. 2007). In contrast, a more recent study by Hao et al. shows the use of rbASCs overexpressing BMP-2, to heal a femoral or tibial critical size bone defect (Hao et al. 2010a).

In addition, our group has also demonstrated that autologous and undifferentiated rbASCs seeded on clinical grade HA disks are a potential treatment for the regeneration of a critical size bone defect in rabbit tibia. The use of untreated rbASCs suggests the ability of these cells to achieve bone regeneration probably through the combined effect of the lesion's microenvironment and the scaffold. This evidence may be useful in a hypothetic one-step bone defect treatment, where mesenchymal stem cells could be purified from the donor tissue, directly seeded on scaffold and immediately implanted into the defect, obviating the necessity of either pre-stimulation or genetic modification (de Girolamo et al. 2011).

In this study, an 8 mm diameter full-thickness bone defect was created bilaterally in the proximal epiphysis of the medial facet of the rabbits' tibia (Fig. 19.6A). The choice to perform circular full-thickness defects instead of a periosteal segmental one was due to the fact that these kind of defects are more difficult to repair, and avoids the use of mechanical stabilization devices, which are known to affect bone healing, permitting to accurately evaluate the ability of autologous rbASCs to improve bone healing in a short follow-up of eight weeks.

HA scaffold constructs were implanted into the defect through a "press-fit" technique in group B and D (Fig. 19.6B, D and E), whereas, a semi-liquid suspension of rbASCs was directly injected in bone defects of group C (Fig. 19.6C and E). Group A was just untreated lesioned bones (Fig. 19.6A and E).

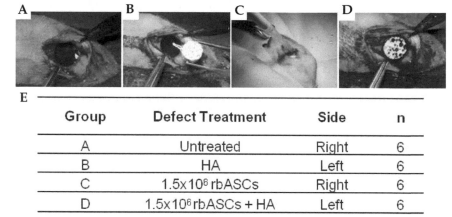

Group	Defect Treatment	Side	n
A	Untreated	Right	6
B	HA	Left	6
C	1.5x10⁸ rbASCs	Right	6
D	1.5x10⁸ rbASCs + HA	Left	6

Figure 19.6 Surgical procedure and constructs implantation. Critical size tibia defect (A), implantation of HA scaffold alone (B) or in association with rbASCs into the defect (D), and filling of the lesion site with a semi-liquid suspension of rbASCs (C). Experimental scheme of the bone defects treatment (E) (modified by de Girolamo et al. 2011).

Gross Appearance Analyses and Radiological Assessment

To evaluate the surgery outcome and the state of bone healing during the study, lateral radiographs were taken immediately after surgery (T_0), 6 wk after intervention (T_6) and post-tibia removal (8 wk, T_8).

At T_0, all the scaffolds were correctly placed and no bone fracture was seen at the lesion site (Fig. 19.7, T_0). A second radiographic analysis was performed 6 wk after implantation and no signs of osteolysis, fracture, or osteopenia were observed: instead, all lesions showed signs of quite an advanced bone remodelling process (Fig. 19.7, T_6).

Eight weeks later, the animals were sacrificed, tibia explanted and the scoring system showed in Table 19.1a was used to evaluate bone integration in terms of good repair for the groups with or without scaffold or integration of the HA disks.

In all the groups, bone defects were satisfactorily filled and, according to the modified Wakitani scale, the stiffness was considered good (Wakitani et al. 1994). Indeed, no significant differences were observed between sham and rbASCs treated group (p>.05, Table 19.1b). In addition, in groups with implanted scaffold no significant differences have been observed between empty or rbASCs seeded ones (p>.05, Table 19.1b).

X-rays bone defects of group A seemed to be partially filled in a non-homogeneous way, whereas the use of rbASCs only (group C) allowed a more complete and homogeneous filling of the lesion. In both groups

| Untreated (A) | HA (B) | rbASCs (C) | rbASCs + HA (D) |

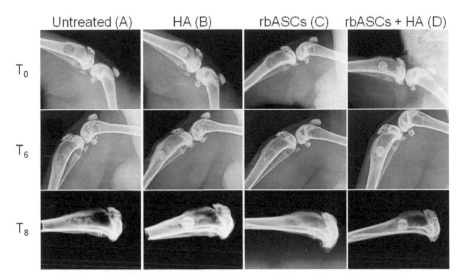

T_0

T_6

T_8

Figure 19.7 X-rays of lesioned bones. Rabbits were X rayed immediately after surgical treatment (T_0), after 6 wk from intervention (T_6), and after tibia removal (8 wk, T_8). Untreated group (group A), hydroxyapatite-alone treated group (group B), only rbASCs treated group (group C) and rbASCs-hydroxyapatite treated group (group D) (with permission from de Girolamo et al. 2011).

Table 19.1 Evaluation of the bone regeneration process by gross appearance analyses.

a		b		
Gross Appearance Analyses		Group	Filling	Stiffness
Filling	Stiffness	A	2.83	2.83
Same level = 3	Same stiffness = 3	B	2.67	2.83
Overgrowth = 2	Softer = 2	C	2.50	2.50
Undergrowth > 1 mm = 1	Very soft = 1	D	2.67	3.00

Filling and stiffness following modified Wakitani scoring scale (a) of the tibia explanted from the four animal groups A, untreated; B, +HA; C, +rbASCs; D, +rbASCs+HA (b). (modified by de Girolamo et al. 2011).

B and D, the scaffolds showed signs of good osteointegration, even if a conspicuous amount of HA was not reabsorbed and was still clearly evident (Fig. 19.7, T_8).

Histology and Immunohistochemistry

Histology and immunohistochemistry analyses were performed with the aim of identifying the nature of new formed bone tissue. Lee et al. have recently reported that the use of different hydroxyapatite supports induces well-recovered cortical bone with newly developed regular bone matrix arranging, concluding that HA may be suitable as a bone substitute in the rabbit tibial defect model (Lee et al. 2010). However, the efficacy of bone substitutes could be ameliorated combining the osteoconductive properties of hydroxyapatite with the osteogenic potential of mesenchymal stem cells. As previously shown, rbASCs are able to colonize scaffolds and move along the interconnected pores, thus contributing to a better bone tissue regeneration. de Girolamo et al. reported that the use of rbASCs alone, administered as a semi-liquid suspension directly injected in the lesion site (group C) induces formation of new bone able to fill the defect, even if a woven matrix conformation indicates an immature bone tissue (Fig. 19.8) (de Girolamo et al. 2011).

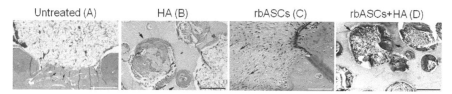

Figure 19.8 Histological analyses of sections from decalcified tibial samples. Haematoxylin-eosin staining of newly formed tissue at the lesion site of the four experimental groups. Untreated bone (A), plus scaffold (B), only cells (C), a bioconstruct rbASCs+HA (D). In group B and D, the new-formed bone within the scaffold pores (arrows) shows a lamellar organization. The reparation tissue in the site of the defect of group C shows a distinct woven matrix (arrows). (A, scale bar—500 µm; B, scale bar—100 µm; C, D, scale bar—200 µm) (with permission from de Girolamo et al. 2011).

The combination of rbASCs seeded on HA disks induces an abundant new bone formation. Pores are filled more homogeneously in comparison to the group treated with only HA, and along the walls new mature bone is observed, as well as connective fibrous tissue in the inner part, indicating an ongoing bone maturation process. Osteoblasts are observed in conjunction with bone trabecules, located at the periphery of the scaffold pores which display a lamellar matrix (Fig. 19.8). These results are also confirmed by OPN and Coll I expressions which are observed in the HA treated bone, just in conjunction with bone trabecules located at the periphery of the scaffold pores, whereas, in rbASCs-HA implanted group, the specific expression is also detected in the middle of the pores (unpublished data). These data indicate that the bone maturation process is more advanced in the cell-treated defects, where lamellar bone is largely represented.

Biomechanical Test

Defects treated with rbASCs show improved mechanical properties, tested by nanoindentation experiments using a NanoTest Indenter System (Micro Materials Ltd., UK) with a diamond Berkovich indenter tip.

Nanoindentation is derived from the classical hardness test, but it is carried out on a much reduced scale. It can be used to determine the hardness of thin layers (non-decalcified samples—thickness ~40 μm) as well as material properties such as elasticity (Er), hardness (H), plasticity, and tensile strength. These measurements involve applying a small force (1mN, 5mN, and 50mN) to a sample using a sharp probe and measuring the resultant penetration depth. The measured values are used to calculate the contact area and hence the particular properties of the sample material.

The rbASCs-HA constructs have an improved capability to bear mechanical loading. This is quite evident at low forces (1mN, corresponding to 200 nm of penetration depth), where it is possible to observe an increase in stiffness and in hardness in the rbASCs-treated group with respect to HA-group. This result might be explained by a higher mineral content present in the lesions treated with rbASCs-HA, thus representing a more advanced step in the healing process (unpublished data).

In conclusion, we suggest the use of ASCs as a safe cellular therapy in future clinical applications where a large bone defect needs to be treated. Indeed, the treatment of bone lesions with autologous rbASCs-hydroxyapatite bioconstructs have improved bone healing, and these results have paved the way for a new study on a large size animal model.

Key Facts

Table 19.2 Key features of ASCs and bone repair.

1. Critical-size bone defects may occur due to trauma, diseases or tumours.
2. The implant of a cells-scaffold tissue engineering construct represents a valid alternative to heterologous or autologous bone.
3. ASCs are optimum candidates in tissue engineering applications for their mesenchymal stem cells features together with their easily availability.
4. Biomaterials, such as hydroxyapatite, are able to osteo-induce ASCs allowing these cells to improve tissue repair.
5. Animal models are required before moving to the clinic.
6. The easy handling and bone similarities between humans and rabbits make this animal a good model in the orthopaedic field.
7. The rabbit tibia regenerated with autologous cells-hydroxyapatite construct leads to a mature tissue similar to the native one.

This table lists the key facts of tissue engineering applied to the regeneration of bone by autologous rabbit ASCs in association with hydroxyapatite. ASCs: Adipose-derived Stem Cells.

Dictionary of Key Terms, Genes, Chemicals or Pathways

- *Adipose-derived stem cells*: mesenchymal stem cells isolated from subcutaneous or omental adipose tissue.
- *Bioconstructs*: combination of progenitor or terminally differentiated cells with either natural or synthetic scaffolds able or not to release active molecules, and to adapt to the wounded tissue.
- *Critical-size bone defect*: the smallest size intra-osseous wound in a particular bone that will not spontaneously heal during the life span.
- *Pre-clinical model*: designed experimental study with the minimum number of animals to determine treatment toxicity and therapeutic effects, and which requires ethical approval.
- *Immunohistochemistry*: technique based on the binding of specific antibodies to cellular antigens used to evaluate the distribution and localization of specific biomarkers in fixed tissues.
- *Nanoindentation*: technique which exploits diamond-coated indenters to measure mechanical properties such as hardness and elasticity of several tissues.

Summary Points

- Fat represents an ideal source of mesenchymal stem cells due to the high number of cells which can be isolated through simple procedures (both surgical and enzymatic). Adipose-derived stem cells (ASCs) are able to self-renew in culture for a long time, are multipotent and low immunogenic, and possess immunomodulatory and anti-inflammatory features.
- ASCs can be associated with scaffolds in tissue engineering approaches.
- Preclinical studies are always required before moving to the clinical phase.
- The rabbit is a good animal model in the orthopaedic field, and rabbit ASCs (rbASCs) can be easily isolated from the interscapular region of New Zealand White rabbits.
- rbASCs rapidly proliferate and show clonogenic ability and osteogenic potential *in vitro*.
- Hydroxyapatite (HA) is largely used in clinic for its osteoinductive property; *in vitro* experiments with rbASCs confirm this material's feature.
- rbASCs-HA bioconstructs have been used to repair a critical size tibia defect.

- X-rays, histology and immunohistochemistry analyses revealed a good quality of the neo-formed bone tissue.
- Nanoindentation tests confirm improved bone elasticity and hardness.

Acknowledgment

The authors sincerely thank Drs. L. de Girolamo, A. Addis, M. Campagnol, C. Domeneghini, A. Di Giancamillo, P. Stortini and D. Carnelli for their valuable work and Finceramica S.p.A. (Faenza, Italy) for providing the scaffolds. This work has been partially supported by the Italian Ministry of University and Research (2006091907_003) and the Italian Ministry of Health (2007-656853).

List of Abbreviations

ALP   :   alkaline phosphatase
ASCs   :   adipose-derived stem cells
BMD   :   bone mineral density
BMP   :   bone morphogenic proteins
BMSCs   :   bone marrow mesenchymal stem cells
CFU-F   :   colony forming unit-fibroblast
CFU-O   :   colony forming unit-osteoblast
Coll I   :   collagen type I
CPC   :   cetylpyridinium chloride
CTRL   :   undifferentiative medium
DMEM   :   Dulbecco's modified eagle's medium
EDTA   :   ethylenediaminetetraacetic acid
Er   :   elasticity
FBS   :   fetal bovine serum
H   :   hardness
HA   :   hydroxyapatite
MHC   :   major histocompatibility complex
MSCs   :   mesenchymal stem cells
MTT   :   (3-(4, 5-dimethylthiazolyl-2)-2, 5-diphenyltetrazolium bromide
OC   :   osteocalcin
ONC   :   osteonectin
OPN   :   osteopontin
OSTEO   :   osteogenic medium
OSX   :   osterix
PA   :   plastic adherence

PBS	:	phosphate-buffered saline
PLGA	:	poly(lactic-co-glycolic acid)
Runx2	:	Runt-related protein 2
SEM	:	scanning electron microscope
SVF	:	stromal vascular fraction
β-TCP	:	β-tricalcium phosphate
TGF-β	:	transforming growth factor β

References

Albrektsson, T. and C. Johansson. 2001. Osteoinduction, osteoconduction and osseointegration. Eur Spine J. 10 Suppl. 2. S96–101.

Arrigoni, E., S. Lopa, L. de Girolamo, D. Stanco and A.T. Brini. 2009. Isolation, characterization and osteogenic differentiation of adipose-derived stem cells: from small to large animal models. Cell Tissue Res. 338: 401–411.

Castaneda, S., R. Largo, E. Calvo, F. Rodriguez-Salvanes, M.E. Marcos, M. Diaz-Curiel and G. Herrero-Beaumont. 2006. Bone mineral measurements of subchondral and trabecular bone in healthy and osteoporotic rabbits. Skeletal Radiol. 35: 34–41.

Choi, E.W., I.S. Shin, S.Y. Park, J.H. Park, J.S. Kim, E.J. Yoon, S.K. Kang, J.C. Ra and S.H. Hong. 2012. Reversal of serological, immunological and histological dysfunction in systemic lupus erythematosus mice by long-term serial adipose tissue-derived mesenchymal stem cell transplantation. Arthritis Rheum. 64: 243–253.

Chu, C.R., M. Szczodry and S. Bruno. 2010. Animal models for cartilage regeneration and repair. Tissue Eng Part B Rev. 16: 105–15.

Constantin, G., S. Marconi, B. Rossi, S. Angiari, L. Calderan, E. Anghileri, B. Gini, S.D. Bach, M. Martinello, F. Bifari, M. Galie, E. Turano, S. Budui, A. Sbarbati, M. Krampera and B. Bonetti. 2009. Adipose-derived mesenchymal stem cells ameliorate chronic experimental autoimmune encephalomyelitis. Stem Cells. 27: 2624–2635.

Dashtdar, H., H.A. Rothan, T. Tay, R.E. Ahmad, R. Ali, L.X. Tay, P.P. Chong and T. Kamarul. 2011. A preliminary study comparing the use of allogenic chondrogenic pre-differentiated and undifferentiated mesenchymal stem cells for the repair of full thickness articular cartilage defects in rabbits. J Orthop Res. 29: 1336–1342.

de Girolamo, L., M.F. Sartori, E. Arrigoni, L. Rimondini, W. Albisetti, R.L. Weinstein and A.T. Brini. 2008. Human adipose-derived stem cells as future tools in tissue regeneration: osteogenic differentiation and cell-scaffold interaction. Int J Artif Organs. 31: 467–479.

de Girolamo, L., E. Arrigoni, D. Stanco, S. Lopa, A. Di Giancamillo, A. Addis, S. Borgonovo, C. Dellavia, C. Domeneghini and A.T. Brini. 2011. Role of autologous rabbit adipose-derived stem cells in the early phases of the repairing process of critical bone defects. J Orthop Res. 29: 100–108.

Di Nicola, M., C. Carlo-Stella, M. Magni, M. Milanesi, P.D. Longoni, P. Matteucci, S. Grisanti and A.M. Gianni. 2002. Human bone marrow stromal cells suppress T-lymphocyte proliferation induced by cellular or nonspecific mitogenic stimuli. Blood. 99: 3838–3843.

Dudas, J.R., K.G. Marra, G.M. Cooper, V.M. Penascino, M.P. Mooney, S. Jiang, J.P. Rubin and J.E. Losee. 2006. The osteogenic potential of adipose-derived stem cells for the repair of rabbit calvarial defects. Ann Plast Surg. 56: 543–548.

Friedenstein, A.J., U.F. Deriglasova, N.N. Kulagina, A.F. Panasuk, S.F. Rudakowa, E.A. Luria and I.A. Ruadkow. 1974. Precursors for fibroblasts in different populations of hematopoietic cells as detected by the *in vitro* colony assay method. Exp Hematol. 2: 83–92.

Gilsanz, V., T.F. Roe, D.T. Gibbens, E.E. Schulz, M.E. Carlson, O. Gonzalez and M.I. Boechat. 1988. Effect of sex steroids on peak bone density of growing rabbits. Am J Physiol. 255: E416–421.

Gonzalez, M.A., E. Gonzalez-Rey, L. Rico, D. Buscher and M. Delgado. 2009. Treatment of experimental arthritis by inducing immune tolerance with human adipose-derived mesenchymal stem cells. Arthritis Rheum. 60: 1006–1019.

Hao, W., J. Dong, M. Jiang, J. Wu, F. Cui and D. Zhou. 2010a. Enhanced bone formation in large segmental radial defects by combining adipose-derived stem cells expressing bone morphogenetic protein 2 with nHA/RHLC/PLA scaffold. Int Orthop. 34: 1341–1349.

Hao, W., L. Pang, M. Jiang, R. Lv, Z. Xiong and Y.Y. Hu. 2010b. Skeletal repair in rabbits using a novel biomimetic composite based on adipose-derived stem cells encapsulated in collagen I gel with PLGA-beta-TCP scaffold. J Orthop Res. 28: 252–257.

Im, G.I. and J.H. Lee. 2010. Repair of osteochondral defects with adipose stem cells and a dual growth factor-releasing scaffold in rabbits. J Biomed Mater Res B Appl Biomater. 92: 552–560.

Kern, S., H. Eichler, J. Stoeve, H. Kluter and K. Bieback. 2006. Comparative analysis of mesenchymal stem cells from bone marrow, umbilical cord blood, or adipose tissue. Stem Cells. 24: 1294–1301.

Lee, M.J., S.K. Sohn, K.T. Kim, C.H. Kim, H.B. Ahn, M.S. Rho, M.H. Jeong and S.K. Sun. 2010. Effect of hydroxyapatite on bone integration in a rabbit tibial defect model. Clin Orthop Surg. 2: 90–97.

Levi, B. and M.T. Longaker. 2011. Concise review: adipose-derived stromal cells for skeletal regenerative medicine. Stem Cells. 29: 576–582.

Li, H., K. Dai, T. Tang, X. Zhang, M. Yan and J. Lou. 2007. Bone regeneration by implantation of adipose-derived stromal cells expressing BMP 2. Biochem Biophys Res Commun. 356: 836–842.

Li, J., Y. Li, S. Ma, Y. Gao, Y. Zuo and J. Hu. 2010. Enhancement of bone formation by BMP-7 transduced MSCs on biomimetic nano-hydroxyapatite/polyamide composite scaffolds in repair of mandibular defects. J Biomed Mater Res A. 95: 973–981.

Li, W.J., H. Chiang, T.F. Kuo, H.S. Lee, C.C. Jiang and R.S. Tuan. 2009. Evaluation of articular cartilage repair using biodegradable nanofibrous scaffolds in a swine model: a pilot study. J Tissue Eng Regen Med. 3: 1–10.

Lian, J.B., G.S. Stein, A. Javed, A.J. Van Wijnen, J.L. Stein, M. Montecino, M.Q. Hassan, T. Gaur, C.J. Lengner and D.W. Young. 2006. Networks and hubs for the transcriptional control of osteoblastogenesis. Rev Endocr Metab Disord. 7: 1–16.

Lovett, M., K. Lee, A. Edwards and D.L. Kaplan. 2009. Vascularization strategies for tissue engineering. Tissue Eng Part B Rev. 15: 353–370.

Najar, M., G. Raicevic, H.I. Boufker, H. Fayyad-Kazan, C. De Bruyn, N. Meuleman, D. Bron, M. Toungouz and L. Lagneaux. 2010. Adipose-tissue-derived and Wharton's jelly-derived mesenchymal stromal cells suppress lymphocyte responses by secreting leukemia inhibitory factor. Tissue Eng Part A. 16: 3537–3546.

Neyt, J.G., J.A. and N.C. Carroll. 1998. Use of animal models in musculoskeletal research. Iowa Orthop J. 18: 118–123.

Oshima, S., M. Ishikawa, Y. Mochizuki, T. Kobayashi, Y. Yasunaga and M. Ochi. 2010. Enhancement of bone formation in an experimental bony defect using ferumoxide-labelled mesenchymal stromal cells and a magnetic targeting system. J Bone Joint Surg Br. 92: 1606–1613.

Pieri, F., E. Lucarelli, G. Corinaldesi, N.N. Aldini, M. Fini, A. Parrilli, B. Dozza, D. Donati and C. Marchetti. 2010. Dose-dependent effect of adipose-derived adult stem cells on vertical bone regeneration in rabbit calvarium. Biomaterials. 31: 3527–3535.

Sarkar, M.R., P. Augat, S.J. Shefelbine, S. Schorlemmer, M. Huber-Lang, L. Claes, L. Kinzl and A. Ignatius. 2005. Bone formation in a long bone defect model using a platelet-rich plasma-loaded collagen scaffold. Biomaterials. 27: 1817–23.

Schaffler, A. and C. Buchler. 2007. Concise review: adipose tissue-derived stromal cells—basic and clinical implications for novel cell-based therapies. Stem Cells. 25: 818–827.

Spencer, N.D., J.M. Gimble and M.J. Lopez. 2011. Mesenchymal stromal cells: past, present, and future. Vet Surg. 40: 129–139.

Torres, F.C., C.J. Rodrigues, I.N. Stocchero and M.C. Ferreira. 2007. Stem cells from the fat tissue of rabbits: an easy-to-find experimental source. Aesthetic Plast Surg. 31: 574–578.

Torroni, A. 2009. Engineered bone grafts and bone flaps for maxillofacial defects: state of the art. J Oral Maxillofac Surg. 67: 1121–1127.

Wakitani, S., T. Goto, S.J. Pineda, R.G. Young, J.M. Mansour, A.I. Caplan and V.M. Goldberg. 1994. Mesenchymal cell-based repair of large, full-thickness defects of articular cartilage. J Bone Joint Surg Am. 76: 579–592.

Wang, L., H. Fan, Z.Y. Zhang, A.J. Lou, G.X. Pei, S. Jiang, T.W. Mu, J.J. Qin, S.Y. Chen and D. Jin. 2010. Osteogenesis and angiogenesis of tissue-engineered bone constructed by prevascularized beta-tricalcium phosphate scaffold and mesenchymal stem cells. Biomaterials. 31: 9452–9461.

Wang, X., J.D. Mabrey and C.M. Agrawal. 1998. An interspecies comparison of bone fracture properties. Biomed Mater Eng. 8: 1–9.

Yanez, R., M.L. Lamana, J. Garcia-Castro, I. Colmenero, M. Ramirez and J.A. Bueren. 2006. Adipose tissue-derived mesenchymal stem cells have *in vivo* immunosuppressive properties applicable for the control of the graft-versus-host disease. Stem Cells. 24: 2582–2591.

Yoshimura, K., T. Shigeura, D. Matsumoto, T. Sato, Y. Takaki, E. Aiba-Kojima, K. Sato, K. Inoue, T. Nagase, I. Koshima and K. Gonda. 2006. Characterization of freshly isolated and cultured cells derived from the fatty and fluid portions of liposuction aspirates. J Cell Physiol. 208: 64–76.

Zhao, M., J. Zhou, X. Li, T. Fang, W. Dai, W. Yin and J. Dong. 2011. Repair of bone defect with vascularized tissue engineered bone graft seeded with mesenchymal stem cells in rabbits. Microsurgery. 31: 130–137.

Zuk, P.A., M. Zhu, P. Ashjian, D.A. De Ugarte, J.I. Huang, H. Mizuno, Z.C. Alfonso, J.K. Fraser, P. Benhaim and M.H. Hedrick. 2002. Human adipose tissue is a source of multipotent stem cells. Mol Biol Cell. 13: 4279–4295.

Transplantation of Human Adipose-derived Stem Cells for Fracture Healing: Multi-lineage Differentiation Potential and Paracrine Effects

Taro Shoji,[1] Masaaki Ii[2] and Takayuki Asahara[3],*

ABSTRACT

Somatic tissues are the most widely used source of stem cells for regeneration research and regenerative medicine and have tremendous therapeutic potential in translational research and clinical applications because they are derived from adult tissues rather than the embryo and are multipotent and relatively safe.

[1]Group of Vascular Regeneration Research, Institute of Biomedical Research and Innovation, 2-2 Minatojima-minamimachi, Chuo-ku, Kobe, Hyogo 650-0047, Japan; and Department of Orthopaedic Surgery, Kobe University Graduate School of Medicine, Kobe, Hyogo, Japan.
E-mail: taro@lexus-sc.jp
[2]Group of Translational Stem Cell Research, Department of Pharmacology, Osaka Medical College, 2-7, Daigaku-machi, Takatsuki, Osaka 569-8686, Japan.
E-mail: masa0331@mac.com
[3]Group of Vascular Regeneration Research, Institute of Biomedical Research and Innovation, 2-2 Minatojima-minamimachi, Chuo-ku, Kobe, Hyogo 650-0047, Japan; and Department of Regenerative Medicine Science, Tokai University School of Medicine, Isehara, Kanagawa, Japan.
E-mail: asa777@is.icc.u-tokai.ac.jp
*Corresponding author

List of abbreviations given at the end of the text.

Although bone marrow stromal cells (BMSCs) are a well-known source of somatic stem cells for regenerative medicine, there is increasing interest in using adipose-derived stromal cells (ADSCs). ADSCs are easily and relatively painlessly isolated from white adipose tissue, can differentiate into multiple mesenchymal lineages including BMSCs and have numerous favourable characteristics, tropic paracrine effects and therapeutic potential for tissue regeneration *in vivo*. Various clinical trials have also demonstrated the regenerative capacity of ADSCs in the fields of orthopaedic, plastic, oral, facial and cardiac surgeries.

Human multipotent adipose-derived stem (hMADS) cells isolated from young donor adipose can differentiate into multiple mesenchymal lineages *in vitro* and *in vivo*, have significant expansion capacity *ex vivo* and promote healing in an immunodeficient rat femur non-union fracture model through the paracrine effects of hMADS-derived osteoblasts and endothelial cells.

Here we summarize the characteristics of ADSCs gleaned from *in vitro* studies, *in vivo* animal studies and recent clinical trials of ADSCs in regenerative medicine.

Introduction

Somatic stem cells are the most widely used stem cells in regenerative medicine. They have promising therapeutic potential and are safe for use in translational research and many clinical applications. Bone marrow stromal cells (BMSCs) have been widely used in regenerative medicine and related research for several decades (Pittenger et al. 1999) but have significant disadvantages, such as the stem cell yield is usually low, the isolation procedures are invasive, painful and bleeding is common in donors and recipients. Adipose tissue, especially white adipose tissue, has recently been identified as an alternative source of adult stem cells; there is a large reservoir of donor tissue, which is also readily accessible, and adipose resident stem cells can differentiate into multiple mesenchymal lineages, including BMSCs (Zuk et al. 2001).

Since the identification of preadipocytes, (Green and Meuth 1974) a number of investigators have characterized adipose tissue-derived stem/progenitor cells, the so-called adipose-derived stromal cells (ADSCs). The therapeutic potential of ADSCs has been widely studied using assays of *in vitro* multipotency and *in vivo* transplantation. These cells have the capacity to differentiate into various mesenchymal cell types such as osteoblasts (OBs) (Rodriguez et al. 2005; Shoji et al. 2010), chondrocytes (Zuk et al. 2001) myocytes (Rodriguez et al. 2005), cardiomyocytes, (Planat-Benard et al. 2004), fibroblasts (Zuk et al. 2001) and adipocytes (Rodriguez et al.

2005). In addition, stromal vascular fraction (SVF) from adipose tissue can differentiate into vascular cell types such as endothelial cells (ECs), smooth muscle cells and circulating blood cells (Planat-Benard et al. 2004).

Of the available ADSC cell lines, we focused on human multipotent adipose-derived stem (hMADS) cells (Rodriguez et al. 2005). hMADS cells from young donors adhere rapidly in tissue culture and have extensive expansion capacity *ex vivo*, continuing to expand to at least over 200 passages. These cells express the mesenchymal cell surface markers CD44, CD49b, CD105 and CD13 but not the hematopoietic and endothelial markers CD34 and CD31. hMADS cells are multipotent, differentiating into adipogenic, osteogenic and myogenic cell types. They possess telomerase activity and are of normal karyotypes; characteristics that lead to long-term engraftment and extensive, diverse regenerative capacity. hMADS cells transplanted into skeletal muscle of the non-immunocompromised X-*linked muscular dystrophy* (*mdx*) mice, a model of Duchenne muscular dystrophy, showed long-term engraftment and efficient regeneration of myofibres expressing human dystrophin (Rodriguez et al. 2005).

The therapeutic potential of hMADS cells in other disease models has not yet been explored; therefore, we tested these cells in a bone fracture healing model on the basis of their capacity for osteogenic and vasculogenic differentiation (Shoji et al. 2010). As 5–10 percent fractures result in non-union, causing serious problematic effects on quality of life, novel therapeutic strategies for non-union healing are clinically warranted. A major cause of non-union is insufficient blood supply and the lack of new bone formation at the fracture site. For example, a subset of distal tibia or scaphoid fractures is clinically observed as hypovascular lesions and at the high risk of non-union. Because neovascularization and osteogenesis are critical in combination for fracture healing, stem/progenitor cell-based therapies are of significant interest as the stem/progenitor cells can differentiate into vascular cell types as well as OBs. Clinical trials testing the regenerative treatment of non-union long bone fractures have used various different cell types, such as BMSCs (Connolly et al. 1991) bone marrow mononuclear cells (Fukui et al. 2011) and endothelial progenitor cells (EPCs) (Mifune et al. 2008). However, a large volume of blood or bone marrow is required to isolate sufficient number of stem cells required for the treatment, and the isolation may involve invasive procedures. In contrast, treatment of non-union fractures with ADSCs may be preferential because large numbers of cells are much more easily obtained.

Here we summarize the characteristics and advantages of ADSCs, focusing on hMADS cells and their multilineage differentiation potentials and beneficial paracrine effects. We also introduce our *in vivo* study of the therapeutic potential of hMADS cells in a rat non-union fracture model.

Characterizations of Adipose-derived Stromal Cells

Although defined cell surface markers of BMSCs remain controversial, most researchers reported that BMSCs express at least typical antigens such as CD13, CD29, CD34, CD44, CD73, CD90, CD105, and CD166. On the other hand, they express neither hematopoietic cell surface markers (CD14, CD16, CD45, CD56, CD61, CD104 and CD106) nor endothelial markers (CD31 and CD144 (De Ugarte et al. 2003). ADSCs also express mesenchymal stem cell markers (CD10, CD13, CD29, CD44, CD54, CD71, CD90, CD105, CD117 and STRO-1) but not hematopoietic or endothelial markers (Romanov et al. 2005). Similar to BMSCs, ADSCs are similar to fibroblasts *in vitro*, exhibiting spindle morphology and rapid adherence (Zannettino et al. 2008). The extensive similarities between BMSCs and ADSCs may be a result of the derivation of ADSCs from blood-borne BMSCs infiltrated into the adipose tissue from blood supply or the existence of a common niche for ADSCs and BMSCs in the peri-microvessels (Zannettino et al. 2008). Chen et al. suggested that pericytes are the ancestors of a subset of BMSCs (Chen et al. 2009), whereas Traktuev et al. defined a pericyte-like subpopulation of ADSCs (Traktuev et al. 2008) and suggested that the majority of CD34$^+$/CD31$^-$/CD144$^-$ ADSCs are resident pericytes that stabilize vasculature via structural and functional interactions with the endothelium (Traktuev et al. 2008).

hMADS cells isolated from young donor's SVF are rapidly adherent in culture (i.e., within 12 hr), have extensive expansion capacity *ex vivo* (>200 passages), express CD13, CD49b, CD44, CD90 (Thy-1) and CD105 but not Flk-1 (VEGF-R2), glycophorin A, CD34, CD15, CD117 (c-Kit), CD133 and STRO-1 (Rodriguez et al. 2005). The differences in the cell surface phenotypes between hMADS cells and other ADSCs may be due to passage number, which may affect CD34 expression in hMADS cells, (Traktuev et al. 2008) or degree of cell differentiation (Romanov et al. 2005). One significant advantage of hMADS cells is evasion of an immune response possibly due to low expression levels of class I HLA and the absence of class II HLA (Rodriguez et al. 2005).

ADSC Multipotency

The multilineage differentiation potentials of ADSCs have been extensively characterized. ADSCs can differentiate into OBs (Rodriguez et al. 2005), chondrocytes (Zuk et al. 2001), myocytes (Rodriguez et al. 2005), cardiomyocytes (Planat-Benard et al. 2004), fibroblasts (Zuk et al. 2001) and adipocytes (Rodriguez et al. 2005) as well as vascular ECs, vascular smooth muscle cells and circulating blood cells (Planat-Benard et al. 2004).

Stem cells have been isolated from the fat of Lewis rats and induced to differentiate into adipogenic and osteogenic lineages *in vitro* as well as *in vivo*, forming bone when subcutaneously implanted into Lewis rats (Lee et al. 2003). Zuk et al. reported the chondrogenic potential of ADSCs from lipectomy-derived human adipose tissue, specifically the fibroblast-like cells from the processed lipoaspirate (PLA) (Zuk et al. 2001). PLA cells could differentiate into adipogenic, chondrogenic, myogenic and osteogenic lineages in the presence of lineage-specific induction factors *in vitro*. Nathan et al. reported that ADSCs from New Zealand White rabbits promoted the healing of defects created in the medial femoral condyle, reconstituting the gross osteochondral defect at a high histological grading through native mechanisms (Nathan et al. 2003). These results demonstrate the *in vitro* and *in vivo* chondrogenic potential of ADSCs.

Various assays have demonstrated the myogenic potential of ADSCs. Transplanted ADSCs repaired skeletal muscle injuries through direct differentiation into skeletal myocytes as well as through fusion with host myotubes (Rodriguez et al. 2005; Sherwood et al. 2004). Rodrigez et al. showed that hMADS cells transplanted into skeletal muscle of non-immunocompromised mdx mice were capable of long-term engraftment and efficient regeneration of myofibres expressing human dystrophin (Rodriguez et al. 2005).

ADSC differentiation into cardiomyocytes and ECs was indicated by the identification of EPCs in SVF isolated from human adipose tissue (Miranville et al. 2004), presence of beating cells in ADSC cultures from rabbit subcutaneous adipose tissue (Rangappa et al. 2003) and presence of cardiomyocytes in primary cultures of ADSCs from murine visceral and subcutaneous adipose tissue (Madonna and De Caterina 2008).

ADSC from human adipose tissue can differentiate into fibroblast-like cells (Zuk et al. 2001). In soft tissue repair assays, ADSCs differentiate into fibroblastic cells that secrete several growth factors essential for wound healing (Kim et al. 2007; Rehman et al. 2004). ADSCs have also been shown to stimulate fibroblast proliferation and migration and type I collagen secretion (Kim et al. 2007; Rigotti et al. 2007), suggesting that these cells promote wound healing by cell differentiation and alteration of the microenvironment through growth factor secretion.

Finally, the plasticity of ADSCs toward ECs was reported by Planat-Benard et al. (Planat-Benard et al. 2004) who showed that cultured human SVF cells transplanted into nude mice differentiate into ECs, incorporate into vessels and promote post-ischemic neovascularization and vessel-like structure formation in nude mice. In addition, they showed that cultured human SVF cells are a homogeneous population of CD34- and CD13-positive cells that spontaneously express the endothelial markers CD31 and von-Willebrand factor when cultured in semisolid medium. They suggested that adipocytes and ECs are derived from a common progenitor. Cousin et al. identified hematopoietic stem cells (HSCs) in several tissues of mesodermal origin and demonstrated the hematopoietic potential of the transplanted SVF from mouse adipose tissue through rescue of lethally irradiated mice (Cousin et al. 2003). Donor cells were detected in circulating blood and hematopoietic tissues of the recipient, indicating that ADSCs could reconstitute the major circulating and tissue hematopoietic lineages. They proposed that adipose tissue could be a useful and convenient source of cells for hematopoietic stem cell therapies.

Paracrine Secretions from ADSCs

It is well known that adipose tissue participates in many endocrine processes through the production of several cytokines and growth factors (Kilroy et al. 2007). Consistent with this, several studies showed that ADSCs secrete a large number of growth factors, including epidermal growth factor (EGF), vascular endothelial growth factor (VEGF), basic fibroblast growth factor (bFGF), keratinocyte growth factor (KGF), platelet-derived growth factor (PDGF), hepatocyte growth factor (HGF), transforming growth factor beta (TGF-β), insulin-like growth factor (IGF) and brain-derived neurotrophic factor (BDNF) (Cai et al. 2007; Chen et al. 2009; Kilroy et al. 2007; Kim et al. 2007; Rehman et al. 2004; Wei et al. 2009). ADSCs also secrete cytokines such as the Flt-3 ligand, granulocyte colony-stimulating factor (G-CSF), granulocyte/macrophage colony-stimulating factor (GM-CSF), macrophage colony-stimulating factor (M-CSF), interleukin-6 (IL-6), interleukin-7 (IL-7), interleukin-8 (IL-8), interleukin-11 (IL-11), interleukin-12 (IL-12), leukaemia inhibitory factor (LIF), tumour necrosis factor-alpha (TNF-α) and hypoxia-inducible transcription factor (HIF) (Kershaw and Flier 2004; Kilroy et al. 2007; Trayhurn and Beattie 2001). We studied HIF secretion by ADSCs under hypoxic stress, and these cells then secrete angiogenic and anti-apoptotic growth factors through HIF autostimulation (Cai et al. 2007; Lee et al. 2009; Rehman et al. 2004). Consistent with this, in our non-union fracture model created by cauterization of the periosteum, there is an ischemic/hypoxic environment around the fracture that may stimulate hMADS cells to secrete HIF and then HIF upregulates intrinsic VEGF and angiopoietin-1 (ANG1) (Shoji et al. 2010). In addition, Ii et al. found that hMADS cell transplants contribute to the recovery of cardiac function following myocardial infarct, reducing infarct size by increasing vascularization through the paracrine effects of VEGF, bFGF and stromal cell-derived factor-1α (SDF-1α) production (Ii et al. 2011).

Therapeutic hMADS Cell Transplantation for Bone Fractures

hMADS cells were isolated from adipose tissue of young donors (age, <7 yr) following surgery at Nice University Hospital (Rodriguez et al. 2005). In brief, adipose tissue (200 mg/ml) was dissociated for 5–10 min in Dulbecco's Modified Eagle's Medium (DMEM) containing antibiotics, 2 mg/ml collagenase and 20 mg/ml foetal bovine serum (FBS). SVF, which contains adipose precursor cells, was pelleted by centrifugation. The SVF pellet was resuspended, seeded on culture plates and cultured in 10 percent FBS/ DMEM. hMADS cells were isolated by collection of the rapidly adherent cell population after 12 hr of culture and expanded by Stem Cell Sciences K.K. (Kobe, Hyogo, Japan).

In our previous study (Shoji et al. 2010) we tested whether hMADS cell transplantation was more effective for fracture healing than conventional treatments because of the ability of these cells to contribute directly to tissue regeneration by differentiation into osteogenic and vasculogenic cells. We also explored the paracrine effect of transplanted hMADS cells on the production of intrinsic pro-angiogenic/-osteogenic cytokines from host cells.

In detail, female F344/N-rnu nude rats (8–10 wk; 150–170 g) were used. Non-healing femoral fractures were induced in all animals by cauterization of the periosteum around the fracture site with a modification of the original method described previously. Immediately after fracture, 1×10^5 hMADS cells or 1×10^5 human fibroblasts (hFB) suspended in 100 ml PBS with 100 ml atelocollagen gel, a bovine-derived bioscaffold, were transplanted into the fracture site. Same amount of PBS (without cells) with 100 ml atelocollagen gel was used as a negative control. Contralateral, unfractured femurs were used as intact controls for histological and functional analyses. Morphological fracture healing was evaluated by radiography and histology 8 wk after surgery. Radiography revealed that a bridging callus uniting the fracture had formed in 9/10 littermates in the hMADS group but only in 4/10 littermates in the hFB group. In contrast, fractures failed to unite in 9/10 littermates in the PBS group (Fig. 20.1). Thus, the frequency of fracture healing judged by morphological criteria was significantly improved in animals receiving hMADS cells (Fig. 20.2).

To evaluate the recovery of local blood flow via neovascularization of the fracture, hind limbs were serially examined by laser doppler perfusion imaging (LDPI) after surgery. The blood perfusion ratios were significantly higher in the hMADS group than in the hFB and PBS groups at 1 and 2 wk after surgery. The enhancment of intrinsic angiogenesis due to the paracrine effects of the transplanted cells was also confirmed by histochemical staining with the endothelial marker isolectin B4 (ILB4). Enhanced neovascularization

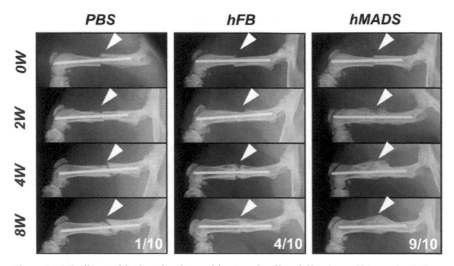

Figure 20.1 Radiographical evaluations of fracture healing following cell transplantation. At week 8, 9 of 10 animals in the PBS group showed no bridging callus and this resulted in non-union. In contrast, fractures in the hFB and hMADS groups were united by bridging callus formation (4 of 10 animals receiving hFB, 9 of 10 animals receiving hMADS cells; red arrowheads, femur fracture sites).

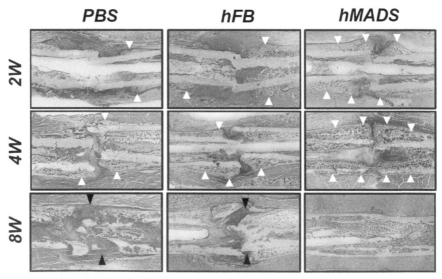

Figure 20.2 Histological evaluation of endochondral ossification by toluidine blue staining. In the PBS group, a thick callus (yellow arrowheads) was observed at week 2, but the fracture remains unhealed at week 4 and the callus was absorbed and a pseudarthrosis was noted (red arrowheads) at week 8. In animals receiving hFB, the callus (yellow arrowheads) was observed at week 4, but the fracture gap (red arrowheads) was not filled with bridging callus at week 8. In the hMADS group, the callus and newly formed trabecular bones were visible in the fracture at week 2, a bridging callus was observed (yellow arrowheads) at week 4 and complete union was visible at week 8.

around the endochondral ossification area was found 2 wk after surgery. Capillary density was also superior in the hMADS group (Fig. 20.3). LDPI analysis during fracture healing showed significant enhancement of blood perfusion via neovascularization in the hMADS group. We explored the mechanism of enhancement of host osteogenesis and angiogenesis by hMADS cell therapy on the basis of expression of pro-angiogenic and -osteogenic cytokines in peri-fracture sites 2 wk after surgery. Using real-time RT-PCR, we found that the relative mRNA expression levels of host VEGF (rVEGF), ANG1 (rANG1) and bone morphogenetic protein 2 (rBMP2) normalized to those of host glyceraldehyde-3-phosphate dehydrogenase (rGAPDH) were significantly higher in the hMADS group than in the control group (Fig. 20.4). These results suggest that the paracrine effects of hMADS cells enhanced intrinsic angiogenesis as well as osteogenesis at the fracture sites via upregulation of pro-angiogenic and -osteogenic cytokines, which resulted in recovery of blood perfusion and rapid fracture healing. Thus, hMADS cells had a greater therapeutic benefit than hFB cells.

The interaction between transplanted hMADS cells and resident cells described above most probably plays a critical role in fracture repair. As described above, adipose tissue can promote tissue regeneration through secretion of various cytokines, such as angiogenic, hematopoietic and pro-inflammatory cytokines as well as HIF under hypoxic conditions (Kershaw and Flier 2004; Khan et al. 2007; Kilroy et al. 2007; Trayhurn and Beattie 2001). As HIF-2a is upregulated by hypoxia *in vitro*, (Khan et al. 2007) hMADS cells might secrete HIF, which then stimulates host and hMADS cells to secrete VEGF and ANG1. This hypothesis is based on the fact that the non-union fracture model used in this study was created by cauterizing the periosteum, leading to an ischemic/hypoxic condition around the fracture sites. This upregulation of endogenous angiogenic cytokines, rVEGF and rANG1, by hMADS-secreted HIF could negate the ischemic/hypoxic conditions found in our non-union fracture model.

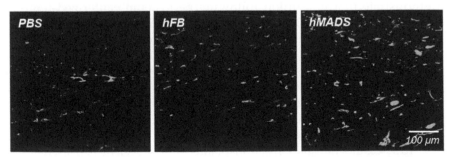

Figure 20.3 Serial improvement of blood flow in fracture sites following cell transplantation. Representative images of vascular staining with isolectin B4 (green) in peri-fracture tissue samples at week 2.

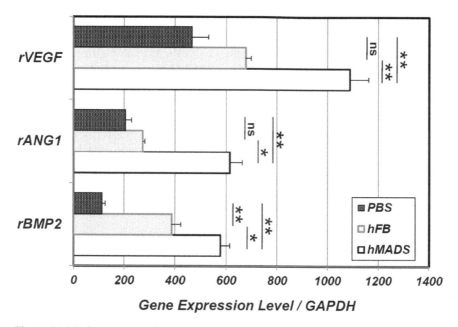

Gene Expression Level / GAPDH

Figure 20.4 **Enhancement of local angiogenesis and osteogenesis around the fracture sites.** RT-PCR analysis of host cytokine expression in peri-fracture tissue 2 wk after surgery. Normalized to expression of rat glyceraldehyde-3-phosphate dehydrogenase (rGAPDH), expression of rat vascular endothelial growth factor (rVEGF), rat angiopoietin-1 (rANG1) and rat bone morphologic protein-2 (rBMP2) was significantly greater in the hMADS group than in the hFB and PBS groups. (*: $p < 0.05$ and **: $p < 0.001$).

The effect of hMADS cells on functional recovery of non-union fractures was assessed by biomechanical three-point bending comparing the fractured and unfractured femurs. All three parameters measured (ultimate stress, failure energy and extrinsic stiffness) were significantly superior in the repaired femurs in the hMADS group. Taken together, our findings demonstrate that non-union following periosteum cauterized-femoral fractures can be successfully repaired by hMADS cell transplantation.

Using an *in vitro* assay, we found that stimulation for 48 hr with BMP-2 or VEGF (dose used: 0, 5, 10, 50 and 100 ng/ml) significantly increased hMADS cell proliferation, peaking at 50 ng/ml of either growth factor. In Transwell culture plates, hMADS cells migrated toward higher concentrations of BMP-2 or VEGF in a dose-dependent manner, suggesting that transplanted hMADS cells may be stimulated *in vivo* by BMP-2 and VEGF in the fracture sites.

To assess the contribution of transplanted hMADS cells to neovasculatures and differentiation into OBs, ILB4, Ulex europaeus agglutinin 1 (UEA1) lectin and antibodies for human osteocalcin (hOC) were applied to the fracture samples 2 wk after surgery; human nuclear antigen (hNA) was also used as a specific marker for human cells. ILB4-positive tubular structures in the granulation tissue surrounding the fracture gap were of host origin, whereas ILB4-negative/UEA1-positive cells within them were human-derived ECs (Fig. 20.5). Human OBs (hNA and hOC double-positive cells) were also detected in the endochondral ossification area of the pericallus (Fig. 20.6), indicating that transplanted hMADS cells had differentiated into two different lineages, ECs and OBs. This was confirmed by RT-PCR, which detected expression of human alleles of the bone markers hOC and collagen 1 alpha 1 (hCol1); EC markers CD31 (hCD31) and vascular endothelial cadherin (hVEcad) and hGAPDH 2 wk after surgery (Fig. 20.7). These results indicate that hMADS cells promote fracture healing through two mechanisms: paracrine effects resulting from upregulation of endogenous pro-angiogenic/-osteogenic cytokines and differentiation into ECs and OBs.

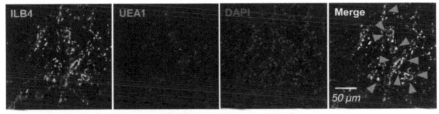

Figure 20.5 Endothelial differentiation of hMADS cells in peri-fracture tissue. Representative images of fluorescent staining for isolectin B4 (ILB4: green), Ulex europaeus agglutinin 1 (UEA1: red) and DAPI (blue) with serial sections of peri-fracture tissue samples. Arrowheads in the merged image indicate ILB4-negative/UEA1-positive cells as differentiated human endothelial cells located within host-derived capillaries detected as ILB4-positive tubular structures in the granulation tissue surrounding the fracture gap.

Color image of this figure appears in the color plate section at the end of the book.

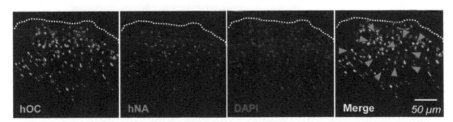

Figure 20.6 Osteoblastic differentiation of hMADS cells during callus formation. Immunofluorescence staining for human-specific osteocalcin (hOC, green), anti-human nuclear antigen (hNA, red) and DAPI (blue) in serial sections of peri-fracture tissue samples. White dotted, curved line indicates the edge of callus. Arrowheads in the merged image indicate hOC and hNA double-positive cells identified as differentiated human osteoblasts (hOBs) in the callus endochondral ossification area receiving hMADS cells.

Color image of this figure appears in the color plate section at the end of the book.

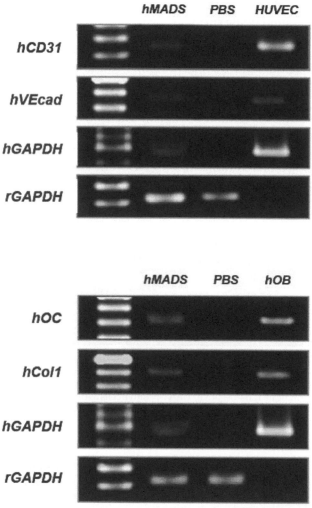

Figure 20.7 RT-PCR analysis of angiogenic and osteogenic markers in differentiated hMADS cells. RT-PCR analysis of tissue from the peri-fracture site showed the gene expression of human-specific endothelial cell markers, human CD31 (hCD31) and human vascular endothelial cadherin (hVEcad); human-specific bone-related markers, human osteocalcin (hOC) and human collagen type 1 alpha 1 (hCol1); human-specific glyceraldehyde-3-phosphate dehydrogenase (hGAPDH); and rat-specific glyceraldehyde-3-phosphate dehydrogenase (rGAPDH). Cultured human umbilical vein endothelial cells (HUVECs) and hOBs were used for positive controls.

Our results highlight the following advantages of hMADS cells in cell therapy: (1) high expansion capacity, (2) differentiation into OBs and ECs, (3) paracrine effects on osteogenesis and angiogenesis in fracture-induced environment, (4) easy cell isolation and (5) long lifespan compared with other adipose-derived cells, such as stromal vascular cells and lipoaspirated cells (Rodriguez et al. 2005; Shoji et al. 2010). Based on these data, we propose that transplantation of autologous hMADS cells is feasible in humans and will soon become a powerful therapy for non-union fractures.

Other Therapeutical Approaches using ADSCs

Various studies have demonstrated tissue regeneration *in vivo* following ADSC transplantation. We conducted *in vivo* studies with ADSCs for the treatment of skeletal muscle disease, osteochondral defects, myocardial infarction (MI) and wound healing.

Transplantation of hMADS cells into the tibialis anterior and gastrocnemius muscles of mdx mice resulted in functional recovery, substantial expression of human dystrophin, long-term engraftment and a high proportion of myofibres expressing human dystrophin (Rodriguez et al. 2005). hMADS cells gave rise to a larger number of myofibres expressing human dystrophin than similar experiments using wild-type primary myoblasts or muscle stem cells. Two mechanisms have been proposed to account for the contribution of hMADS cells to muscle regeneration: (i) de novo generation of muscle-specific cells from hMADS cells and (ii) changes in gene expression after direct fusion of hMADS and host cells.

Osteochondral defects of the rabbit left medial femoral condyle have been successfully repaired by direct reconstitution of the injured osteochondral tissue by transplanted ADSCs from New Zealand White rabbit adipose tissue (Lee et al. 2003). The histological grading score in the ADSC-treated animals was superior to animals treated with periosteum-derived stem cells, and the repaired tissues had the solidity of intact cartilage and were superior to osteochondral autografts.

Ii et al. examined the therapeutic effect of hMADS cells in a rat MI model and demonstrated functional myocardial recovery by direct and indirect effects (Ii et al. 2011). hMADS cells transplanted into the ischemic myocardium of nude rats immediately following MI lead to superior improvement in cardiac function assessed by echocardiography compared with the animals receiving PBS; infarct sizes were significantly reduced and capillary densities were increased in the peri-infarct myocardium of hMADS-treated animals. *In vitro*, hMADS cells produced multiple pro-angiogenic growth factors and chemokines, such as VEGF, bFGF and SDF-1α. In ischemic myocardium, SDF-1α from transplanted hMADS cells recruited circulating bone marrow-derived EPCs to the infarct, suggesting that myocardial repair involved synergy between the transplanted hMADS cells and endogenous EPC recruitment. However, only small numbers of hMADS cells differentiated into cardiac or vascular lineage cells in this study.

The effects of ADSCs on skin wound healing have been examined by Kim et al. (Kim et al. 2007) who also analyzed their effects on fibroblast activation/proliferation, collagen synthesis and migration. ADSC transplants promoted human dermal fibroblast (HDF) proliferation through both direct contact and paracrine mechanisms. *In vitro*, HDFs upregulated collagen types I and III as well as fibronectin and downregulated matrix metalloproteinase-1 (MMP-1) when co-cultured with ADSCs or grown in ADSC-conditioned medium (ADSC-CM). ADSC-CM also stimulated HDF migration in *in vitro* wound healing models. In animal models, ADSCs significantly reduced wound size and accelerated re-epithelialization from the edges of skin wounds. These data indicate that ADSCs are therapeutically suitable for dermal wound healing because of their secretory behaviour and activation of host HDFs.

Recent Clinical Trials with ADSCs

Based on the evidence described above, clinical trials for regenerative medicine using ADSCs are on-going. ADSCs are commonly used specifically in plastic surgery, i.e., for soft tissue repair, such as craniofacial and breast tissue.

Cell-assisted lipotransfer is a promising technique for breast reconstruction or augmentation; however, graft survival is low because of necrosis (Yoshimura et al. 2008). The post-operative atrophy and necrosis of injected fat tissue has been successfully minimized by inclusion of ADSCs in lipoinjection, and breast augmentation was to the satisfaction of patients.

Rigotti et al. treated radiotherapy tissue damage with lipoaspirate transplantation (Rigotti et al. 2007). Patients with severe or irreversible functional damage due to radiotherapy were transplanted with purified autologous lipoaspirates from healthy donors by repeated low-invasive computer-assisted injections. Treatments were assessed using a symptom classification scale, cytofluorimetric characterization and ultrastructural evaluation of the targeted tissue. Lipoaspirate transplantation significantly improved the symptom classification score, and progressive regeneration of the target tissue ultrastructure including neovessel formation and improved hydration were observed. These clinical outcomes led to systematic improvements or remission of symptoms in all patients. They concluded that lipoaspirate transplantation has therapeutic potential for the repair of microangiopathies due to radiotherapy, and it may be potentially useful for other forms of microangiopathy.

Ongoing randomized trials in Europe are addressing the need to treat acute MI and chronic myocardial ischemia with ADSCs freshly isolated using Celution™ (Cytori Therapeutics Inc.) (Sanchez et al. 2010) these are APOLLO (AdiPOse-derived stem ceLLs with ST-elevation myOcardial infarction) and PRECISE (adiPose-deRived stEm & regenerative Cells In the treatment of patients with non-revaScularizable ischEmic myocardium) in Madrid, Rotterdam and Copenhagen. PRECISE is a prospective, double-blind, randomized, placebo-controlled and sequential dose-escalation trial for 36 patients with end-stage coronary artery disease not amenable for revascularization and moderate–severe left ventricular (LV) dysfunction. ADSCs were delivered via transendocardial injection after LV electromechanical mapping using a NOGA XPTM delivery system (BDS). In APOLLO, freshly isolated ADSCs were delivered through coronary arteries after appropriate infarct-related artery repair with stent implantation to 48 patients suffering from AMI and LV ejection fraction impairment.

Other clinical trials include the repair of critical calvarial bone defects, treatment of chronic fistulas in Crohn's disease (Garcia-Olmo et al. 2005) and plastic surgery for soft tissue augmentation. Trials of ADSC therapies for osteochondral disease or injury, such as bone fracture repair, pseudoarthritis and cartilage defects, are not currently underway but are likely in the near future.

Conclusion

ADSCs are a promising and increasingly prevalent therapeutic tool and will be applied safely in translational research and clinical trials for regenerative medicine. hMADS cells isolated from young donors are rapidly adherent in culture, are of normal karyotypes and have extensive expansion capacity. ADSCs isolated by other methods and from other tissue types are also multipotent, have favourable paracrine effects and are capable of long-term engraftment to varying degrees. Taken together, although ADSCs are a promising tool for autologous cell transplantation, further optimization of ADSC isolation and selection of the most appropriate adipose tissue are required before efficient organ-specific tissue regeneration in clinical settings is a reality.

Key Facts of hMADS Cells

- hMADS cells are somatic stem cells isolated from adipose tissue discards after surgery of young patients.
- Among the several types of ADSCs, we characterized hMADS as the rapidly adherent cells (<12 hr) in primary culture of adipose tissue.
- hMADS cells have stronger expansion capacity *ex vivo* than other ADSCs.
- hMADS cells can differentiate into multiple mesenchymal lineages and vascular endothelial cells *in vitro* and *in vivo*.

- The therapeutic potential of hMADS cells in fracture healing, muscle disease and ischemic heart disease has been demonstrated *in vivo* using animal models.

Dictionary

- *hMADS cells*: A subset of ADSCs. They are the rapidly adherent cells (<12 hr) isolated from adipose tissue of young donors, multipotent within mesenchymal lineages and have extensive expansion capacity *ex vivo*.
- *Pseudarthrosis*: Non-union of a bone fracture due to inadequate healing.
- *Non-healing femoral fracture model*: An animal fracture model exhibiting non-union. This model is induced in animals by cauterization of the periosteum around the fracture site.
- *Hypoxia-inducible transcription factor (HIF)*: A factor secreted in response to hypoxia. HIF is an upstream regulator of VEGF and angiopoietin-1.
- *Endothelial progenitor cells (EPCs)*: A population of cells in the blood having the potential to differentiate into ECs. They can support neovascularization.

Summary Points

- ADSCs are the resident stem cells from a large tissue reservoir and are easily collected. They have similar multipotency to BMSCs, differentiating into multiple mesenchymal lineages.
- hMADS cells are a type of ADSCs that are isolated from young donors. hMADS cells exhibit rapid adherence in culture, have greater extensive expansion capacity than other ADSCs and are of normal karyotypes.
- hMADS cells are multipotent, exert favourable paracrine effects and have long-term engraftment potential.
- hMADS cell transplantation promotes fracture healing through direct differentiation into OBs and ECs as well as paracrine effects.
- Various *in vivo* studies and clinical trials using ADSCs have demonstrated favourable results. ADSCs, especially hMADS cells, are likely to be widely used in clinical settings in the near future.

List of Abbreviations

ADSC   :   adipose derived stromal cell
ADSC-CM :   conditioned medium of ADSCs

ANG1	:	angiopoietin-1
BDNF	:	brain-derived neurotrophic factor
bFGF	:	basic fibroblast growth factor
BMP-2	:	bone morphogenetic protein-2
BMSC	:	bone marrow stromal cell
DMEM	:	Dulbecco's modified Eagle's medium
ECs	:	endothelial cells
EPC	:	endothelial progenitor cell
FBS	:	fetal bovine serum
GAPDH	:	glyceraldehyde-3-phosphate dehydrogenase
G-CSF	:	granulocyte colony stimulating factor
GM-CSF	:	granulocyte/macrophage colony stimulating factor
hCD31	:	human CD31
hCol1	:	human collagen 1 alpha 1
HDF	:	human dermal fibroblast
hFB	:	human skin fibroblasts
HGF	:	hepatocyte growth factor
HIF	:	hypoxia-inducible transcription factor
hMADS cells:		human multipotent adipose-derived stem cells
hNA	:	human nuclear antigen
hOC	:	human osteocalcin
HSC	:	hematopoietic stem cell
hVEcad	:	human vascular endothelial cadherin
IGF	:	insulin-like growth factor
IL	:	interleukin
ILB4	:	isolectin B4
KGF	:	keratinocyte growth factor
LDPI	:	laser doppler perfusion imaging
LIF	:	leukemia inhibitory factor
LV	:	left ventricular
M-CSF	:	macrophage colony stimulating factor
MI	:	myocardial infarction
MMP-1	:	matrix metalloproteinase-1
OB	:	osteoblast
PDGF	:	platelet-derived growth factor
PLA	:	processed lipoaspirate
SDF-1	:	stromal cell-derived factor-1
SVF	:	stromal vascular fraction
TGF-β	:	transforming growth factor-beta
TNF-α	:	tumor necrosis factor-alpha
UEA1	:	ulex europaeus agglutinin-1
VEGF	:	vascular endothelial growth factor

References

Cai, L., B.H. Johnstone, T.G. Cook, Z. Liang, D. Traktuev, K. Cornetta, D.A. Iingram, E.D. Rosen and K.L. March. 2007. Suppression of hepatocyte growth factor production impairs the ability of adipose-derived stem cells to promote ischemic tissue revascularization. Stem Cells. 25: 3234–43.

Chen, C.W., E. Montelatici, M. Crisan, M. Corselli, J. Huard, L. Lazzari and B. Peault. 2009. Perivascular multi-lineage progenitor cells in human organs: regenerative units, cytokine sources or both? Cytokine Growth Factor Rev. 20: 429–34.

Connolly, J.F., R. Guse, J. Tiedeman and R. Dehne. 1991. Autologous marrow injection as a substitute for operative grafting of tibial nonunions. Clin Orthop Relat Res. 259–70.

Cousin, B., M. Andre, E. Arnaud, L. Penicaud and L. Casteilla. 2003. Reconstitution of lethally irradiated mice by cells isolated from adipose tissue. Biochem Biophys Res Commun. 301: 1016–22.

De Ugarte, D.A., K. Morizono, A. Elbarbary, Z. Alfonso, P.A. Zuk, M. Zhu, J.L. Dragoo, P. Ashjian, B. Thomas, P. Benhaim, I. Chen, J. Fraser and M.H. Hedrick. 2003. Comparison of multi-lineage cells from human adipose tissue and bone marrow. Cells Tissues Organs. 174: 101–9.

Fukui, T., T. Matsumoto, Y. Mifune, T. Shoji, T. Kuroda, Y. Kawakami, A. Kawamoto, M. Ii, S. Kawamata, M. Kurosaka, T. Asahara and R. Kuroda. 2011. Local Transplantation of Granulocyte Colony Stimulating Factor-Mobilized Human Peripheral Blood Mononuclear Cells for Unhealing Bone Fractures. Cell Transplant. 21: 707–721.

Garci-Olmo, D., M. Garci-Arranz, D. Herreros, I. Pascual, C. Peiro and J.A. Rodriguez-Montes. 2005. A phase I clinical trial of the treatment of Crohn's fistula by adipose mesenchymal stem cell transplantation. Dis Colon Rectum. 48: 1416–23.

Green, H. and M. Meuth. 1974. An established pre-adipose cell line and its differentiation in culture. Cell. 3: 127–33.

Ii, M., M. Horii, A. Yokoyama. T. Shoji, Y. Mifune, A. Kawamoto, M. Asahi and T. Asahara. 2011. Synergistic effect of adipose-derived stem cell therapy and bone marrow progenitor recruitment in ischemic heart. Lab Invest. 91: 539–52.

Kershaw, E.E. and J.S. Flier. 2004. Adipose tissue as an endocrine organ. J Clin Endocrinol Metab. 89: 2548–56.

Khan, W.S., A.B. Adesida and T.E. Hardingham. 2007. Hypoxic conditions increase hypoxia-inducible transcription factor 2alpha and enhance chondrogenesis in stem cells from the infrapatellar fat pad of osteoarthritis patients. Arthritis Res Ther. 9: R55.

Kilroy, G.E., S.J. Foster, X. Wu, J. Ruiz, S. Sherwood, A. Heifetz, J.W. Ludlow, D.M. Stricker, S. Potiny, P. Green, Y.D. Halvorsen, B. Cheatham, R.W. Storms and J.M. Gimble. 2007. Cytokine profile of human adipose-derived stem cells: expression of angiogenic, hematopoietic, and pro-inflammatory factors. J Cell Physiol. 212: 702–9.

Kim, W.S., B.S. Park, J.H. Sung, J.M. Yang, S.B. Park, S.J. Kwak and J.S. Park. 2007. Wound healing effect of adipose-derived stem cells: a critical role of secretory factors on human dermal fibroblasts. J Dermatol Sci. 48: 15–24.

Lee, E.Y., Y. Xia, W.S. Kim, M.H. Kim, T.H. Kim, K.J. Kim, B.S. Park and J.H. Sung. 2009. Hypoxia-enhanced wound-healing function of adipose-derived stem cells: increase in stem cell proliferation and up-regulation of VEGF and bFGF. Wound Repair Regen. 17: 540–7.

Lee, J.A., B.M. Parrett, J.A. Conejero, J. Laser, J. Chen, A.J. Kogon, D. Nanda, R.T. Grant and A.S. Breitbart. 2003. Biological alchemy: engineering bone and fat from fat-derived stem cells. Ann Plast Surg. 50: 610–7.

Madonna, R. and R. De Caterina. 2008. *In vitro* neovasculogenic potential of resident adipose tissue precursors. Am J Physiol Cell Physiol. 295: C1271–80.

Mifune, Y., T. Matsumoto, A. Kawamoto, R. Kuroda, T. Shoji, H. Iwasaki, S.M. Kwon, M. Miwa, M. Kurosaka and T. Asahara. 2008. Local delivery of granulocyte colony stimulating factor-mobilized CD34-positive progenitor cells using bioscaffold for modality of unhealing bone fracture. Stem Cells. 26: 1395–405.

Miranville, A., C. Heeschen, C. Sengenes, C.A. Curat, R. Busse and A. Bouloumie. 2004. Improvement of postnatal neovascularization by human adipose tissue-derived stem cells. Circulation. 110: 349–55.

Nathan, S., S. Das De, A. Thambyah, C. Fen, J. Goh and E.H. Lee. 2003. Cell-based therapy in the repair of osteochondral defects: a novel use for adipose tissue. Tissue Eng. 9: 733–44.

Pittenger, M.F., A.M. Mackay, S.C. Beck, R.K. Jaiswal, R. Douglas, J.D. Mosca, M.A. Moorman, D.W. Simonetti, S. Craig and D.R. Marshak. 1999. Multilineage potential of adult human mesenchymal stem cells. Science. 284: 143–7.

Planat-Benard, V., C. Menard, M. Andre, M. Puceat, A. Perez, J.M. Garcia-Verdugo, L. Penicaud and L. Casteilla. 2004. Spontaneous cardiomyocyte differentiation from adipose tissue stroma cells. Circ Res. 94: 223–9.

Rangappa, S., C. Fen, E.H. Lee, A. Bongso and E.K. Sim. 2003. Transformation of adult mesenchymal stem cells isolated from the fatty tissue into cardiomyocytes. Ann Thorac Surg. 75: 775–9.

Rehman, J., D. Traktuev, J. Li, S. Merfeld-Clauss, C.J. Temm-Grove, J.E. Bovenkerk, C.L. Pell, B.H. Johnstone, R.V. Considine and K.L. March. 2004. Secretion of angiogenic and antiapoptotic factors by human adipose stromal cells. Circulation. 109: 1292–8.

Rigotti, G., A. Marchi, M. Galie, G. Baroni, D. Benati, M. Krampera, A. Pasini and A. Sbarbati. 2007. Clinical treatment of radiotherapy tissue damage by lipoaspirate transplant: a healing process mediated by adipose-derived adult stem cells. Plast Reconstr Surg. 119: 1409–22; discussion, 1423–4.

Rodrigues, A.M., D. Pisani, C.A. Dechesne, C. Turc-Carel, J.Y. Kurzenne, B. Wdziekonski, A. Villageois, C. Bagnis, J.P. Breittmayer, H. Groux, G. Ailhaud and C. Dani. 2005. Transplantation of a multipotent cell population from human adipose tissue induces dystrophin expression in the immunocompetent mdx mouse. J Exp Med. 201: 1397–405.

Romanov, Y.A., A.N. Darevskaya, N.V. Merzlikina and L.B. Buravkova. 2005. Mesenchymal stem cells from human bone marrow and adipose tissue: isolation, characterization, and differentiation potentialities. Bull Exp Biol Med. 140: 138–43.

Sanchez, P.L., R. Sanz-Ruiz, M.E. Fernandez-Santos and F. Fernandez-Aviles. 2010. Cultured and freshly isolated adipose tissue-derived cells: fat years for cardiac stem cell therapy. Eur Heart J. 31: 394–7.

Sherwood, R.I., J.L. Christensen, I.M. Conboy, M.J. Conboy, T.A. Rando, I.L. Weissman and A.J. Wagers. 2004. Isolation of adult mouse myogenic progenitors: functional heterogeneity of cells within and engrafting skeletal muscle. Cell. 119: 543–54.

Shoji, T., M. Ii, Y. Mifune, T. Matsumoto, A. Kawamoto, S.M. Kwon, T. Kuroda, R. Kuroda, M. Kurosaka and T. Asahara. 2010. Local transplantation of human multipotent adipose-derived stem cells accelerates fracture healing via enhanced osteogenesis and angiogenesis. Lab Invest. 90: 637–49.

Traktuev, D.O., S. Merfeld-Clauss, J. Li, M. Kolonin, W. Arap, R. Pasqualini, B.H. Johnstone and K.L. March. 2008. A population of multipotent CD34-positive adipose stromal cells share pericyte and mesenchymal surface markers, reside in a periendothelial location, and stabilize endothelial networks. Circ Res. 102: 77–85.

Trayhurn, P. and J.H. Beattie. 2001. Physiological role of adipose tissue: white adipose tissue as an endocrine and secretory organ. Proc Nutr Soc. 60: 329–39.

Wei, X., Z. Du, L. Zhao, D. Feng, G. Wei, Y. He, J. Tan, W.H. Lee, H. Hampel, R. Dodel, B.H. Johnstone, K.L. March, M.R. Farlow and Y. Du. 2009. IFATS collection: The conditioned media of adipose stromal cells protect against hypoxia-ischemia-induced brain damage in neonatal rats. Stem Cells. 27: 478–88.

Yoshimura, K., K. Sato, N. Aoi, M. Kurita, T. Hirohi and K. Harii. 2008. Cell-assisted lipotransfer for cosmetic breast augmentation: supportive use of adipose-derived stem/stromal cells. Aesthetic Plast Surg. 32: 48–55; discussion, 56–7.

Zannettino, A.C., S. Paton, A. Arthur, F. Khor, S. Itescu, J.M. Gimble and S. Gronthos. 2008. Multipotential human adipose-derived stromal stem cells exhibit a perivascular phenotype *in vitro* and *in vivo*. J Cell Physiol. 214: 413–21.

Zuk, P.A., M. Zhu, H. Mizuno, J. Huang, J.W. Futrell, A.J. Katz, P. Benhaim H.P. Lorenz and M.H. Hedrick. 2001. Multilineage cells from human adipose tissue: implications for cell-based therapies. Tissue Eng. 7: 211–28.

Index

About the Editors

Victor R. Preedy BSc, PhD, DSc, FSB, FRCPath, FRSPH FRSC is a Professor at King's College London and also at King's College Hospital. He is attached to both the Diabetes and Nutritional Sciences Division and the Department of Nutrition and Dietetics. He is also Director of the Genomics Centre and a member of the School of Medicine. Professor Preedy graduated in 1974 with an Honours Degree in Biology and Physiology with Pharmacology. He gained his University of London PhD in 1981. In 1992, he received his Membership of the Royal College of Pathologists and in 1993 he gained his second doctoral degree, for his outstanding contribution to protein metabolism in health and disease. Professor Preedy was elected as a Fellow to the Institute of Biology in 1995 and to the Royal College of Pathologists in 2000. Since then he has been elected as a Fellow to the Royal Society for the Promotion of Health (2004) and The Royal Institute of Public Health (2004). In 2009, Professor Preedy became a Fellow of the Royal Society for Public Health and in 2012 a Fellow of the Royal Society of Chemistry. In his career Professor Preedy has carried out research at the National Heart Hospital (part of Imperial College London) and the MRC Centre at Northwick Park Hospital. He has collaborated with research groups in Finland, Japan, Australia, USA and Germany. He is a leading expert on the science of health. He has lectured nationally and internationally. To his credit, Professor Preedy has published over 570 articles, which includes 165 peer-reviewed manuscripts based on original research, 90 reviews and over 50 books and volumes.

Dr. Vinood B. Patel, PhD is currently a Senior Lecturer in Clinical Biochemistry at the University of Westminster and honorary fellow at King's College London. Dr. Patel obtained his degree in Pharmacology from the University of Portsmouth, his PhD in protein metabolism from King's College London in 1997 and carried out post-doctoral research at Wake Forest University School of Medicine, USA where he developed novel biophysical techniques to characterise mitochondrial ribosomes. He presently directs studies on metabolic pathways involved in fatty liver disease, focussing on mitochondrial energy regulation and cell death. Dr.

Patel has published over 150 articles, including books in the area of nutrition and health prevention.

Dr. Rajkumar Rajendram, AKC BSc. (hons) MBBS (dist) MRCP (UK) FRCA, graduated in 2001 with a distinction from Guy's, King's and St. Thomas Medical School, in London. As an undergraduate he was awarded several prizes, merits and distinctions in pre-clinical and clinical subjects. This was followed by training in general medicine and intensive care in Oxford, during which period he attained membership of the Royal College of Physicians (London, MRCP) in 2004. Dr Rajendram went on to train in anaesthesia in the Central School of Anaesthesia, London Deanery and became a fellow of the Royal College of Anaesthetists (FRCA) in 2009. He has completed higher training in regional anaesthesia, pain medicine and intensive care. He returned to Oxford in his current position as a Locum Consultant in General Medicine at the John Radcliffe Hospital, Oxford. He teaches on several courses for post-graduate examinations and has authored several research papers, review articles, textbook chapters and books.

Color Plate Section

Chapter 1

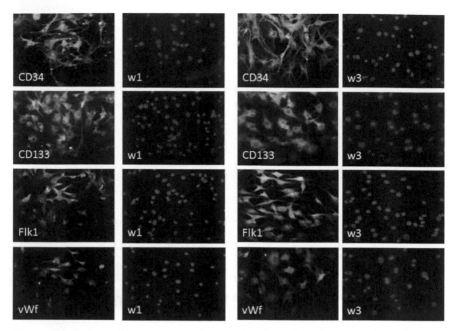

Figure 1.1 EPC surface marker. Characterization of EPCs at 1 and 3 wk using immunohistochemistry staining. The cultured EPCs appeared to have a cobblestone shape with strongly positive CD133, and spindle shape with strongly positive CD34 and Flk1, with both shapes having weakly positive vWF staining at 1 and 3 wk (unpublished data).

Chapter 6

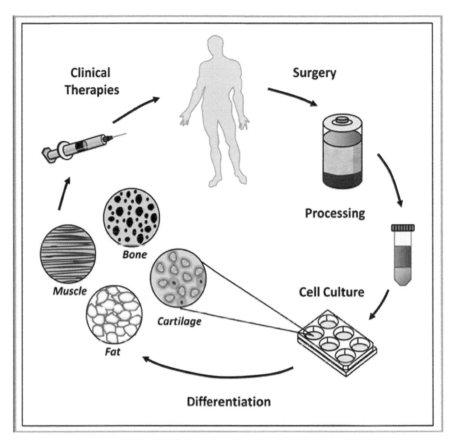

Figure 6.1 Schematic outline of differentiation potential of ASCs isolated from lipoaspirate tissues. Human lipoaspirate is an abundant source of ASCs that under appropriate culture conditions can differentiate along different lineages.

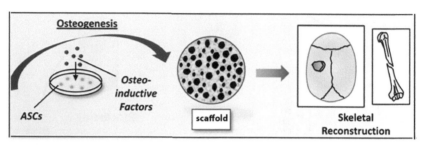

Figure 6.2 Schematic outline of ASCs isolation from human liposaspirate and their use for bone regeneration. Human lipoaspirate in a sterile container is processed for isolation of ASCs. Plated ASCs are grown and subsequentially treated with osteoinductive factors (e.g., BMPs, FGF-2, VEGF), seeded on scaffolds and implanted on injured bones to promote skeletal reconstruction.

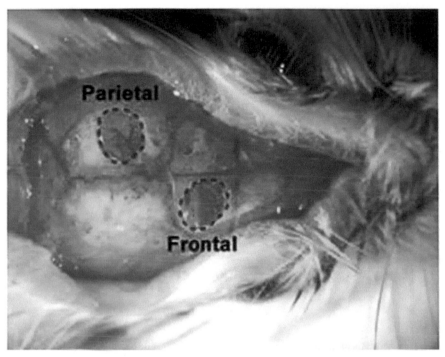

Figure 6.3 Model of murine calvarial bone defect employed to assess the ability of ASCs to regenerate bone. An example of murine calvarial defects created in the right frontal and left parietal bones of adult post-natal day 60 mice with meticulous care to avoid injury to the dura mater. Dotted circles mark the defects. Note the continuous blood vessels, indicating intact dura mater.

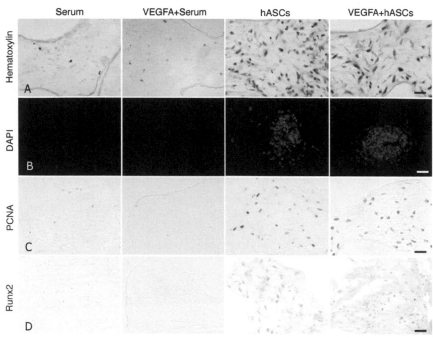

Figure 6.4 hASCs attachment to coral-apatite scaffolds and initiation of *in vivo* **osteogenic differentiation at 48 hr post implantation on calvarial defects.** (A) staining for Hematoxylin and (B) DAPI revealed hASCs attachment *in vivo* after 48 hr. (C) immunohistochemistry for the proliferation cell nuclear antigen PCNA to monitor cell proliferation of untreated and VEGFA treated hASCs. (D) immunohistochemistry for Runx2 detecting the presence of osteoprogenitors in both VEGFA treated and untreated hASCs. In contrast, no staining is observed in the control groups. Scale bars: A, C, D 50 μm, B, 100 μm (Modified from Behr et al. 2011 Stem Cells).

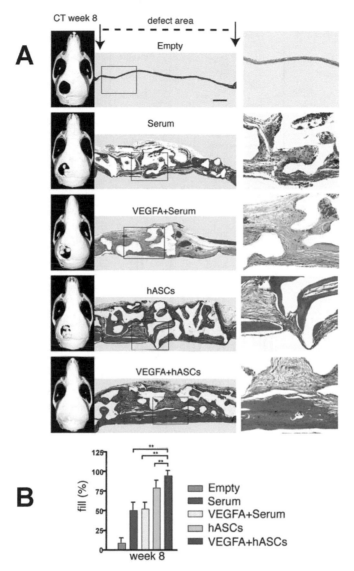

Figure 6.5 Repair of critical sized calvarial defects with VEGFA treated hASCs. (A) Micro-computed tomography (µCT) and corresponding H&E staining of critical-sized calvarial defects in parietal bones of nude mice at postoperative week 8, to assess bone healing. Defects were treated with coral apatite scaffolds loaded with either 2 mg VEGFA-treated or VEGFA-untreated hASCs. Treatment of defects loaded with serum or VEGFA + serum were controls, as well empty defect. Boxed areas are enlarged in the right column. Scale bar: 400 µm. (B) Quantification of defect healing according to µCT results revealed significantly increased healing of defects with VEGFA-treated and VEGFA-untreated ASCs loaded coral scaffolds. (Modified from Behr et al. 2011 Stem Cells).

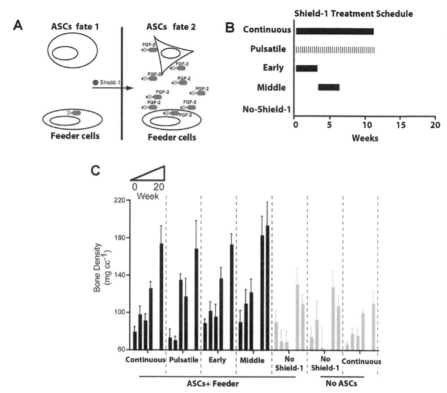

Figure 6.6 Chemical tunable control of FGF-2 release for calvarial bone defect regeneration with ASCs. (A) schematic of feedercells (DD-FGF-2 cells) engineered tosecrete a FGF-2 fusion protein under a regulation system driven by the presence of the synthetic ligand (Shield-1). These DD-FGF-2 cells induce paracrine/osteogenic inductive signals on ASCs loaded on scaffold placed into mouse calvarial defects. (B) schematic of *in vivo* treatment groups. Treatment groups (*n* = 4–6 mice/group) consisted of CD-1 nude mice with calvarial defects implanted with scaffolds on which were seeded DD-FGF-2 cells and ASCs or scaffolds loaded only with DD-FGF-2 cells. Each group was administered with intraperitoneal injections of Shield-1 (6 mg/kg) every other day for first 12 wk (**Continuous**), every 4th day for the first 12 wk (**Pulsatile**), every other day for the first 3 wk (**Early**), or every other day during wk 4–6 (**Middle**). (C) Quantification of defects healing according to μCT results obtained from treatment and control groups at wk 0, 4, 8, and 20. (Modified from Kwan et al. 2011, 286 J Biol Chem).

Chapter 8

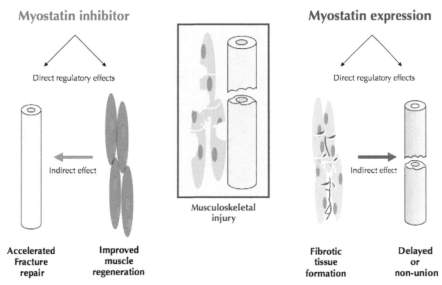

Figure 8.5 General model summarizing effects of myostatin inhibition (left) and myostatin expression (right) on musculoskeletal repair and regeneration. Blocking myostatin signaling following injury of muscle and bone is likely to enhance bone healing in two ways (left side of figure): directly, because suppression of myostatin activity will enhance chondrogenesis and indirectly, because enhancing muscle regeneration is likely to restore the normal secretion of paracrine, osteogenic trophic factors (e.g., IGF-1) from muscle. Myostatin expression following musculoskeletal injury inhibits bone healing in two ways (right side of figure), first by suppressing chondrogenesis directly and second by increasing fibrosis in injured muscle thereby reducing secretion of osteogenic paracrine factors from muscle tissue.

Chapter 14

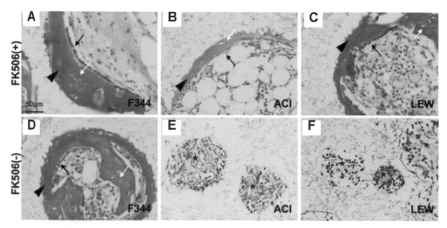

Figure 14.3 Histological analyses of transplants 4 wk after transplantation by H&E staining. Ceramic sections of transplants from FK506 treated rats (A-C) or non-treated rat (D-F). Scale bar: 50μm. Black arrows: osteoblasts; white arrows: osteocytes; arrowheads: bone matrix; asterisk: inflammation (from Kotobuki et al. 2008; with permission from Cognizant Communication Corporation).

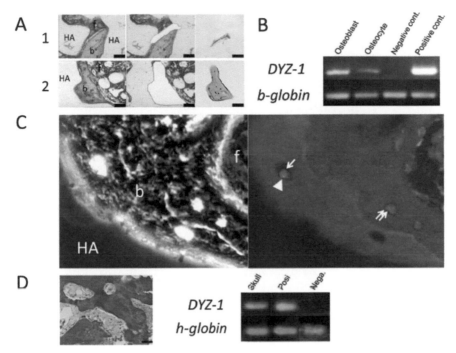

Figure 14.8 Analyses of the retrieved constructs and skull fragment. (A) Areas of the osteoblast lining (1) and osteocytes with bone matrix (2) were dissected from the original section using laser-assisted microdissection. The images show the original sections (left), after dissection (middle), and retrieved samples (right). The retrieved samples were used for detection of the Y chromosome. New bone (b) as well as fibrous tissue (f) can be seen within the pores of the ceramic (HA). Bar = 50 μm. (B) Detection of male-specific DYZ1 on the Y chromosome using PCR analysis. *β-globin* was used as an internal control. Osteoblast: DNA extracted from the retrieved osteoblast sample, as shown in A1; Osteocyte: from the retrieved osteocyte sample, as shown in A2; Negative cont.: negative control (female DNA); Positive cont.: positive control (male DNA). (C) Phase contrast (left) and fluorescence (right) images of FISH analysis of the retrieved construct. The osteocytes have the donor's (XY) and the patient's (XX) chromosomes. Nuclei are stained with DAPI (blue). The green and red dots indicate the Y chromosome (arrowhead) and the X chromosome (arrows), respectively. New bone (b) and fibrous tissue (f) were observed within the ceramic (HA). Original magnification ×400. (D) Decalcified section of the skull fragments (left). H-E staining, Bar = 100 μm. DNA samples from the other skull fragments were used for detection of the donor DYZ1 sequences using PCR analysis (right). Skull: extracted DNA sample from the skull fragment; Posi: positive control (male DNA); Nega: negative control (female DNA) (from Tadokoro et al. 2009, partially modification; with permission from Elsevier Inc.).

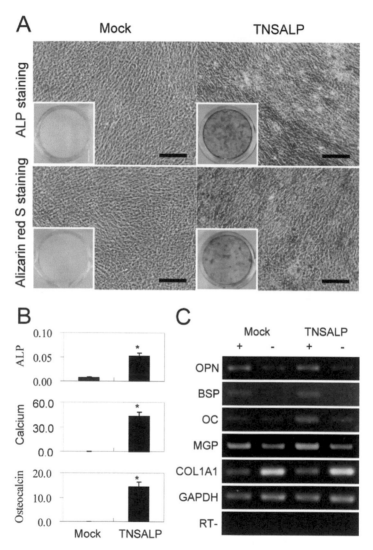

Figure 14.9 *In vitro* **osteogenic differentiation potential of** *TNSALP*-transduced MSCs. (A) ALP activity and mineralization. Cells were stained by ALP and alizarin red S staining. 100× magnification, bar: 100 μm. Inset: Observation of whole areas of stained culture plate. (B) Biochemical analysis. ALP activity (μmol/30 min/μg DNA), calcium content (μg/well), and osteocalcin (OC) content (ng/well) are shown. Values are expressed as averages and standard deviations (n = 4). Asterisks indicate statistically significant differences between mock- and *TNSALP*-transduced MSCs (*t*-test: *P* < 0.01). (C) Bone-related gene expression of MSCs. Mock- and *TNSALP*-transduced MSCs were cultured in osteogenic differentiation medium (+) or control medium (−) (from Katsube et al. 2010; with permission from Nature Publishing Group).

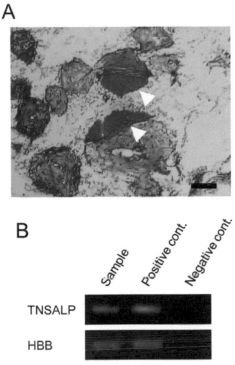

Figure 14.10 *In vivo* bone formation ability of *TNSALP*-transduced MSCs. (A) Histological appearance of osteogenic constructs using *TNSALP*-transduced MSCs. The harvested sample was decalcified, sectioned, and stained with hematoxylin and eosin. Arrowheads indicate newly formed bone. The newly formed bone areas in the section were harvested by laser-assisted microdissection. (B) PCR analyses of DNA samples of newly formed bone harvested by microdissection. The dissected samples (sample), human MSCs with *TNSALP* expression vector (positive control), and rat MSCs (negative control) were used to detect the transduced *TNSALP* gene (TNSALP) and human β-globin (HBB) (from Katsube et al. 2010; with permission from Nature Publishing Group).

Chapter 18

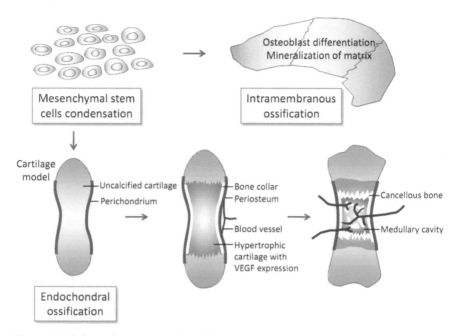

Figure 18.1 Schematic representation of the two types of osteogenesis. 1) Intramenbranous ossification involves bone formation through the direct osteoblastic differentiation; 2) endochondral ossification generates bone via transition from an avascular hypertrophic cartilage to a vascularized and mineralized matrix.

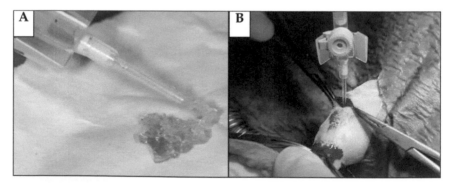

Figure 18.3 Application of hydrogels as injectable bone substitute materials. (A) Hydrogels ejection from syringe through a needle cannula. (B) Injection of a hydrogel into a 1.8mm-bone defect of a guinea pig tibia model.

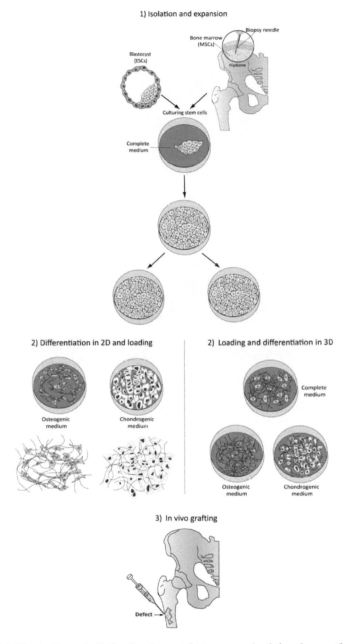

Figure 18.4 Illustration of all the fundamental steps required for the synthesis and application of a cell-material system for bone tissue regeneration. (1) cell isolation and expansion, (2) cell seeding and differentiation, and (3) *in vivo* grafting.

Chapter 20

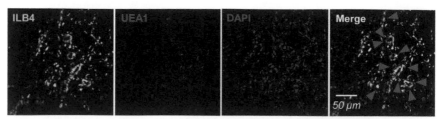

Figure 20.5 Endothelial differentiation of hMADS cells in peri-fracture tissue. Representative images of fluorescent staining for isolectin B4 (ILB4: green), Ulex europaeus agglutinin 1 (UEA1: red) and DAPI (blue) with serial sections of peri-fracture tissue samples. Arrowheads in the merged image indicate ILB4-negative/UEA1-positive cells as differentiated human endothelial cells located within host-derived capillaries detected as ILB4-positive tubular structures in the granulation tissue surrounding the fracture gap.

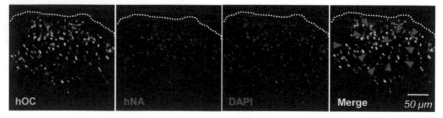

Figure 20.6 Osteoblastic differentiation of hMADS cells during callus formation. Immunofluorescence staining for human-specific osteocalcin (hOC, green), anti-human nuclear antigen (hNA, red) and DAPI (blue) in serial sections of peri-fracture tissue samples. White dotted, curved line indicates the edge of callus. Arrowheads in the merged image indicate hOC and hNA double-positive cells identified as differentiated human osteoblasts (hOBs) in the callus endochondral ossification area receiving hMADS cells.

T - #0340 - 071024 - C14 - 234/156/19 - PB - 9780367380397 - Gloss Lamination